Pioneering in Outer Space

A course of lectures on selected topics in modern physics and space flight

HERMANN BONDI, F.R.S.

Professor of Applied Mathematics, King's College, London and Director-General of the European Space Research Organisation

SIR MARK OLIPHANT, F.R.S.

Emeritus Professor, Fellow of the Australian National University, Canberra

LEE B. JAMES

Director of Lunar Operations, George C. Marshall Space Flight Centre, Huntsville, Alabama

G. E. MUELLER

Vice-President, General Dynamics Corporation, Washington, D.C.

G. HAGE

Vice-President for Development, Boeing Company, Seattle, Washington

Heinemann Educational Books
London and Edinburgh

Heinemann Educational Books Ltd

LONDON EDINBURGH MELBOURNE TORONTO
SINGAPORE JOHANNESBURG AUCKLAND IBADAN
HONG KONG NAIROBI NEW DELHI

IN THE SAME SERIES

Nuclear Energy Today and Tomorrow

ISBN 0 435 68281 4

First published in Great Britain 1971

Published by Heinemann Educational Books Ltd
48 Charles Street, London W1X 8AH
Printed in Great Britain by
Biddles Ltd., Guildford, Surrey

PREFACE

The contributions comprising this book were delivered in 1970 at the thirteenth International Science School organized annually by the Science Foundation for Physics within the University of Sydney. These Schools are attended by students selected from Australia and New Zealand in addition to twenty students, chosen for their outstanding ability, from America, Britain, and Japan. The aim is to stimulate and develop science consciousness in Australia and throughout the world.

On behalf of the Foundation we wish to take this opportunity of thanking Professor Hermann Bondi, F.R.S., Mr. George Hage, Colonel Lee B. James, Dr. George Mueller and Professor Sir Mark Oliphant, F.R.S., for having given so generously of their time and effort.

Sydney, *August*, 1970 H. MESSEL and S. T. BUTLER

INTERNATIONAL SCIENCE SCHOOL SERIES

Edited by

H. MESSEL

B.A., B.Sc., Ph.D.
Professor of Physics and
Head of the School of Physics
University of Sydney

S. T. BUTLER

M.Sc., Ph.D., D.Sc.
Professor of Theoretical Physics,
University of Sydney

THE SPONSORS

The Science Foundation for Physics within the University of Sydney gratefully acknowledges the generous financial assistance given by the following group of individual philanthropists and companies, without whose help the 1970 International Science School for High School Students and the production of this book would not have been possible.

FULL SPONSORS

Ampol Petroleum Limited

The James N. Kirby Foundation

The Nell and Hermon Slade Trust

The Sydney County Council

W. D. & H. O. Wills (Aust.) Limited

PART SPONSORS

A. Boden, Esq.

Philips Industries Pty. Ltd.

CONTENTS

Europe's Space Effort and Gravitation

U.S. Space Flight

Science and Mankind

Europe's Space Effort

and

Gravitation

by

Hermann Bondi

Professor H. Bondi,
Director-General of the European Space Research Organisation.

CHAPTER ONE

Is Space Only for the Big Ones?

Nowadays we are so deeply impressed by the gigantic achievements of the two great Powers, such as putting a man on the Moon or sending probes to Venus and Mars, that one wonders whether there is room for anybody else in space. Is there any possibility or any desirability of going into space for anybody else? Would they be able to catch up with the great space powers? Is it possible to do something in a modest way that is useful and advantageous? Alternatively, we can put the question—what do you lose if you are not in space? One can perhaps analyse these questions best if one divides the gains from space into three classes. What can you gain from exploring, using space as a medium? In other words what can science, in the narrow sense of the word, learn through engaging in space research? Secondly, we can ask what can any other user gain by using space as a means? How much money is there to be made out of space in this sense? And the third question is, how much benefit does industry gain by working for space? How much better is a company that has built a satellite in the struggle for markets in certain fields than a company that has not built satellites? It is clear from what has been said that there must be three partners in space, universities and research institutions on the one hand, the government on the second and industry on the third.

Let us look at the benefit to scientists. In the first instance there is the possibility of actually getting apparatus or a man to the object to be studied. The most famous example here of course is the man on the Moon but equally there are the probes that have visited Mars and Venus. It is, of course, ideal to proceed to a soft landing and a major success here was the Soviet Venus probe. Soon, the Mars orbiter and lander in the American programme will greatly increase our knowledge and equally there are of course plans for visits to other more distant planets by automatic equipment. But it is not only the solid objects in space that can be directly visited, but simply the material in interplanetary space. To get out of the area of the Earth and make an in-position examination of the material that exists between the Earth and the Moon is of the highest importance. We have here a unique possibility to study what the scientist calls a plasma (that is an ionized medium) and indeed this is an ionized medium that is very much more tenuous than could be obtained in the best laboratory vacuum. In such a medium electric currents flow virtually without resistance. This leads to behaviour that is still ill-understood and that needs a lot more investigation. In this connection one might also mention one of the most famous experiments of Apollo 11 when Dr. Geiss, of the University of Berne, devised a sheet that was put up on the Moon by the astronauts and caught the particles of the solar wind. The sheet was then returned to Earth and analysed for the composition of this exceedingly tenuous stream of particles coming from the Sun. This kind of *in situ* work also extends to the outermost atmosphere with satellites orbiting not too far away, a few hundred kilometres above the surface of the Earth, there traversing the outer ranges of the atmosphere and the ionosphere and capable of carrying out investigations of great interest and value. A second way of using space platforms is to get out of the Earth's atmosphere. As is well known, the atmosphere of the Earth is opaque to very many kinds of radiation. Indeed it is transparent in only two windows, the region of the visible light and a particular range of radio frequencies. That our eyes and those of other animals are essentially tuned to a transparent range of the frequency medium is of course a direct consequence of evolution. Therefore, we have the idea that the atmosphere is extremely transparent. Scientifically a great deal of use has been made of this through the centuries by the astronomer who has been watching the stars in the visible light with telescopes on the ground. In more recent times radio-astronomy has used the other window amongst the radio frequencies and in this field Australia has of course been in the foreground during the last quarter of a century and you know what tremendously important and exciting results have come from this, such as the quasars and the pulsars. But the astronomer is just as interested in other frequencies. Indeed one of the matters that radio astronomy has revealed is how vast is the amount of information that we do not get if we do not look at frequencies other than visible light. Another salutary lesson is this: Radio astronomy gave results of the greatest interest from the day it was started with serious means in the years just after the war, but it needed a decade and a half of this fruitful scientific work before the first really quite outstanding discovery was made, the discovery of the quasars. This just shows that even in our time, when progress seems to be so rapid, long periods must elapse before the technology of a new subject is sufficiently well understood and before one starts looking for what might be the most exciting discovery of all. Thus new frequencies form a clear field of space research that is to supplement the efforts of the optical and the radio astronomers in space from outside the atmosphere. The astronomer can then look at the celestial radiation in any wavelength he pleases, gamma-rays and x-rays at the short end, infra-red or ultra-violet, extremely long radio waves or anything else that his apparatus might be capable of receiving. We are only at the very beginning of these researches but they have already shown great interest. Again we must follow here the lesson of radio astronomy and before it optical astronomy. Astronomy is by its nature an inexhaustible subject. The fact that somebody has made certain measurements in the ultra-violet doesn't mean that it is stupid for somebody else to take measurements in the ultra-violet. I want to remind you here that the opening of the 200-inch telescope in California did not make other ground based optical telescopes obsolete. On the contrary, it ushered in the period of the most rapid development of telescopes a little or substantially smaller than itself. It was never possible to say that it was useless to use other instruments because the 200-inch could do more. The 200-inch was extremely busy and not suited necessarily to all types of astronomical work. It will be noticed that for this kind of operation it is not necessary to send a spacecraft very far, it has got to be above the atmosphere, that is all. There is no need to travel as far as Mars or Jupiter. Perhaps it might be a great thing to put a radio telescope on the far side of the Moon, not because there are things you can observe from there that you can't observe from here, but merely because on the Earth we generate in our lives, from our motor cars, from all the communication equipment, etc., so much interference that it certainly makes the radio astronomer's life difficult. On the back of the Moon he is shielded from all this. Yet it is sometimes difficult to avoid the impression that maybe by the time a radio telescope is built on the back of the Moon that area will not be so radio quiet either. There will be lots of package-tour tourists sent there by travel agencies on the Earth roaming about in vehicles that could make quite a lot of radio noise. A third line of space research is to examine the Earth itself from out in space. The advantage of a space platform here is that if it is orbiting it surveys a great deal of the Earth in the course of time and can, for example, gather information from automatic stations in rather inaccessible positions—say, buoys on the oceans which would enable us to learn a great deal more about oceans. It is of course well known how much meteorologists gain by viewing cloud patterns from above and how this has led to the possibility of forecasting the movement of hurricanes. The study of Earth's resources is another field in its infancy. The further development in meteorology beyond the study of cloud patterns will, I am certain, be enormously fruitful.

Next, let us consider the application of space. What is space good for as a medium to be exploited industrially for the benefit of all of us? What services can we get from space that we can't get otherwise? The best known of course is telecommunication. We know that the short-wave radio links that alone can house the enormous amount of information we like to transmit, cannot travel round the Earth. In order to jump the oceans therefore it is either necessary to use submarine cables or, and this seems to be a lot cheaper, go via a space platform. There are enormous possibilities here for the future.

Communication needs are constantly growing. There will be as much use made of any communication facility as can be provided. The market is virtually infinite. The possibility of satisfying needs, however, is always limited. Available frequencies are in constant demand. Technically, one goes to higher and higher frequencies to get wider and wider bands, to transmit not just tens of thousands of telephone messages but television in colour, if necessary, and to link distant computers for the exchange of data. However, when one comes down to wave-lengths of the order of a centimetre or so, the atmosphere ceases to be wholly transparent; particularly, if there is heavy rain there is a lot of attenuation. One can go some way with more powerful transmitters in the satellites and on the ground, with bigger antennae on the satellites and on the ground.

There is a long way to go before the use of space telecommunications is exhausted, but one can already see that there will be limits. Perhaps later on it will become possible to go to still shorter wavelengths and overcome some of the atmospheric problems. Perhaps in the visible. But we do know what a terrible obstacle, even to the sun's powerful radiation, clouds are. Whatever these more distant limitations, the exploitation of space for telephone and television and data transmission is certain to be in tremendous demand, extremely useful and highly profitable. It is, of course, not just something for the future but something which has been used for many years now.

In the same general area, the control of air traffic over oceans offers a real field for space exploitation. To be able to locate an aircraft over the oceans, far out of the range of ground radar stations from a satellite; to be able to communicate with it, without any doubt at any time, to avoid collisions; to enable them to follow the best route. All these are matters where there are great gains to be had from space. The advantages that meteorology can expect are clearly also going to have great economic consequences. Indeed, hurricane warnings from space have already led to considerable benefits. The study of the resources of the Earth, whether of mineral type or underwater, or the state of harvests—all this is a great field for space exploitation. Beyond it lies the exciting field of space manufacture. It is very much on the cards that certain industrial techniques that are extremely difficult and extremely expensive on the Earth will be performed so much more easily in space, that the whole operation may be cheaper. Certain techniques of welding, production of ball bearings and so on are all in this class.

Much further in the future, but just as much real, is long distance transport through space. The supersonic aircraft is not yet in civil use, but we are confidently looking forward to cutting the large travel times on the Earth substantially. I firmly believe that this will lead to a great intensification of industrial and commercial co-operation between far distant countries, like Japan and Europe, Australasia and Europe, South America and North America, and so on. The time taken by travel and its strain now certainly put limitations on the willingness of private industry to co-operate inter-continentally, particularly if business is of the small and medium type. But the supersonic aircraft will not be the end point of travel evolution. To use a transport that takes off from the ground, flies through space at tremendous speed and can reach any other point on the surface of the Earth with a good quiet landing in perhaps two hours, would make a further vast difference to global industrial, commercial and political co-operation. I realize that this is still twenty years or so away, but it will undoubtedly be achieved.

To come now to the third side, industrially. What does industry gain from being occupied in space activities? The technological gain is the gain that comes from every very demanding task, an improvement in capability. To a certain extent, this is straightforwardly technical. A knowledge of how to handle new materials; an ability to design and manufacture the most complicated electronic gadgets; an ability to design for the hostile space environment with its vacuum and its sharp temperature fluctuations; and, equally, for the similarly hostile launching environment with its tremendous vibrations. Beyond this is the task of designing for reliability and long life for unserviced automatic equipment. This should, in due course, lead to a better understanding of reliability engineering to the benefit of us all in our use of equipment every day of our lives. Beyond this are the management techniques perfected in the demanding field of space. To make sure that all the vast number of bits of a space enterprise fit together, work together, are tested in a coherent way and in no way conflict with each other, is an enormous task for management. A task, above all, that must insist that all the people who devise this and devise that, work together sufficiently, and talk to each other often enough. In this sense, space is intensely human because it requires the co-operation of a vast number of human beings. It is this field of applied psychology that can be called management. It has to get over certain human reluctances and difficulties; above all, our shyness and our tendency to be absorbed in a particular task. Who hasn't heard a man say, "How awful, I'm just engaged in my job which is fascinating, interesting and difficult, and here I am called away to a meeting to waste my time!" Well this is not what work in a big enterprise, like space, involves. In a big enterprise, the meeting *is* nine-tenths of the work. To make this bit and that bit is not half as difficult, not one-tenth as difficult, as to ensure that the one who makes the one bit and the one who makes the other *talk* to each other enough about their work, so that these two pieces can work together in the end.

But this long prospectus for these space advantages goes only some way to answer my original question. What is there in space for the small man? For the country not participating on the scale of the United States or the U.S.S.R.? I think the answer is clear now. Space is not just a prestige matter. Space is not just a matter of getting there first. There were many who followed Columbus after his discovery of America. There are vast numbers who split the atom since Rutherford did it first. Many manufacturers make a lot of money from producing motor cars, not only the one whose car wins the Monte Carlo Rally. To exclude one's industry from such an enterprise would seem to be very rash. The really difficult question governments have to debate is to what extent, and at what time, they want to go into these matters. Space is expensive or, to put it differently but meaning the same thing, it requires many people to work on it and a large percentage of them must be highly qualified. Any space installation, even the smallest one, must contain certain essential features so that one cannot make them arbitrally small and cheap. These are public enterprises of real magnitude but not perhaps of the overwhelming magnitude sometimes ascribed to them.

Perhaps a few figures will help. When one speaks of public expenditure, that is expenditure by us all jointly, one comes very soon to millions and billions, and one loses one's understanding of what this means. I would like to have a general rule that public expenditure should always only be expressed on any and every occasion in expenditure per head of population. Then we can come to figures that we can compare with what we, ourselves, spend in our private capacity and it will be much easier for us to see what is reasonable and what is not reasonable. We know rather well the figure for the United States civil space programme, NASA. This comes to approximately 20 dollars per head per year. Per head here of course means every man, woman or child. Four billion dollars a year may sound a staggering sum that leaves us completely unable to comprehend it, but 20 dollars a year is the sort of sum one has an understanding for. In a family of four that would, of course, be 80 dollars a year. It's real money, the sort of money one notices but not a major factor in one's expenditure. Perhaps it is worth saying that the women of America spend annually at the hairdressers a sum slightly larger than NASA spends on space. To put it differently, the number of people who deal with women's hairstyles, whether as hairdressers or producing the materials to be used, or even as owners of the buildings in which the shops are located and they receive rent for; all these together must be more people than work on the space programme. To talk of other public expenditures, the United States' expenditure on defence is 20 times as large as the NASA expenditure. That is to say, 400 dollars per head per year, 1,600 dollars for an average family of four. Now that is a lot of money, that is a substantial proportion of what that family spends on its private concerns. Even sending a man to the moon is thus not a colossal enterprise; it is a significant enterprise, it is more, shall I say, than a flea-bite, but a perfectly bearable cost.

I have been much concerned with what Europe should or could be doing in this field, but perhaps it is worth saying why this should be government expenditure. After all, when the industry of the country makes motor cars, then the cost of this is not met out of taxation, quite the contrary. Why should space be supported by the government, when motor car manufacture is not? The answer is simply that, in space, some of the gains—the scientific ones—are the cultural type that every government supports through its support for science, while others are of the industrial type (and every government is interested to make its industry as modern and competitive as possible). Finally in the straight industrial side, for communications for example, again some initial pump priming is needed because the returns come through many years after the putting in of money, more years than in most industrial enterprises. Therefore this is a longer time than banks and industry are normally willing to finance themselves. Hence the government is involved and where this leads the European countries to I will discuss in the next chapter.

CHAPTER TWO

Europe in Space

When we look at the world today, industrially, then we have in the very front rank the United States and the U.S.S.R. Behind them is what one might call the second division, including in Europe most of the countries of Northern and Western Europe and East Germany, Poland and Czechoslovakia. Outside Europe, there is of course Australia itself, and, perhaps leading the whole second division, Japan. And, overshadowed by her neighbour, there is Canada. There are quite a few countries on the borderline of the second division and I will not be concerned with a very precise definition here. But this second division has a number of characteristics: All the countries concerned are rich in highly qualified manpower both in science and technology. They each have firms working in many of the most sophisticated fields of modern technology. The proportion of their national production that lies in advanced fields is very significant, yet they are all small in numbers of total population compared with the two Super Powers with their 200 million plus population each. Even Japan has only half that and West Germany a third. The others vary from big countries like Britain or France, to quite small ones like Switzerland or the Netherlands. Yet this huge variation in size leads to less internal difference in this group in industrial capabilities than there exists between them and the United States. Even a rather small country like the Netherlands has some of the biggest and most successful companies in the world, like "Philips", "Shell" or "Unilever". Switzerland is a tiny country, yet there are many fields of engineering where it is quite outstanding. Nevertheless, the interest that these countries have shown in space so far is relatively modest. Even in the country with the highest per capita expenditure amongst this group, France, the space expenditure per head is only two dollars per year, roughly one-tenth of the expenditure in the United States. Thus, quite apart from the fact that these countries do not have the population base of the United States, they spend vastly less per head and this is a question of deliberate choice. Of course, it is not poverty. These second division countries are amongst the very richest in the world, and generally have per capita incomes between half and two-thirds of that of the United States. Why are they spending so much less on space? What can they achieve with such, much lower expenditure? These are real questions of public expenditure. That is, they are the very meat of politics. And for a number of reasons, some clear, some not so clear, space has not caught on in these countries as it has in the United States and the Soviet Union.

I think in the first instance there is a lack of self-confidence; a feeling that, because of the smaller size, one couldn't achieve anything outstanding like the two Super Powers, and that therefore it's not worth doing anything unless there is direct, patent and immediate justification. Then why do these countries not join together to achieve a base as large as the United States? Indeed, the ten countries of Europe joined in ESRO, the European Space Research Organisation, have a combined population distinctly larger than that of the United States, and a combined gross national product not very much smaller than that of the United States. But, of course, combining is not very easy. It is essential in our day for the not so huge countries to learn to work together; but international collaboration is a difficult technology in its own right. A technology that has so far been learnt only very partly. A technology in which it will take many years before we reach perfection. Why is this so difficult? It seems, when one first hears about it, that internationally doing something should be quite a simple thing. We all put our money together and then we spend it in the best way possible. But this is not how nation States are organized.

The modern State is a very complicated machinery and the first rule is that the State is sovereign. It does what it thinks best, subject, of course, to such treaties as it may freely have entered. Any such treaty which diminishes its sovereignty is a painful and serious matter. Just think, to mention a particular example, of a financial or economic crisis. The standard cure is for the Minister of Finance to look at the budget and to slash things here, impose a stand-still order there, and order a reduction in a third place. It is very awkward in this kind of operation if he comes across a figure that he is not allowed to change because it is internationally agreed. And this means that the savings that are considered necessary fall that much harder on that which remains. At all times he spends the taxpayers' money and he is responsible to them for what he does with it. He must spend it in the best way and make sure that it is properly used. To spend it in the best way, he has to get advisers for the many different special fields. To make sure that it is properly used, he must have it checked by the Government Audit and order any particular investigations when he sees fit. However, if the money is spent internationally it is, to some extent, outside his control. Conversely, imagine that the Minister of Finance is rich at a particular time, and says: "Let us make an advance in this promising field." What a pity that it cannot be done because the matter has to be internationally settled, and one or two of the other countries concerned happened to be in financial straits at that time and are not willing to go further.

Thus it is natural and reasonable for countries to be very careful about undertaking international commitments, and to endeavour to safeguard themselves against these commitments not being fruitful one way or the other. One safeguard consists of insisting, in as many places as possible, on unanimity amongst the participating countries, so that nothing can be done against you. But, of course, even a unanimous international decision reduces sovereignty. It cannot be taken back, as a national decision can. Another safeguard is not to put all your eggs into the international basket. If you do things internationally, then do a little bit of the same activity nationally too. Of course, this splits the money voted and is bound to lead to a reduction in efficiency. But perhaps one of the greatest difficulties lies in the problem of advice. The subjects concerned are always difficult, and advice is not easy to give. On particular questions, different perfectly reasonable people may come to different conclusions, because it is hard to decide on them. In most such questions, many interests are involved and the advice is naturally left to a committee on which all these interests are represented. The representative of one field may have a very forceful personality, and the representative of another the opposite. It is, therefore, clear that each government will get advice from its own group of advisers different from what another government gets, not because they belong to different nations, not because one is more competent than the other, but simply because they are different people. Different governments may, therefore, get contradictory advice. The governments are bound to take their advisers' view even if it was reached by a decision of 51% to 49%. Governments are paid to take decisions and the government must fully represent this view. Two governments may, therefore, be in conflict—and you have an international problem.

I do not wish to sound too pessimistic. On the contrary, I am certain that the problems are being overcome and will be overcome more fully in future. But it is just a plain fact that there are serious problems and it's no use being impatient with them. What, then, have we achieved in Europe? There have been appreciable national efforts in the four biggest European countries—Germany, U.K., France and Italy. France, the leading country in the field, has built its own launchers and its own satellites, some of them launched by its launchers, some by American launchers. The U.K. is working on a launcher and has had several satellites launched by the Americans; as has Italy and, most recently, Germany.

It is perhaps important at this stage to put launchers and satellites into their relative classes. A launcher is, invariably, an expensive device to develop, but when it is developed then you can produce it in appreciable numbers at a more reasonable cost and use it to put bits into orbit; generally at a cost of a few thousand dollars per pound. Scientific satellites are generally not repetitive, or only in part repetitive. Each has to be developed, tested and built. Enormous reliability has to be ensured because, owing to the high launch costs, it has to work first time. This "making sure" is immensely expensive in manpower and, therefore, immensely expensive. Roughly speaking, a satellite costs about fifty to a hundred times as much as the same weight in gold.

The stimulus for the international effort in Europe came from space science, from European scientists who knew of, or were involved in, the American effort, and who from 1960 made strong efforts to involve their governments in a joint endeavour; to be modelled on the immensely successful high-energy physics joint European enterprise of CERN at Geneva. After various initial steps, the European Space Research Organization came into being in the spring of 1964. We have ten member countries—the four big European nations, Germany, U.K., France and Italy, and six of the smaller ones, Denmark, Sweden, The Netherlands, Belgium, Switzerland and Spain. The joint purpose of these countries getting together was to advance in space research and technology. That is, they wanted to give opportunities both to their scientists to carry out experiments in space, and to their industries to acquire space technology. For this purpose, the organization of ESRO was set up and given considerable technical strength and management powers. We have a large technical centre, ESTEC, at Noordwijk in The Netherlands. Nearly 700 people work there now. Their main task is the leadership of industry in our projects. In a big satellite project, we act as the design authority, but the work is carried out in industry under a prime contractor whom we select. Moreover, to ensure an equitable distribution of work between the member countries, our efforts have led to the formation of Europe-wide consortia of great firms co-operating on our major projects. In some ways, it is one of ESRO's chief achievements to have brought European industry together in this remarkable way in which industry has really learned to co-operate in great detail and with great success.

ESRO's activities are not confined to space research by satellite. There is also space research by sounding rockets, which carry highly complex payloads above the atmosphere for a few minutes and either radio their results back or the scientific payload descends by parachute and is recovered. ESTEC constructs some of these payloads itself but most of them are made by industry under ESTEC leadership. As far as both, satellites and sounding rockets,

Figure 2-1. General view of ESRO's large technical centre, ESTEC at Noordwijk in The Netherlands.

Figure 2-2. ESRO Sounding Rocket Range (ESRANGE) in Kirura, Sweden. Main building and antenna arrays.

Figure 2-3. Preparation of a Skylark payload at ESTEC in Holland.

are concerned, ESTEC is in charge of testing them and of laying down the extremely complex qualification procedures that alone can ensure that all the highly complex equipment survives the extreme vibrations during launch and functions perfectly in the hostile environment of space. A third task of ESTEC is to prepare Europe for the more complex projects of the future, by stimulating technological research, largely through placing, monitoring and devising suitable technological research contracts with industry,

Figure 2-4. REDU Station in the Belgian Ardennes, 40 km south-east of Dinant. On the left, main building of the station. The smaller buildings farther away house the array of the interferometer antennae.

partly by doing work itself. A fourth task of ESTEC is to keep in constant touch with the scientists who build the experiments for our spacecraft, and to make sure that during the development, during the qualifications and the launch and thereafter, all the scientific demands are met as fully as possible, and to eliminate any possible conflict between the demands of different experiments. Moreover, ESTEC is much concerned, under the leadership of the Directorate of Planning and Programmes at my headquarters in Paris, to advise on the feasibility of projects, their likely problems and cost.

Next, it is a huge task to get down the information from a satellite in orbit. This is so particularly if the satellite is in a low orbit because it then passes over any point on the ground at irregular intervals for a very short time. To get down, on command, the information is a major undertaking. So we also have our own world-wide network of telemetry stations in the

Falkland Islands, Spitzbergen, in Belgium and Alaska. We also co-operate with other similar networks—the huge network of NASA and the network of the French organization CNES. This work is all directed and controlled from our operations centre at Darmstadt in Germany to which all the communication links run, and from where our satellites are controlled and where the data is evaluated in big computers. It can thus be seen how ESRO spreads throughout the world, and I have not yet mentioned our sounding-rocket range, at the northernmost point of Sweden, for the investigation of the Aurora, nor the fact that we launch rockets from other places, notably an Italian range in Sardinia and, when our astronomers want to look at southern skies, from Woomera. Another support for our space work is a Plasma Research Institute in Italy.

What emerges from all these labours? We have now launched more than 100 sounding rockets, and four satellites have been launched for us by NASA, three of them in 1968. They are modest in size, weighing around 200 lbs. to 250 lbs. each, but highly complex for their size. For example, the number of commands that can be given to the spacecraft to switch on or off different equipment to put it into a different mode to get the apparatus on the spacecraft to turn it to face in a different direction, varies between 30 and 50 on the three spacecraft. Two of them are in low orbit—one has as its chief task the investigation of the Aurora, these curious lights on the Northern and Southern Hemispheres produced at the tail-end of the radiation belts and arising through a highly complex interaction between trapped particles, that is, particles trapped by the Earth's magnetic field in the outermost atmosphere. Another one is essentially devoted to the study of the relationship between the Sun and the Earth, in particular to the radiation arising from the Sun during solar outbursts. A third one (HEOS-1) is a real space probe that goes deep into space to a distance more than half-way to the Moon to investigate the inter-planetary medium and, in particular, the border between where the magnetic field of the Earth is in control and the area beyond it which is governed by the solar wind. Many of these disturbances are very insufficiently understood so far and indeed the whole structure of the medium out there is not well known. Sometimes, one can use this medium as a sort of laboratory to study how ionized particles behave in the area which is more tenuous than the best vacuum can produce on Earth. In spring 1969, this spacecraft released some 45,000 miles above the Earth a little package that, after it had separated from the spacecraft sufficiently to ensure the spacecraft's safety, exploded and spread particles that were rapidly ionized by the Sun to make a cloud whose movement charted the movement of the interplanetary medium. This radiant cloud was visible from the Earth, from observing stations spread throughout North and South America, for nearly half an hour and was quite spectacular.

These three spacecraft function extremely well. They have been in orbit now for around two years and almost all the equipment has survived far longer than it was planned to. The Auroral investigation satellite, which was deliberately put into a low orbit for this investigation, is now nearing the end of its life by re-entering the atmosphere, and burning up. We had a little less good luck with a repeat of the Auroral satellite which was launched late in 1969, again deliberately into a very low orbit that would give it a short life. Accidentally, as it turned out, the orbit was even lower than intended and the life even shorter than expected—a mere two months. Moreover, one of the most awkward bits of equipment on it didn't work for very long, namely, the tape-recorder which stores information. Thus, with rather infrequent interrogation of the satellite, almost all information should still have been brought down. Though after this failure only real time data were obtained, the satellite produced in its short life much of considerable scientific value. This satellite brought us up to the real complexities of space work.

We have five further scientific satellites in various stages of development. One is another inter-planetary probe which this time will visit the particularly interesting and totally unexplored region where the magnetic fields of inter-planetary space and the Earth exactly neutralize. This is likely to lead to hitherto not-at-all understood electrical phenomena which our spacecraft, due to be launched at the end of 1971, will be the first to explore. A big satellite of more than 1,000 lbs. is in preparation for launch into a low orbit in 1972 to investigate ultra-violet and particle radiation which cannot be observed on the surface of the Earth. In particular, a sky survey will be made—that again will be completely novel. Not long afterwards, a smaller satellite will be launched that, in particular, will investigate the constitution of the uppermost atmosphere. Then, in 1974 (complex satellites take long to prepare, particularly the experimental part) we will put up a satellite that will do astronomy in the cosmic ray area. It will be a satellite totally devoted to gamma ray astronomy and will study the highly penetrating radiation in its direction of arrival and energy distribution, and should give results of the utmost interest. We are in the early stages of preparation for a scientific satellite at geo-stationary height, again investigating the magnetic area surrounding the Earth and, in particular, linking observations there to Auroræ observations lower down.

In recent years, the greatest interest centred on extending Europe in space from pure science to applications. We are in the active stages of achieving agreement on two different kinds of application —telecommunication within Europe, both for telephony and television by satellite, which is likely to be a long programme but one of the utmost importance—and we are in contact with the Americans on the possibility of controlling the air traffic over the Atlantic from satellites to ensure constant location of the aircraft on this very busy route, and instant communication with them to avoid any and all possibility of collision. A little later, we hope to go into problems of meterology. So you see that our ambitions are large. Our achievements are already quite real and we are well on the way to giving Europe a presence in space—not a gigantic presence because the means voted to us are not of that magnitude, but something that is useful, interesting, inspiring and helps us all forward.

CHAPTER THREE

Newtonian Gravitational Theory

Newton's theory of gravitation is one of the most successful scientific theories known. Not only was it the first one and led the way to man's whole concept of what a scientific theory should be, but after 300 years it is still the standard method of calculating orbits in the solar system, with only quite minute corrections to be applied in a very few cases owing to the change in the theoretical picture that has occurred in this century.

Newton's theory rests on three basic assumptions (in addition to the three laws of dynamics that are used to describe the behaviour of matter under *any* force).

(i) The principle of Galileo that all bodies fall equally fast.
(ii) The notion that gravitational forces add linearly.
(iii) The inverse square law.

Each of these assumptions merits some discussion. According to Newton's second law of dynamics, when a force acts on a body it accelerates. Applying the same force (e.g., that of a stretched spring) to different bodies, the *direction* of acceleration is the same for all, but the amount is different. Moreover, if the accelerations of two bodies in response to one and the same force are in the ratio R, then their accelerations in response to any other force applied equally to both bodies will also be in the ratio R. Thus a property can be assigned to bodies called their *inertia* and measured by their *inertial mass* which characterizes their resistance to being accelerated, so that force equals inertial mass times acceleration. This quantity is a *scalar,* i.e., it in no way distinguishes between directions in space, since the direction of acceleration in response to any one force is the same for all bodies.

So far great stress has been laid on the *same* force being applied to different bodies resulting in different accelerations. Since all bodies, however, fall equally fast, clearly the gravitational force acting on different bodies must be different, being strictly proportional to their inertial masses. This mass will then cancel on the two sides of the equation "force equals mass times acceleration" leading to Galileo's principle. How can bodies react so differently (i.e., by experiencing a different force) to a field of gravitation? Consider a magnetic field. A piece of iron will be strongly affected, a piece of glass hardly at all. There is something in the piece of iron to which the magnetic field can hook on which is absent in the piece of glass. If we call what the gravitational field can hook on to in a body its *passive gravitational mass*, then Galileo's principle states that inertial and passive gravitational masses are strictly proportional to each other. Note that they can be measured quite independently. Measuring the acceleration in response to a stretched spring tells one the inertial mass of a body, while weighing a body with a spring balance measures its passive gravitational mass. The universal constancy of this ratio was established with great accuracy by Eotvös some 60 years ago and with even greater accuracy (one part in 10^{11} or so) by Dicke in recent years.

By Newton's third law ("action equals reaction") a body attracted by the Earth must also attract the Earth. Not only do bodies have their passive gravitational masses responding to gravitation, they must also have *active gravitational masses* generating gravitation. By Newton's third law the active and passive gravitational masses must be equal. In modern physics one insists that quantities are always defined in terms of the way they are measured. Active gravitational mass is measured by the gravitation producing effect of a body, e.g., the orbit of the Earth or any other planet tells us what the active gravitational mass of the Sun is.

Three quite different types of measurement thus yield the three masses, inertial, passive and active gravitational. Our current theories, based on experiments, tell us that they are all equal (if suitable units are used), but it is as well to appreciate that they are logically distinct in case yet more accurate experiments or novel theoretical considerations suggest that the equality is not perfect.

The linearity of gravitation seems plausible, but linearity is not in fact very common. What we mean by it is, e.g., that the combined attraction of Sun and Earth on the Moon is equal to the sum of that of the Sun, with the Earth absent, and that of the Earth, with the Sun absent. In many areas this just is not the case. If we put a dam across a river, the forces (and the flow pattern of the water!) are not the sum of those obtaining if the half of the dam nearer one side of the valley were present, with the other half absent, plus those obtaining with the other half present and the first half absent. Indeed non-linearity is extremely common, but the student sometimes gets the opposite idea since the simplicity of linear problems leads to the few occurring in nature being selected for teaching purposes. It was of course only the assumption of linearity that enabled Newton to calculate the attraction of a spherical Earth on the Moon by adding the contribution of all the particles of the Earth.

As regards the inverse law, it is well known how it leads to the definition of a gravitational potential V, whose derivatives (the gradient) describe the acceleration of particles in the field. This field is source-free in empty space, and it is only matter which is its origin. Mathematically this is expressed by stating that a certain linear combination of the second derivatives of V (the Laplacion) is proportional to the density of matter (Poisson's equation).

Having stated the main principles of Newton's gravitational theory it may be helpful to criticize a common interpretation of its applications. It is said (as stated above) that its aim is to some extent the calculation of the *acceleration* of particles in the field and thus the evaluation of the gradient of the potential V. But how is this acceleration measured? One's natural reaction is to say "relative to the solid Earth". But not only do the Earth's rotation and revolution introduce serious corrections, but of course a universal theory like gravitation must be applicable far from the Earth. How can we measure acceleration in empty space, when any comparison body accelerates equally fast? Indeed it is now common knowledge that in free fall, in an orbiting spaceship, gravity disappears and a condition of weightlessness exists. Only the circumstance that we live on an Earth solid and firm enough to resist its own gravitational pull produces the feeling of weight here. But then is gravity only observable in those very special circumstances (rather unusual in the universe) in which we live? Would a civilization living entirely in orbiting spaceships be unaware of gravity? Clearly not, for they would see the relative motions of the planets, satellites and themselves, vastly differing from the relative motions of bodies not under the influence of any

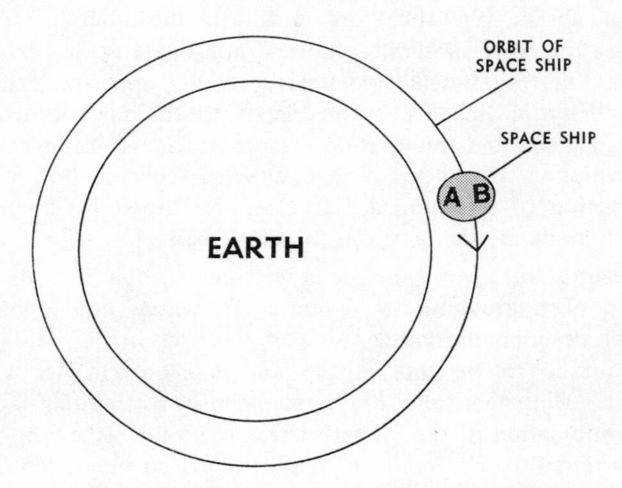

Figure 3-1. Tidal forces on spaceship in orbit.

force since such bodies, by Newton's first law, move in straight lines with constant velocity. But then, this would mean that gravity was only observable *globally* and not *locally*, a possibility conceivable but not agreeable. However, if we think about the spaceship rather more carefully, the local observability of gravitation returns. For a spaceship is not a point but a body of finite size. At any one moment there will be a side, *A*, nearest the Earth and a side, *B*, furthest from the Earth. (*Figure 3.1.*) The acceleration of free fall will be greater at *A* than at *B*, while the centripetal acceleration of the orbit of the whole ship will be a compromise between the two. Thus the spaceship is falling too slowly, as seen by a particle at A, and too fast, as seen by a particle at B. Hence there are forces trying to elongate the spaceship along the A-B axis, and any dust will tend to accumulate at A and at B. A sufficiently accurate instrument will measure these forces and so allow the astronauts to discern the existence of gravitation *locally*. What they are measuring is the difference in the gravitational acceleration at neighbouring points or, to put it mathematically, the local variations in the first derivatives of V, and thus they find the second derivatives of V. There are six such second derivatives which therefore constitute the *intrinsic* gravitational field. Note that it is a linear combination of three of them that, through Poisson's equation, is equal to the local density of matter. In a perfectly uniform gravitational field these intrinsic quantities vanish but such fields are rather special and very rare in nature.

The forces trying to elongate the spaceship are made use of in some artificial satellites to orient the space craft permanently towards the Earth through *gravity gradient stabilization*. The effect is enhanced by fitting a long boom to the spacecraft. The forces trying to orient this boom along the spacecraft-centre of the Earth line lead to pendulum-like oscillations which are gradually reduced to zero by dissipating their energy in the sloshing of a viscous liquid carried in a closed container on board the spacecraft.

The forces here described are very familiar on the Earth, which is after all in a free falling orbit about the Sun and about the Moon. (N.B.: In each case the motion is about the common centre of mass of the two bodies concerned, which in the Earth-Sun system is well inside the Sun, in the Earth-Moon system just inside the Earth.) Thus there will be forces trying to elongate the Earth along the Earth-Moon line and along the Earth-Sun line. The oceans will flow accordingly, generating the *tides*. It is clear from this argument that there must be *two* tidal bulges of the lunar tide (on the sides of the Earth pointing towards and away from the Moon) and similarly two solar tides (on the sides of the Earth pointing towards and away from the Sun). The Sun is vastly more massive than the Moon but also nearly 400 times as far away. Since the tide raising forces are due to the local variation in force, those due to a single body go with the derivative of the inverse square law and are thus proportional to the inverse cube of the distance. The upshot is that the solar tide-producing forces are a little less than half of those due to the Moon. Thus the lunar tides predominate, but are powerfully reinforced when the directions of the Sun and Moon coincide (spring tides at full Moon and new Moon) and greatly weakened when they are at right angles (neap tides at first and last quarter).

The Earth performs its daily rotation in the presence of these forces and turns, as it were, under the tidal bulges of the oceans. The highly complex shapes of the shoreline lend to an enhancement of the tides in some regions (British Isles, Nova Scotia, etc.) and to very modest tides generally in low latitudes. Although the tides of the ocean are the most spectacular, it is clear that both the atmosphere and the solid Earth are subject to the same forces. The atmospheric tides lead to just measurable barometric variations, and the deformation of the solid Earth is also discernible with very sensitive modern equipment.

The rotation of the Earth under the tidal bulges of air, water and solid Earth means that any frictional forces in the motion of these substances will tend to slow down the rotation of the Earth. The angular momentum of the Earth-Moon system must of course be maintained through an acceleration of the Moon driving it further from the Earth. The mechanism by which this occurs is readily understood. The friction pulls the tidal bulges forward in the sense of the Earth's rotation, and the gravitational pull of these bulges on the Moon results in its acceleration. The friction of the atmospheric tides has negligible effect, the friction of the oceans, largely occurring in the shallow seas, is of only minor importance, but the friction in the solid Earth results in a measurable slowing down of the Earth's rotation. This is friction against the process of repeated compression and extension which dissipates mechanical energy as heat. The process is familiar from motor car tyres where the contraction-expansion cycle of the rubber in fast running leads to a very marked heating. It is possible to infer from the measured slowing down of the rotation and from some other arguments that in its early youth, four billion years ago or so, the Earth rotated perhaps three times as fast as now, so that the "day" lasted only eight hours.

Gravitational fields in empty space have a mathematically rather beautiful and simple character, since the potential V satisfies Laplace's equation. In particular it turns out that outside an isolated body V must necessarily consist of a sum of terms, each term varying with distance from an origin within the body like a negative integral power of r, r^{-n-1} as it is normally written. The first term, $n = 0$, gives a potential varying inversely with r and therefore a field of force (its derivative) that is the familiar inverse square law. This part of the field is the same in all directions, and its magnitude is simply proportional to the mass of the body. The next term, $n = 1$, must vary with direction so that it changes smoothly from maximum in one direction to an equal and opposite magnitude in the opposite direction. Its coefficient is called the dipole moment which thus has direction (that of the maximum field) as well as magnitude. It is most familiar from electricity where it is due to a dipole, neighbouring equal and opposite charges. In gravitation where we have no negative masses, it is most easily imagined as due to a displacement. Instead of having a mass at point P, we have an equal mass at point Q and may regard this as due to the superposition of a mass at P and of a dipole having positive mass at Q and a negative mass at P just cancelling the mass there. Indeed in a gravitational system the dipole moment divided by the mass is simply the displacement of the centre of mass from the origin from which r is reckoned. By making this origin coincide with the centre of mass, the dipole term can always be reduced to zero.

The next term, $n = 2$, is however of intrinsic importance. Its variation with angle is more complex and is best discussed in the axially symmetric case. There is an "equatorial" plane of symmetry. The term reaches a maximum at right angles to this plane (equal in both directions) and has a minimum of opposite sign and half the magnitude of the maximum on this plane. This so-called quadrupole term may be imagined as being due to a spheroidal structure of the source. If the source is a prolate spheroid rather than a sphere, the mass term $n = 0$ will be accompanied by a positive quadrupole term leading to an enhanced field along the axis of the quadrupole and a diminished field in its equatorial plane. For an oblate spheroid the opposite will be true.

The higher terms, $n = 3, 4, 5 \ldots$ need not concern us in detail. To describe the structure of the source with more precision. Note that the larger n, the faster the effect diminishes with distance. Very far away any source has virtually the same field as a sphere of the same mass. As one comes nearer one discerns first the location of the centre of mass through the dipole, then any spheroidal distortion through the quadrupole moment, etc.

The identification of the coefficients of the field terms with quantities relating to the local structure of the source has a particular significance for $n = 0$ and $n = 1$. The coefficient of the $n = 0$ term is the active gravitational mass, which, through its equality with inertial mass, is subject to the same conservation law. Thus it cannot change. The dipole term describes the location of the centre of mass, the ratio of the dipole term to the mass term being the distance of the centre of mass from the origin of co-ordinates. By the law of conservation of momentum, the centre of mass of an isolated body can have no acceleration. Thus the dipole term can depend, at most, linearly on the time.

Note that the laws of conservation of mass and momentum have their origin not in Newton's gravitational theory, but in his dynamics. Nevertheless, when applied to his gravitational theory, they give the result that the time variation of the first two terms in the expansion of the field of an isolated source is severely restricted. By contrast all the remaining terms are completely free in their time-dependence. A body can change its shape in any way it pleases provided its mass and the velocity of its centre of mass remain fixed. For example, a spherical body may change its shape to that of an oblate spheroid. This will increase its attraction in its equatorial plane and diminish it along its axis of symmetry, both changes in force varying like r^{-4}. At large distance where the mass term ($n = 0$) predominates (since its acceleration varies like r^2) there is only a negligible change in attraction.

There is thus a certain reciprocity. The only intrinsic part of the gravitational field is the tidal force which tends to change a sphere into a prolate spheroid and more complex deviations from a sphere. Conversely, the only freely variable parts of the field are those due to spheroidal and more complex changes of shape. The interaction of these effects and their relation to energy may be studied by means of an example.

Consider two celestial bodies far from any others and inhabited by engineers of such a level of competence that they can change the shapes of their bodies at will but always maintaining an equatorial plane of symmetry and sticking to spheroidal shapes with axes of symmetry at right angles to the orbital plane. The energy required for a change comes on each body from an internal electrical storage battery which in turn can be charged if the change yields energy. Of course these changes of shape affect the mutual attraction of the bodies and therefore their orbits. (We assume that the orbital plane is the equatorial plane of both bodies.) To avoid this nuisance, the engineers of body *A* ("Tweedledum") persuade the engineers of body *B* ("Tweedledee")

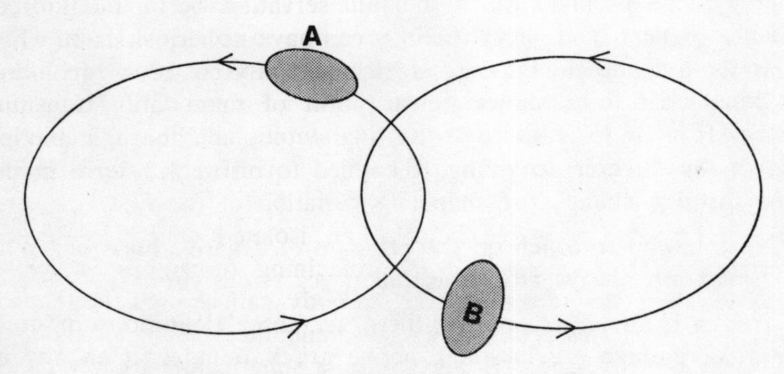

Figure 3-2. Tweedledum and Tweedledee.

always to counterbalance this effect. Thus whenever *A* becomes oblate, thereby increasing attraction in its equatorial plane, *B* becomes prolate and diminishes it so that all the time the attraction is the same as if both bodies were perfectly spherical. Their mutual orbits are therefore Keplerion, and we shall assume them to be ellipses of high eccentricity. (*Figure 3-2.*) Thus there are marked and periodic changes in the distance between the bodies, leading, through the r^{-4} law, to very large periodic fluctuations in the tidal forces they exert on each other. The tidal force would like to transform the bodies into prolate spheroids with the axes of symmetry pointing at each other and thus in the orbital and hence equatorial plane. This is not allowed by the symmetry. The best thing that the tidal forces are allowed to do is to make the bodies oblate spheroids which will extend them towards each other and diminish their polar axes (as the tidal forces try to do) but will also extend them in the equatorial plane at right angles to the line joining them, instead of contracting them there as the tidal forces try. But a gain on two out of three directions is still a gain, as one can show by detailed calculation. Now the cunning engineers of *A* allow their body to go oblate when *B* is near, gaining a lot of energy for their battery in the process since the strong tidal forces are pulling in this way. When *B* is far, they go spherical. Since the tidal forces are then weak, little energy is used up. The poor engineers on *B*, however, have, by treaty, to use a lot of energy to go prolate against the pull of the powerful tidal forces when *A* is near, and gain little in reverting to spherical shape when *A* is far and the tidal forces are weak. Thus in each revolution *A* gains and *B* loses, showing clearly how pure gravitation transmits energy from *B* to *A* through the tidal forces and through making use of the freedom to change the third (quadrupole) term in the field through changes of shape.

Next I want to touch on the problem of negative mass or: Why is gravitation always attractive and never repulsive?

We have already seen that there are three definitions of mass (inertial, passive gravitational, active gravitational). Can any or all of them be of the opposite sign to the usual one?

Negative inertial mass would be curious indeed. Such a body would run away if pulled, and approach if pushed. How serious this would be is readily seen by the example of electric charge. Normally, since like charges repel and unlike attract, accumulations of one charge disperse and are neutralised by attracting the opposite charge. Thus energy is needed to charge up bodies, and energy can be gained by allowing them to discharge. With charge carriers of negative inertial mass the opposite would be the case. Huge charges would build up and would be constantly increased by being joined by like charges carried by negative inertial masses, since the repulsion between the charges would make them approach. All such evidence as we have is against such spontaneous electric charging processes.

If then we accept that inertial mass is always positive, the same must hold for passive gravitational mass unless Galileo's principle is wrong, and some bodies fall up. The Eotvös-Dicke experiments suggests that there is not the slightest admixture of such anomalous material in terrestrial matter, but of course it may exist elsewhere in the universe. Such strict separation could only be considered likely if there were a separation mechanism at work. Suppose then that Newton's third law, equating action and reaction, was valid even for such matter. Then active gravitational mass would be negative too. The result as regards the interaction of positive and negative masses would be exactly the opposite of electrical interactions; like masses would attract each other, unlike masses would repel. Thus separation would naturally occur. If a large region of the universe contained at one time an equal mixture of the two types of mass, any accidental excess of one type would attract more of its own kind and repel the opposite kind which would tend to agglomerate in other regions. Eventually, full separation would be achieved.

In spite of its agreeable features, this theory is not very credible, due to a property of gravitation not so far mentioned, its weakness. It is true that a strong force holds us to the Earth, but then the Earth is a very large body. What force should be compared with gravitation? It is perhaps best to go to elementary particles. An electron and a proton attract each other electrically, with an inverse square-law dependence, and, knowing their masses, we can calculate their gravitational attraction. Since this depends on distance in the same manner, the ratio is independent of their separation, and turns out to be close to the enormous number 10^{40}. This means that gravitation is very, very much weaker than the forces that hold an atom together. The forces between atoms that are responsible for chemical bonds and the solidity of matter also originate in the electrical character of electrons. Though a little weaker than the forces holding atoms together, they are still enormous compared with the gravitational attraction between atoms. Thus if in the day of a mixed universe a positive mass hydrogen atom joined a negative mass hydrogen atom to form a hydrogen molecule, the feeble gravitational repulsion would be totally unable to break the bond. It would be even more true that an atom formed of a proton of one kind of matter and an electron of the opposite type could not possibly be torn apart by the gravitational repulsion. Thus the separation mechanism would not be perfect. But we know from the Eotvös-Dicke experiment that separation is very complete indeed. This strongly suggests that negative mass of the kind postulated does not in fact exist.

One can imagine other kinds of negative mass according to the signs given to each of the three kinds of mass, but no very plausible matter seems to emerge. Hence it seems likely that gravitation is indeed universal and that Galileo's principle that all bodies fall equally fast has no exception.

The fact that gravitation, in spite of its extreme weakness, is observable at all, is due to its additivity, that is, it is always of the same sign. A proton on the surface of the Earth is attracted gravitationally by *all* the particles in the Earth, but the enormously greater electrical attraction of the electrons of the Earth is so perfectly balanced by the electrical repulsion of the protons that no electrical effect remains. Thus the very fact that all gravitation has the same sign overcomes its extreme weakness.

CHAPTER FOUR

General Relativity

For some 50 years now, another theory of gravitation has superseded Newton's theory, namely Einstein's theory, known by the rather misleading name of General Relativity. It is a remarkable fact that this theory was created for reasons of intellectual discontent with Newton's theory rather than because of experimental evidence, and today is regarded as our best theory of gravitation, mainly on intellectual grounds, the experimental evidence for it, while quite satisfactory in so far as it goes, being somewhat thin.

There are two main facets of this discontent with Newton's theory, one that it says nothing (or at least nothing tenable) about high velocities and in particular about light, and secondly that it takes Galileo's principle as an extraneous fact and thus gives to such a fundamental law the status of an accidental coincidence. (A third difficulty of Newton's theory will be discussed later.)

It might be advantageous to put the matter of velocity in a historical context. Modern physics started with the work of Galileo and of Newton on dynamics. Their most important break with their predecessors was to regard *acceleration*, and not velocity, as the basic quantity requiring explanation. In essence they said (as for example in Newton's first law of dynamics) that it is silly to ask for the force responsible if a body is moving with *any* constant velocity. Only change in velocity, that is acceleration, is related to force. Thus all observers moving themselves with constant velocity vector will be dynamically equivalent. They are called *inertial observers* as they find the law of inertia (Newton's first law) to be correct when the motion of others is referred to themselves. The velocity of an inertial observer, however large, is totally irrelevant for its equivalence in dynamical matters with any other inertial observer.

This picture led to no difficulties until, 200 years after Newton, the theory of the propagation of light and electromagnetism was founded by Maxwell. Now a basic velocity, the velocity of light entered. Could differently moving inertial observers, equivalent dynamically, also be equivalent optically? For while the different constant velocity vectors of different inertial observers did not matter for dynamics where only acceleration is relevant, it was thought that it must matter for light, whose velocity is itself of paramount importance. This notion of the possible optical non-equivalence of inertial observers led to conflict with experiment and to serious theoretical worries brilliantly resolved by Einstein with his Special Theory of Relativity, which extended the equivalence of Newton's inertial observers from dynamics to all of physics. The problem of the supposed non-equivalence was resolved by pointing out that each inertial observer used his own clock for measuring purposes. Time was shown to be not a universal quantity, but since each observer measured it with his own clock, each observer had his own time. This allowed them all to observe the same velocity of light, thus making all inertial observers truly equivalent.

The discrepancies between calculations based on Newton's universal time and those based on the private times of Einstein's observers are negligible at velocities small compared with the speed of light, but become very significant as this speed is approached. As dynamics is normally applied at low speeds (even a jet airliner travels at less than one-millionth of the speed of light), these discrepancies were not noticed for over 200 years. Nowadays when we accelerate elementary particles to enormous speed in big accelerators and investigate cosmic radiation containing very fast particles, the difference can be very large indeed. (It is even appreciable in the paths of the electrons in a television tube.) In countless experiments there has been excellent agreement with the calculations based on special relativity.

One of the most important results of special relativity is that nothing can be accelerated past the speed of light. Whatever we do to speed up a particle originally slower than light, its speed will never attain, let alone exceed, the speed of light. This is indeed clear from the fact that at any instant an inertial observer moving momentarily at the speed of the particle will see light travelling with the speed of light. While relativity does not wholly exclude the possibility of the existence of entities travelling faster than light, they would have such strange properties that we must count it as fortunate that none have ever been discovered. It thus follows that information cannot travel faster than light.

Another important consequence of relativity is that energy has mass. This was shown by Einstein in a beautifully simple ideal experiment for the case of radiant energy. First it should be pointed out that light exerts a pressure. This is clear from Maxwell's theory. Consider any electro-magnetic wave (such as light or ultraviolet or X rays or infra red or, easiest to think about for the present example, radio waves). If it hits a sheet of metal, i.e., a reflecting surface, the electric and magnetic fields of the incident wave generate electric currents in the sheet which themselves generate fresh electro-magnetic waves. In the forward direction this secondary wave just cancels the continuation of the incident wave creating a *shadow*, while in the rearward direction this secondary wave is the reflected wave. (An absorbing, i.e., black surface, similarly has electric currents in it which, however, only generate a shadow but no reflected wave.) In either case a force, the radiation pressure, is bound to arise between these currents in the surface and the electric and magnetic fields of the waves just outside it. Although the radiation pressure is small for ordinary intensities of light, it can easily be measured in the laboratory. Where light intensity is large, as near the Sun, small dust particles are blown away fast by radiation pressure.

To come now to Einstein's ideal experiment, imagine a long hollow box, black on its inside, lying at rest on a smooth horizontal table. Near one end is a flashbulb, radiating equally forward and backward, and activated by a battery through a switch. When the switch is closed, a flash of light is emitted by the bulb. The radiation hitting the near end is absorbed there; the radiation pressure gives the box a push and sets it in motion, while a similar amount of radiation takes time to travel to the far end of the box. When it hits this end and is absorbed, the pressure of the radiation gives the box an equal and opposite push, bringing it to rest again. To an external observer, the box, originally at rest, moved for a short time (the travel time of the light in the box) and then came to rest again, without any external force acting on it. By the law of conservation of momentum the whole system can never have had any momentum and thus its centre of mass must have stayed put. But as the box has shifted, there must have been an internal shift of (inertial) mass from the battery end to the far end. Indeed after the flash the battery end has less energy, since of the energy that left the battery, only part has been absorbed (and so turned to heat) at the near end. By contrast, the far end has gained energy by the absorption. So this shift of energy has implied a shift of mass, proving that energy has mass, whether it is in the form of chemical energy in the battery or heat energy in the ends of the box, or indeed in any other form. Indeed, in any normal material, the energy locked up in the nuclei is a significant fraction (several tenths of one per cent, differing substantially between different materials) of the total mass. Thus the Eotvös-Diche experiment would readily have discovered if the relation between the inertial and passive gravitational mass were different for the mass of energy than for any other mass. Thus energy responds to gravitation and, by the law of action and reaction, must generate gravitation (active gravitational mass).

Some of the defects of Newtonian theory are now evident. It says nothing about any interaction of gravitation and light (which after all is a form of energy). It does not allow for the relativistic behaviour of fast particles, etc. But highly unsatisfactory though this is intellectually, measurement is difficult. Gravitational accelerators are so small that they significantly change appreciable velocities only after acting for some time. If, as in a great proton-synchroton, protons chase round a circuit a couple of hundred metres or more in length at near the speed of light, even a hundred circuits take them less than 0.1 milleseconds during which time an internally horizontally moving body only falls barely 5×10^{-6} cm, far less than its deflection by stray electric or magnetic fields. With lesser lengths, the effect of gravitation on fast particles is even less significant. However, there is one effect of gravitation on radiation where measurement has become possible. This is the famous

Einstein gravitational shift of spectral lines, first theoretically demonstrated by Einstein through one of his great ideal experiments about 1910, and first measured in 1960 by Pound and Rebka.

In the ideal experiment one imagines a high tower. (*Figure 4-1.*) An endless chain of buckets runs over pulleys at the top and at the bottom. Each bucket is filled with the same number of atoms of the same element, but all the buckets on the left are filled with atoms in the ground state (state of lowest energy), all the buckets on the right with atoms that have more energy (they are said to be in an excited state). Atoms can be put from the ground state to an excited state by letting them absorb light of a particular frequency, and they emit light of the same frequency when they change from the excited state to the ground state. Thus the excited

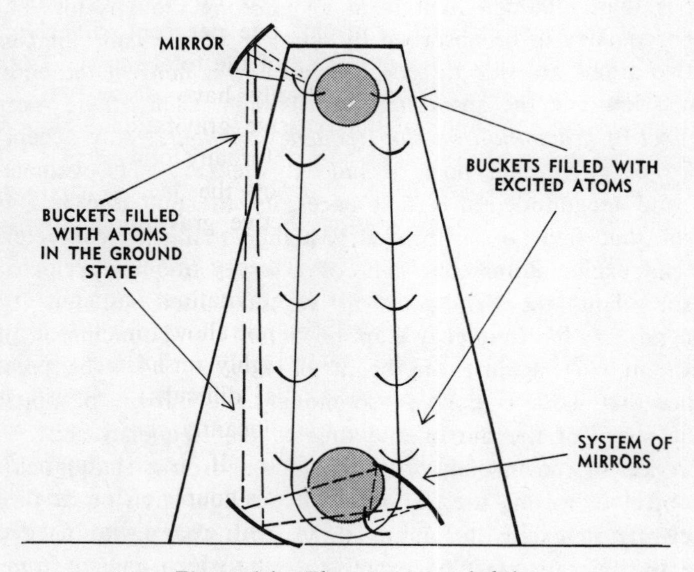

Figure 4-1. The gravitational shift.

atoms contain stored energy and, as the energy has mass, they have more mass than atoms in the ground state. Accordingly the buckets on the right are heavier than the buckets on the left, and so the chain will begin to move, the right side descending and left side ascending. We now arrange it so that when a bucket from the right side reaches the bottom, the atoms in it revert to the ground state, emitting light of the characteristic frequency. We have a system of mirrors to catch this light and concentrate it on the buckets just reaching the top of the tower. These contain atoms in the ground state which, on absorbing light of the characteristic frequency, go into the excited state. Thus we ensure that the buckets on the right always contain excited atoms and the buckets on the left always atoms in the ground state. Thus the chain will keep moving and, if a generator is attached to one of the pulleys, we will have a power station consuming nothing. This is a real *perputuum mobile*, a way of getting something for nothing. This is not possible (law of conservation of energy). Thus there must be a mistake in our argument. The only possible fault is that, although light from an emitting atom is just of the right frequency to be absorbed by an atom of the same kind when the two atoms are side by side, this no longer holds if the emitting atom is low and the absorbing atom is high. Thus there must be *an effect of gravitation on the frequency of light*. Any acceptable theory of gravitation must include this effect. To evaluate its sign and magnitude, it is first necessary to state that it is well known that light of a frequency higher than the characteristic one can excite atoms, but light of a lower frequency can never do so. Thus the light must arrive red-shifted (of too low a frequency). Its frequency can be raised by bouncing it off a mirror moving against the beam of light. The light pressure implies that work is done in so moving the mirror, proportional to the speed of the mirror and thus to the frequency shift. If a succession of moving mirrors (e.g., mounted on a shaft) achieves the object of raising the frequency of the light arriving at the top sufficiently to excite the atoms there, then the power needed to drive the mirrors must be exactly equal to that gained from the chain since otherwise energy could not be balanced. The shift of frequency can be calculated from this equation. It is always small; even for a tower 90 m. high, the shift in frequency is only one part in 10^{14}, but this has been measured. There are much larger shifts on astronomical bodies (two parts in a million for the Sun, appreciable for White Dwarf stars), but other effects make the measurement of these very difficult.

To sum up this section: An acceptable gravitational theory must be *relativistic* (i.e., compatible with Special Relativity). This will automatically enable it to comprehend the interaction of gravitation with fast particles and light, including the Einstein gravitational shift of frequency.

To come to the question of Galileo's principle that all bodies fall equally fast, this is not an organic part of Newton's theory of gravitation, but only an incidental rule, viz., inertial mass equals passive gravitational mass. Moreover, it is logically virtually incompatible with Newton's law of inertia. For suppose you were asked to test this law that a body on which no forces act moves in a straight line with constant velocity. You would carefully select a non-magnetic body so that it would be unaffected by any magnetic field, make sure that it was electrically uncharged to avoid the effect of electric fields, make certain that there were no ropes pulling it, that it was dense enough and of such a shape that accidental air draughts would not affect it, etc. But you simply could not exclude the gravitational force on it, since no choice of material can avoid it. Thus Newton's law of inertia, the foundation of his dynamics, is untestable. This is profoundly unsatisfactory. Nor is it any relief to advise this experiment to be carried out far away from any gravitating body. For we need the law here and not far out in space which furthermore seems to contain enough stars and galaxies for it to be impossible to remove the experiment arbitrarily far from any matter.

Einstein's resolution of this difficulty, like several other of his contributions, has a fierce simplicity. Instead of regarding inertial and passive gravitational mass as "accidentally" equal, he regards them as *one and the same quantity since inertia and gravitation are the same phenomenon.* This requires a reformulation of the law of inertia. Newton's law may be divided into two parts: A: There exists a standard motion which matter follows when it is not acted upon by a force. B: This standard motion is motion in a straight line with constant velocity.

Since gravitation is the same as inertia it cannot be a force. A force is only something whose effect on a body can be switched on or off at will (like an electrostatic pull present if the body is charged, absent if it is discharged). Something unavoidable and universal like gravitation must be accepted as part of the background, as necessarily present. Thus part A of the law of inertia can be retained, but the standard motion is now not, as in B, particularly simple, but is the set of possible trajectories under inertia and gravitation. Thus, at the price of complicating B, inertia and gravitation have been unified. The equality of inertial and passive gravitational mass is no longer a law, it is a tautology.

But how are these motions to be described? In the previous lecture it was shown how the intrinsic gravitational field is described by the relative acceleration of neighbouring particles. But we now use our chance to go relativistic by describing the relative acceleration of neighbouring particles whether their relative velocity is small (the Newtonian case) or large (speed of light). Of course this is a richer situation requiring more specifications. Where Newton could do with six quantities (the second derivatives of a single potential), 20 quantities (involving the second derivatives of 10 potentials) are needed to describe these relative accelerations in their dependence on relative position and relative velocity. It will be remembered that in Newtonian theory a linear combination of some of these six quantities equalled the source of the field, the density of matter. In full analogy, 10 linear combinations of the 20 quantities equal the sources of the field in relativity. Why 10 sources? Because now not only the density of matter, but its motion also generate gravitational fields. Since matter in motion gains passive gravitational mass through its kinetic energy, the equality of action and reaction require motion to add also to the active gravitational effect of matter. Thus the sources are roughly speaking the density and the density multiplied by each of the three components of velocity and by their squares and products, making 10 sources in all.

There is a further gain in all this complexity. Whereas in Newtonian theory of gravitation the laws of the conservation of mass and momentum were extraneous constraints on the source of the field, now they are consequence of the field. For the 10 linear combinations of the field (which equal the 10 sources) are not mathematically independent of each other but satisfy mathematically four certain identities.* These must be satisfied equally by the sources, and these identities are the laws of conservation of mass, momentum and energy.

Thus tremendous physical steps have been achieved. Inertia has been unified with gravitation, and the law of gravitation with the laws of conservation of matter and momentum. Gravitational theory has been made compatible with Special Relativity. All this has admittedly led to awkward mathematical complexity. But the real test of a physical theory is neither its unifying accomplishments nor its mathematical form, but agreement with experimental evidence. What is the score here?

First it is an astounding result for such a novel theory that for slowly moving bodies of no undue size or density the same orbits emerge as for Newton's theory. In view of the countless tests of Newton's theory in the solar system this is a necessary result for the acceptability of the theory, but for a theory with such a different starting attitude it is remarkable. Real discrepancies between Newton's theory and Einstein's occur in three circumstances only:

(i) when relative velocities are high;

(ii) when the mass density corresponding to pressure energies is comparable with the mass density of matter;

(iii) when the gravitational potential energy is comparable with the mass.

The first of these follows clearly from the relativistic character of Einstein's theory compared with the non-relativistic character of Newton's theory. The second cause is of the same kind, for pressure arises from the random motion of molecules, and the two

---

* To give a trivial example to show what is meant, suppose field variables x, y, z are connected with sources u, v, w by the relations

$$y - z = u$$
$$z - x = v$$
$$x - y = w$$

It follows from the structure of these equations that the sources satisfy $u + v + w = 0$.

densities become comparable when the velocity of random motions becomes comparable to the speed of light. The third one is of a slightly different kind. On the surface of any gravitating body one can define the velocity of escape, the minimum velocity necessary for the ejection of a particle which is not to fall back on the body. The third condition is essentially that the velocity of escape becomes comparable to the speed of light.

It is only discrepancies of the first kind that are testable so far. One effect, the Einstein shift of frequencies, has already been discussed, and indeed the theory gives the right shift. A second one also concerns light, in this case the lateral deflection of light rays passing close to a massive body. The only such body accessible is the Sun. The observations can only be carried out during a total eclipse, and the results, at the verge of the measurable, are in reasonable agreement with the theory. Finally, for fast moving bodies there should be a small deviation from the Newton-Kepler orbits. The difference is best measurable for the fastest planet, Mercury, and is in excellent agreement with the theory. Thus the experimental evidence, as far as it goes, speaks for the gravitational theory of Einstein.

It is our, no doubt fortunate, fate to live in a part of the universe where bodies massive enough to cause appreciable gravitational fields do not move fast, do not have enormous pressure, do not have colossal velocities of escape. But the theorist is not bound in his imagination by what exists in our neighbourhood. He can torture the theory by seeing what answers it gives in circumstances that are not self-contradictory, but happen not to occur in our neighbourhood. Such work certainly tests and illuminates our theory. And who knows that such circumstances might not exist somewhere in our incredibly complex universe? Thus we are led to the problem of gravitational waves.

CHAPTER FIVE

Gravitational Waves

In the Newtonian theory of gravitation, the problem of the speed of gravitation does not arise. The link between the source and its field is immediate and instantaneous. The gravitational field is attached firmly to matter like an aura surrounding it. Wherever the source goes and however it moves, its field of gravitational attraction goes with it.

It has already been pointed out that the validity of Newtonian theory is limited to cases where relative velocities are small compared with the speed of light. What happens if this restriction does not hold? We evidently have to use Einstein's theory of gravitation, as this is a relativistic theory and thus capable of dealing with high velocities. However, before one can ask any theory to deal with a problem, it is necessary to formulate precise and sensible questions. In the case of a theory of the immense mathematical complexity of General Relativity, it is moreover desirable to consider what the possible answers may look like, so that they can be analysed reasonably.

The first task must be to set up an ideal experiment that would test questions concerning the speed of gravitation. We have already met ideal experiments in the previous lecture, but perhaps an ideal experiment should be defined here as an experiment that can be described in a self-consistent way but, for reasons of cost or technical difficulties, cannot be performed. (It has been said that we do not perform experiments on, say, the constitution of the stars by building stars because of the cost.) The weakness of the critical faculty of man makes it essential to be doubly severe in criticizing any ideal experiment, as the snags that a realization would show up have to be found intellectually rather than by experience.

How, then, would one devise an ideal experiment to discover the speed of gravitation? In its most primitive form the question could be put in the form: How soon would the Earth leave its orbit if the Sun suddenly ceased to exist? Would it happen instantaneously or some eight minutes later (when we would *see* the disappearance of Sun, as light takes some eight minutes to reach the Earth from the Sun) or at some other time? Since the law of conservation of mass is an essential addition to Newton's theory and an integral consequence of Einstein's theory, this question is seen to be nonsensical. The disappearance of the Sun is incompatible with our theories of gravitation.

A slightly more refined question is: How soon would the Earth leave its orbit if the Sun suddenly started to move off at right angles to the plane of the Earth's orbit. Again the question is nonsensical, since the law of conservation of momentum is linked to our theories of gravitation exactly like the law of conservation of mass. (It will be recollected that in an earlier lecture it was pointed out that the first two coefficients in the expansion of the gravitational potential could not be varied at will.)

A sensible question, however, does result if we ask how soon the orbit of the Earth would be affected if the Sun changed its shape from spherical to, say, spheroidal and prolate, leaving aside whether such a development is compatible with the internal constitution of the Sun (a question that has nothing to do with the theory of gravitation). The only limitations on this development apparent from our theories are that the mass must be unchanged, the position of the centre of mass must be unchanged, and the velocity of no particle of the Sun may reach or exceed the speed of light. Though this last restriction is not trivial, it is none too serious. It simply means that a major change in the shape of the Sun must take a few seconds (light can travel a distance equal to the Sun's diameter in a bare five seconds). Since we are looking for transmission delays of under eight minutes, a blurring of the signal by a few seconds is unimportant.

From the ideas of Special Relativity it emerges that information cannot travel faster than light. The orbital behaviour of the Earth would reveal changes in the shape of the Sun. Thus a relativistic theory would require that the Earth cannot leave its orbit before we can *see* the change in the shape of the Sun.

Transmission of information, however, opens a whole new chapter. It is an essential tenet of modern physics that information cannot be transmitted without the transmission of energy. For any apparatus receiving the information must abstract some energy from what is arriving in order to function. The amount of energy involved is not clearly specified, but *some* energy must be transmitted. Transmission of energy is well known from electromagnetic theory where one distinguishes between two kinds of transmission, inductive and radiative. Inductive transfer involves generally the near field, but its most essential characteristic is that the transmitter only loses energy if there is a receiver to take it up. An electrical transformer is a good example. The two coils are close to each other, and the primary current experiences a resistive component of impedance (other than the, for our purposes irrelevant, Ohmic loss) only if the secondary circuit is closed through a resistance. If the secondary is open-circuited, no energy is taken from the primary.

Electromagnetic radiative transfer of energy is quite different. The load on a radio transmitter is totally independent of how many people have switched on their receivers. It radiates energy just as perfectly if it is surrounded only by empty space as if there are absorbers. How can this occur? Where does the energy go? There is a clear link between the finite speed of propagation of radiation and radiative loss of energy. In induction there is, for all practical purposes, an instantaneous interaction between receiver and transmitter. At any moment, the energy is divided between transmitter and receiver. Whatever energy is not in the transmitter, is in the receiver. There is too little journey time between the two for any but a negligible amount of energy to be on the way from one to the other. In the radiative case, on the other hand, there is a significant time lag between transmission and reception. Thus a good deal of energy must be on the way in transit in any given time. Thus this energy has left the transmitter. It cannot yet "know" that there exists a receiver. If it is in transit for a while, it might be in transit for ever, as far as concerns the transmitter. Therefore the loss of energy of the transmitter is fixed, whether an absorber exists or not.

There are some very puzzling features in this. In electromagnetic theory itself, as in most of physics, there is no *direction of time*. It makes no more difference whether, in evaluating the equations, we let time advance or go backward than whether we let the x-co-ordinate increase or decrease. This is just the same as in dynamics. If a clockwise orbit is possible for the Earth, so is an equal anti-clockwise orbit. If a ball can move from A to B, then it can move from B to A. Yet in radiation there is a sense of time. The receiver receives *after* the transmitter radiates. Why is there this sudden intrusion of a particular direction of time, when up to this point the theory could not care less whether time went one way or another? This is an obscure and difficult point in our understanding of physics. There seem to be three, and only three, points in physics where the direction of time matters, namely in radiation theory, in the theory of heat (hot and cold water mix by themselves to make lukewarm water, but lukewarm water does not of its own accord separate into hot and cold water) and in cosmology, where distant objects show a red shift of the spectra, not a blue shift. There is good reason to believe that the three phenomena are linked. In particular, it can be argued that the red shifts mean that all radiation is always absorbed at very large distances, if it has not been absorbed nearby. Thus the red shifts imply a direction of time which shows itself in radiation and, through radiation, in the theory of heat. The connections are not too well understood, but the essential point for our purposes is that, as soon as one comes to radiation, one must select a direction of time such that transmitters emit *before* receivers absorb.

If we want to apply the ideas of electromagnetic theory to the gravitational case we must first surmount a hurdle, the universal character of gravitation. In the electromagnetic description of a TV station there is a great deal that can be omitted because it has no electric charge or current. The buildings, the announcer or actors, the furniture, all this is electrically neutral and irrelevant. Where the news comes from that the announcer reads, how the play was written or how the manuscripts were brought to the TV studio, what the actors ate, all this can be neglected for sure, since it does not produce electromagnetic radiation. Thus there are free variables, to put it mathematically, in electromagnetic theory. A description of the system can obviously be self-consistent without being complete. One can describe all the electronics, TV cameras, antennae, etc., of a TV station without this in any way specifying the newsreader or the news he reads. Indeed the electric system is compatible with *any* news reader and *any* news. This is a vital aspect of any useful physical theory. It must always have a free input, an interface with reasonably arbitrary behaviour. For our knowledge can of necessity only be partial. A theory that gives a useful answer only if initially *everything* is known with perfect accuracy is a useless theory. There must be room, plenty of room, for ignorance. A useful theory can tell us a fair amount from partial initial knowledge.

In gravitational theory the situation is not so clear. Everything has mass and thus contributes to the gravitational field. One cannot leave anything out of the description and know for sure that it could not have affected the situation. To come back to our example of the Sun changing suddenly from a sphere to a prolate spheroid, the chief question was how long it took the *news* of this change to reach the Earth in its orbit. But was it true news? Was the change not necessarily pre-programmed, as it were, in the structure of the Sun? Could a sufficiently good computer on the Earth not have predicted, from its knowledge of the preceding gravitational field of the Sun, when and how this change would occur? But if there is no news and thus no fresh information in the change of the Sun's shape, the Earth's orbit will not tell us about the speed of gravitation. I think that probably this fear is unfounded. There could be a time bomb in the Sun, as powerful as you like, triggered by an arbitrarily small watch and hence of arbitrarily small gravitational influence, or even by a random event, a peak of noise for example. Then nobody, inside or outside the Sun, could have forecast *when* the change of shape would occur. Thus there would be true *news* in the change of shape. Thus the rules and regulations of special relativity apply, and the Earth cannot possibly leave its orbit *before* one can detect by looking at the change of shape of the Sun. It might, however, be later.

Thus the delay, the one-sidedness of time, must be as compatible with gravitational theory as with electromagnetic theory. But if gravitation conveys information it must transport energy. If it transports energy radiatively, the loss of energy of the transmitter is independent of the existence of the receiver. Furthermore, such a loss of energy implies a loss of mass, since all energy has mass. Thus gravitational radiation must diminish the mass of the transmitter. This shows how complicated the theory of gravitation must be. We started with the Newtonian theory and its linearity. This meant that the mass A_0, the dipole moment A_1, the quadrupole moment A_2, and the higher moments A_n were all independent of each other. Moreover A_0 was conserved by conservation of mass, while A_1 was restricted by conservation of momentum, since an isolated body could only move with constant velocity, so that dA_1/dt had to be constant. A_2 and the higher moments were arbitrary functions of the time. Note, however, that for an unchanging body in uniform motion A_2 varied quadratically with the time, so that d^2A_2/dt^2 had to be constant. The arbitrariness of the time variation of A_2 for a body of changing shape would reveal itself therefore through d^3A_2/dt^3 not vanishing.

In relativity the situation is far more complex. Let us forget for the moment the difficulties of defining mass and moments in a truly relativistic (i.e., rapidly changing) situation. Emission of radiation will occur if the quadrupole or higher moments change. This can only lead to a mass loss, never to a mass gain. Thus the derivative of the mass must depend *quadratically* (to ensure constant sign) on the variations of the moments. Thus the theory must be non-linear. The mass must be affected by the other moments, and in a uni-directional way. Dimensional considerations readily lead to an important formula. Using the speed of light and the constant of gravitation for conversion, mass, length and time may all be expressed in the same units. Thus we try to find dA_0/dt. This is a non-dimensional quantity, since A_0 (mass) and t (time) have the same dimensions. Restricting ourselves to the quadrupole moment we must therefore construct a non-dimensional quantity from A_2. The only such quantity is d^3A_2/dt^3, which we have already seen to be the true measure of variation of A_2. Thus we are led to the formula

$$dA_0/dt = -k(d^3A_2/dt^3)^2$$

There k is a numerical constant like $1/4\pi$. The full non-linearity is now self-evident. Moreover the frequency dependence is made clear. If A_2 oscillates with frequency ω, the rate of mass loss and therefore the radiated power are proportional to ω^6. (For each derivative yields a multiplying factor ω.)

For low frequencies the radiated power will therefore be extremely low, for high ones extremely high. As an example consider the solar system looked at from far outside it. Due to the motion of the planets (dominated by Jupiter with its large mass), the quadrupole and higher moments of the system as a whole vary, with the period of Jupiter's revolution (a little over 10 years) most important. If we work out the gravitationally radiated power, it is a mere 1½ kW, a ridiculously low amount for this huge system, a mere flea-bite compared with the electromagnetic radiation of the Sun whose power is nearly 4×10^{23} kW. Is thus gravitational radiation something totally unimportant in our universe? Perhaps not. It would be parochial to think that the ponderous motion of Jupiter round the Sun is characteristic of all regions of the universe. Are there anywhere massive systems in faster motion? In the relatively near regions of our galaxy we know many pairs of stars revolving round each other almost in contact. A pair of stars like our Sun in contact would revolve round each other in a few hours. Their gravitational radiation would then be vastly increased by the sixth power of the frequency, compared with the Jupiter case. Allowing for the differences in separation and mass, one arrives at a gravitational radiation power a few percent of the electromagnetically radiated power. No longer a flea-bite, but not perhaps very impressive. If we could only get the stars still closer together, they would revolve faster and the effect would be greater. To get them closer, their radii must be smaller, i.e., they must be denser. We know very dense stars, the White Dwarfs. Their masses are close to that of our Sun, but their radii are smaller by a factor near 100. We do not know of a pair of White Dwarfs revolving round each other. But let us imagine such a pair in near contact. We know nothing that makes this impossible, except that the gravitationally radiated power would be so huge (millions of millions of times the electromagnetically radiated power) that this loss of energy would make the system collapse in an astronomically very short period (100,000 years or less). Thus stars as we know them could be enormously powerful radiators of gravitational waves. Moreover, our part of the galaxy is certainly very peaceful compared with the region near its centre. We know little about it, but we know for sure that very violent processes go on there.

To sum up then: while we cannot positively identify any stellar system as a powerful gravitational radiator, it seems more probable than not that such radiators exist, particularly in the central region of our galaxy.

So far we have only spoken of the transmission of these waves. How could they be received? What would they do? Do they travel with the speed of light or more slowly? On this last point the theory shows, not unexpectedly, that the bulk of the wave travels with the speed of light. The wave, like any gravitational influence, must express itself as a relative acceleration of neighbouring particles. The wave is completely transverse, i.e., the effect is at right angles to the direction of propagation. A simple wave travelling along the z axis might at one moment tend to drive apart test particles in the x direction while making them approach each other in the y direction. A little later this might be reversed. Such a simple wave is called polarized. There exists another polarization in which the stretch-compress directions are at 45° to those of the wave mentioned. A theoretically simple detector might consist of a freely falling rough stick on which two massive spheres can slide, but are constrained in their motion by springs. An incident wave would make the spheres change their distance apart. The friction of this motion would lead to a heating that might be measured. A practical detector has been built in recent years by J. Weber of the University of Maryland. By an extreme refinement of measurement technique, he has made apparatus that can measure changes in the length of cylinders about 1 m. long when these changes are only about 10^{-14} cm.! Even with the best insulation, such changes will be produced by local traffic, effect of wind on the ground, etc. He has therefore two identical such systems placed nearly a thousand miles apart. Any changes in length detected that are different for the two are ascribed to local perturbations but he has also detected changes occurring simultaneously on the two. These he ascribes to gravitational waves. Moreover, there is some evidence that the effects occur most when the apparatus is aligned by the Earth's rotation to be sensitive to disturbances originating in the centre of our galaxy.

The evidence is perhaps still near the margin of error but it is highly suggestive. It is more likely than not that 1969 will go down in the history of science as the year when gravitational radiation was detected. The new subject of gravitational wave astronomy will join optical astronomy, radio astronomy, X-ray astronomy, etc.

It would be wrong to end this chapter without referring to the real difficulties that still exist in our theoretical understanding, due largely to the extreme mathematical complexity. As has been stressed, under certain assumptions, general relativity reduces to Newtonian theory. For our purposes where extremely high potentials and pressures need not be considered, this similarity essentially relates to static (or slowly moving) state of the system. Radiation, on the other hand, occurs when the system is in a dynamic state with rapid internal motions. The Newtonian case we understand well. In particular we know how to relate the characteristic features of the field at large distance (expansion in r^{-1}) such as A_0, A_1, A_2 . . . to the structure of the source, in particular to the distribuiion of mass. We can do this because we can carry our mathematical analysis from far regions to near ones and in fact cover all space. In the relativistic dynamic case we are prevented by mathematical complexity from looking at the situation except from far away. It is likely that the impossibility of inferring some characteristics of the structure of the source from the far field is due not only to mathematical obstacles but to the fact that the complexity of the far field is due to the whole history not only of the source, but of the whole field.

In any case the physical identification of the characteristic parameters of the distant field is clear only in the static case. The ideal way to investigate the dynamics of the field and to link it to the known static situation is to suppose that an isolated source has been static for all time past, then goes through various changes of shape, and finally settles down in a static state again. Such a "sandwich" of dynamic meat between static bread is indeed desirable, but not attainable. It is rather easy to devise theoretically a static situation and to describe its transition into a dynamic state. It has not so far been possible to get the system back into a static state. For some time it was thought that the difficulty was purely technical, but then Newman and Penrose discovered a curious relation which throws some light on this matter. They found that certain combinations of the coefficients describing the far field are necessarily constant, irrespective of whether the situation is static or dynamic. In the simple case of axial symmetry and equatorial symmetry (quite sufficient for our purposes) this combination equals, *in the static case* only, the product A_0A_2. Thus if, e.g., the system is initially spherically symmetrical so that $A_0A_2 = 0$, this must apply at all times. But it is physically clear that an initially spherical body may, after some time, settle down to a prolate or oblate state. If this is so, but A_0A_2 vanishes in the static state because of the constancy of A_0A_2, then the only possibility of reconciling these statements is that the field *can never become static again*. How could this be explained? It is a fact that it is the rule rather than the exception for waves not to be confined to the wave front. Thus, normally there is a "tail" of disturbance following the front. It is true that this is absent in the propagation of light waves or sound waves in three dimensions in a homogeneous medium (Huygens' principle) but in many other cases its presence is well known (dispersive media, cylindrical waves, etc.). Thus for example if we set off an explosion along a line in the air we would hear not just a bang, but a bang followed by a rumble. However, in all cases known in physics so far, this rumble eventually dies out. It seems as if in gravitational waves the rumble goes on permanently. Thus, after a dynamic phase of wave emission, the field apparently *never* returns to a truly static situation, but keeps "ringing" in some way. If the field is never static we can never again identify the Newman-Penrose parameters at large distance with the mass distribution of the source, and the difficulty of reconciling the shape of the source with the constancy of this parameter disappears.

I hope I have made it clear that this is very much an unfinished subject. A lot more work is needed theoretically to help our understanding, and experimentally to extend Weber's observations. But I also hope to have shown you that it is a fascinating subject.

U.S. Space Flight

by

G. Hage

L. B. James

G. E. Mueller

Mr. G. Hage,
Vice-President for Development,
Boeing Company, Seattle, Washington.

Colonel L. B. James,
Director of Lunar Operations,
George C. Marshall Space Flight Centre,
Huntsville, Alabama.

Dr. G. E. Mueller,
Vice-President,
General Dynamics Corporation,
Washington, D.C.

CHAPTER ONE

Origins and Building Blocks of Manned Space Flight

by G. Hage

From about July 16 to July 24, 1969, the Sydney Australia Sun, Daily Mirror, Sunday Telegraph and Morning Herald, along with the rest of the world Press, headlined one of the greatest news stories of all time (*Figure 1-1*). Two men from planet Earth, members of the Apollo 11 astronaut crew, had set foot on the Moon for the first time in the history of mankind. This was a colossal achievement, not only for America but for all of the free world people, who because of their participation shared in the afterglow of this once-in-a-lifetime success story. Australia made its important contribution to the success of Apollo 11 by furnishing facilities that helped provide the all-important communications link-up between the Apollo 11 crew and Earth, via radio and TV.

But space firsts were not always this jubilant for the free world. On October 4, 1957, Soviet Russia scored the then most important first in space science: Sputnik I, the first artificial satellite ever to be placed in orbit by man. Sputnik I was soon followed by space feats of even greater magnitude. It was then that the Iron Curtain countries enjoyed the glory of great space accomplishments, and revelled in the world-wide publicity that ensued. In America, a cloud of gloom settled over the land. Practically everyone felt frustrated and mortified at having been bested by a rival nation. For years, America and Russia had been pitted against each other in the post-World War II "cold war". By moving ahead of the United States in what became known as the space race, the Soviet Union gained invaluable and unassailable prestige. Even

LATE FINAL EXTRA

No. 19,527 MONDAY, JULY 21, 1969 Five cents

THE SUN

Telephone 2-0944. Jones Street, Broadway. Letters to Box 506, G.P.O., Sydney, 2001.

ON TH MOON

● CITY: Colder tomorrow. ● Lotteries: New Jackpot 302, P. 27; Special 1772, P. 31 ● Finance, P. 28 ● TV,

MAN ON MOON

HISTORIC WORDS:

"The Eagle has landed"

MAN'S centuries-old dream came true today forty seconds after 6.17 a.m. (Sydney time) the U.S. astronauts Armstrong and Aldrin dodged a huge crater and landed their lunar vehicle gently on the rock-studded moon in a cloud of lunar dust. Full report pages 2, 3, 4, 5 and

FIRST STEP FIVE HRS. EARL

ASTRONAUT Neil Armstrong opened the hatch of the lunar module soon after noon (Sydney time) today—and was just 10ft away from the incredible moment when man first sets foot on the moon.

The two astronauts, exhilarated by the successful desce had urged space officials to let them begin the historic mo walk nearly five hours earlier than planned.

Figure 1-1. History is made.

more dismaying was the realization that if Russia had developed rockets powerful enough to orbit artificial satellites, the time would soon come when she would have the capability to deliver rockets containing deadly nuclear warheads to all parts of the world.

Now, I am not going to suggest that Sputnik I and Russia's follow-up space probes were wholly responsible for America's Apollo programme. But it would be hypocritical not to admit that the Soviet space effort and achievements provided one of the most important motivations behind America's concentrated space effort that ultimately led to the Apollo 11 lunar landing mission.

Prior to Sputnik I, America, to be sure, did have a space programme. But it little resembled the one that followed shortly afterward. That is what I am going to talk about today — the U.S. pre-Sputnik space effort and the stepped up, concentrated space programme that followed on the heels of Sputnik I, up to, but not including, the Apollo programme. The Apollo and post-Apollo programmes will be covered in subsequent talks by me, by Lee James, and by the man who headed NASA's manned space flight programme during the Gemini and Apollo periods, through to Apollo 12, Dr. George Mueller.

It is not without good reason that discussions on early rocket technology pay tribute to Konstantin Tsiolkovsky of Russia, Herman Oberth of Rumania and Germany, and Robert Goddard of America. Years before Sputnik I was even contemplated, all three of these rocket pioneers had significantly contributed to the development of the rocket that, at a later day, was destined to liberate man from the grip of Earth's gravitational pull.

In America, as early as 1914, Dr. Robert Goddard was granted two patents for his ideas of a multi-stage rocket and liquid propellants. In static laboratory tests, he proved the theory that a rocket can perform in a vacuum and is therefore capable of operating in outer space.

Dr. Goddard worked in rocketry unceasingly from 1917 until his death in 1945. In 1920, he experimented with liquid-fuel propulsion, trying out an idea for a hydrogen and oxygen fuel supply he had conceived as early as 1909. In March, 1926, he accomplished the first liquid-fuel rocket flight in history.

During the 1930s, Dr. Goddard flew numerous rocket flight tests in the state of New Mexico, as he continued to develop the science of rocketry. Testimony to his lifetime of successful efforts in this field are over 200 patents in his name, covering all of the fundamentals of successful rocket flight, from fuels, multi-stage design, and guidance and control, to payloads.

But even as recently as World War II, Goddard's work went virtually unnoticed in the United States. In Germany, however, liquid-fuel rocket study had proceeded during the 1930s and, by 1944, the German V-2 ballistic rockets were being launched from Germany against Britain. In basic design, the supersonic V-2, with its 200-mile range, was almost identical to Dr. Goddard's much smaller liquid-fuel rocket.

The potential of large rockets demonstrated by the German V-2 fostered post-war aspirations for the exploration of space with instrumented, and eventually manned, rocket vehicles. The U.S. armed services pursued several rocket projects in the post-war period, exploiting captured V-2 components and adaptations of the V-2 engines, as well as other engine developments. One such V-2 experiment, launched from White Sands, New Mexico, in 1946, rose 55 miles high to acquire the first ultra-violet spectrum of the Sun above the density of Earth's atmosphere, where ultra-violet rays, for the most part, are absorbed. By 1949, the Navy had developed the more powerful Viking rocket, for high altitude atmospheric probes. By 1952 the National Advisory Committee for Aeronautics (NACA) was studying the prospects of future manned space flight. That same year, NACA and the U.S. Air Force jointly began the X-15 rocket-powered airplane research project (*Figure 1-2*). The U.S. intercontinental ballistic missile programme commenced in 1954, leading to the development of a new generation of multi-stage rocket systems endowed with sufficient thrust to transport military payloads.

In 1956, Project Vanguard, the first United States Earth satellite programme, was initiated under the management of the Naval Research Laboratory. Project Vanguard aimed at developing a satellite-launching vehicle and tracking system, and at placing at least one satellite in orbit with an experimental payload during the International Geophysical Year which began on July 1, 1957.

Figure 1-2. X-15 airplane dropping away from a B-52 in flight.

But before a Vanguard rocket and satellite payload could get off the launch pad, Sputnik I was in orbit. And to further depress the sagging morale of the American people, a bigger and heavier Sputnik II, carrying the dog Laika, was orbited the following month, on November 3.

But the United States immediately began to move into action in the field of space science. An earlier proposal, which authorized the U.S. Army Ballistic Missile Agency to provide a backup rocket/satellite system, named Explorer, for the Vanguard, was resurrected.

The launch of the first Vanguard rocket with potential orbit capabilities was attempted on December 6, 1957. The first stage engine lost thrust two seconds after ignition; and the vehicle burned up on the pad. The launch of the first U.S. satellite, Explorer I,

Figure 1-3. A Jupiter-C rocket.

was accomplished by the Army Ballistic Missile Agency on January 31, 1958 (*Figure 1-3*). Explorer I detected a belt of radiation about 500 miles above the equator that was subsequently confirmed by Explorer III, launched on March 26, 1958, and named the Van Allen radiation belt. Meanwhile, after a second unsuccessful Vanguard launch attempt in February of 1958, the first successful Vanguard satellite was launched into orbit on March 17, 1958. Analysis of the orbit of Vanguard I revealed that the Earth is pear shaped, rather than bulging slightly at the equator and being somewhat flattened at the poles, as previously believed (*Figure 1-4*).

Following Russia's success in orbiting Sputnik, President Eisenhower and the U.S. Congress carefully reviewed the national competence and potential in America's missile and space development.

Figure 1-4. Vanguard I is launched.

Figure 1-5. Mariner IV sets off on its Mars probe.

Early in 1958, the President's Advisory Committee on Government Organization recommended, and the President approved, that a civilian space agency be established, patterned after the successful National Advisory Committee for Aeronautics (NACA), and responsible for all non-military space activities in an integrated programme. The President's Science Advisory Committee also urged national action to develop space technology. A space agency bill was soon forwarded by the President to the Congress for consideration. The National Aeronautics and Space Act of 1958, which evolved from the bill, was passed by Congress and signed by the President on July 29, 1958, creating the National Aeronautics and Space Administration (NASA).

In addition, the Space Act of 1958 spelled out the national objectives in space as follows:

(1) the expansion of human knowledge;

(2) improvement of aeronautical and space vehicles;

(3) development and operation of space vehicles;

(4) long-range studies for peaceful and scientific use of aeronautics and space;
(5) international co-operation;
(6) effective utilization of resources.

The new space agency was formed around a nucleus of former NACA personnel. Eight thousand NACA employees, including scientists, engineers, and technicians, were transferred to NASA. NASA also absorbed NACA research facilities—the Langley, Ames, Lewis, and Edwards Research Centres—with their 40-year legacy of NACA aeronautical, rocket propulsion and missile research.

Army, Air Force, Navy and Department of Defence non-military space projects were also transferred to NASA, including their space probes, satellites and rocket engine programmes. Among the projects and personnel acquired were the Project Vanguard scientific satellite programme and 200 highly qualified scientific and technical personnel from the Naval Research Laboratory.

By the end of 1958, the Jet Propulsion Laboratory of the California Institute of Technology, previously under contract to the Army, was brought under NASA direction. At the same time, the Army Ballistic Missile Agency at Huntsville, Alabama, was made responsive to NASA requirements, as well as the large liquid-fuelled Saturn rocket programme at Huntsville, initiated in 1958 under Department of Defence auspices. Then, in mid-1960, a group of rocket experts of the Army Ballistic Missile Agency's Development Operations Division, and their facilities, were transferred to NASA. With this transfer, the George C. Marshall Space Flight Centre was established in Huntsville to provide large booster support for advanced manned flights.

The NASA Launch Operations Centre at Cape Canaveral, Florida, was established in 1960 and became a NASA Centre in July, 1962. It was renamed the John F. Kennedy Space Centre in November, 1963, concurrent with the redesignation of Cape Canaveral as Cape Kennedy.

Specific capabilities for both manned and unmanned non-military space flights were created by this organizational realignment. Efforts previously fragmented were brought together under the new space agency, each contributing to the total national programme. Key personnel from the Langley Research Laboratory, comprising the nucleus of the Project Mercury team, were transferred to the Manned Spacecraft Centre at Houston, Texas. The Project Vanguard team formed the nucleus of the new Goddard Space Flight Centre in Greenbelt, Maryland, which concentrated on unmanned satellites and spacecraft, and created the basis for the world-wide Goddard tracking and communications network. This network was later expanded and refined to support both unmanned and manned space flights.

At this point I would like to stress that an intimate interrelationship exists between the NASA unmanned and manned programmes. Later, I shall focus attention on manned flights, particularly those precursors of the Apollo programme, Projects Mercury and Gemini. But it is important to keep in mind the complementary role of the unmanned missions. The unmanned scientific investigations of the Earth, Moon, Sun, planets, stars, galaxies and outer space invariably support advances in the manned programmes. Indeed, some unmanned probes were essentially forerunners of specific manned flights. The unmanned probes that I will next discuss were conducted concurrently with the manned flights to be discussed later.

Much of the emphasis of NASA's unmanned satellite programme has been on learning in detail about the Earth and its relationship with the Sun, on exploring the Moon, Venus and Mars, together with interplanetary space, and on providing practical applications.

Study of the Earth focuses on the atmosphere, the ionosphere and the magnetosphere. To learn about the Sun, instruments were pointed directly at the Sun to observe its corona, chromosphere and centre, with particular attention devoted to flares, solar storms and sunspots.

In the exploration of the solar system, NASA conducted flights past the two planets nearest to Earth, Venus and Mars, as well as several flights through interplanetary space. In 1962, after a journey of 109 days, Mariner II passed within 21,500 miles of Venus, transmitting data about that planet from 36 million miles away. The most important finding about Venus was that its

Figure 1-6. The Tiros IV weather satellite.

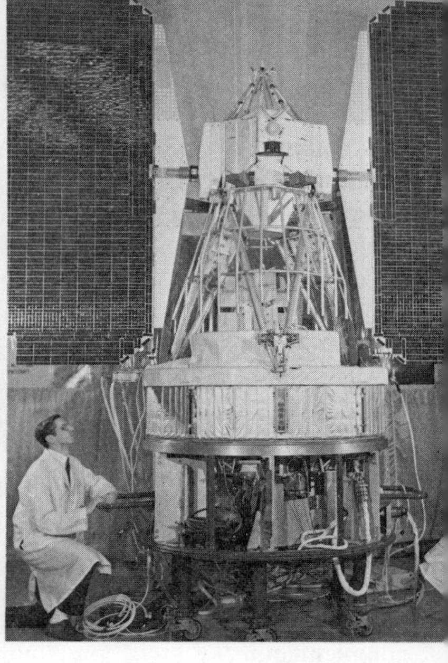

Figure 1-7. The Nimbus B spacecraft.

atmosphere was hotter than previously believed. Temperatures were recorded as high as 600 degrees Fahrenheit.

In 1965, Mariner IV travelled 228 days and passed within 7400 miles of Mars. Over a distance of 134 million miles, Mariner IV transmitted to Earth 21 clear photographs, indicating that the surface of Mars, like that of the Moon, is cratered. Radio signals sent when Mariner passed behind Mars demonstrated that the planet's atmosphere is thinner than expected. The data transmitted failed to supply evidence of a dust belt, a magnetic field or radiation belts (*Figure 1-5*).

Spacecraft in the Pioneer series have measured particles and fields in interplanetary space up to distances of more than 50 million miles.

Closer to home, unmanned spacecraft orbiting the Earth have demonstrated practical applications of space technology to Earth sciences, particularly weather, communications and navigation.

In meteorology, there have been two unmanned spacecraft programmes, TIROS and NIMBUS. The TIROS series consisted of 10 satellites, orbited between 1960 and 1965, which photographed cloud cover, obtained infra-red measurements and transmitted the results to Earth. Since most weather originates over the five-sixths of the Earth's surface covered by water—where weather stations, for the most part, are non-existent—satellite photographs of a developing storm system or typhoon are especially valuable for providing warnings that save lives and property. In addition, the infra-red readings enable scientists to understand the Earth's energy balance. Since 1967, the U.S. Department of Commerce has operated a system of weather satellites similar to those in the TIROS series (*Figure 1-6*).

But TIROS suffers from two limitations. Since its orientation in space is constant, it points toward Earth during only half of Earth's orbit. In addition, TIROS satellites do not pass over the polar regions where some of the most interesting weather is found. To compensate for this, we are now flying the larger NIMBUS satellites, which are placed into near-polar orbits and point constantly toward the Earth (*Figure 1-7*).

As many of you may know, a station exists at the University of Melbourne to receive TIROS and NIMBUS photographs, which should make a very real contribution to the advance of meteorology here in Australia. Two other stations have been constructed at Darwin and Perth.

In communications, a number of experiments have been carried out successfully, beginning in 1960 with the Echo balloon satellite, and later with Telstar, Relay and Syncom satellites. In principle, a communications satellite acts as a tall antenna tower, which enables us to transmit to distances beyond the horizon as seen from the Earth's surface. The Syncom (synchronous communications) satellite demonstrated that it is possible to station a transmitting antenna at an altitude of 22,300 miles. At this altitude, the spacecraft maintains its position relative to a fixed point on the equator.

A synchronous orbit is also used by the Early Bird satellites of the Communications Satellite Corporation, the first commercial venture in space, which is now supplying telephone and television communications between the United States and Europe.

A third application of space technology now in operation is navigation by satellite. The U.S. Navy Transit satellites supply all-weather navigation services to units of the fleet.

Now I will discuss at length the first two NASA manned space flight projects—Mercury and Gemini.

The recommendation to place a man in orbit around Earth and recover him was approved by the NASA Administrator on October 7, 1958, and Project Mercury was immediately set in motion. The basic objectives of Project Mercury were to orbit a man around Earth, observe his physical and mental reactions in the space environment, and recover the man and spacecraft; thus, a closely integrated relationship between government agencies, contractors, and the scientific community was essential.

The NASA Space Task Group was organized to oversee integration and control of the Mercury Programme. Beginning late in 1958, aeromedical personnel from the Army, Navy and Air Force were assigned to the Space Task Group to work with NASA personnel and a special committee on life sciences. This group established an astronaut selection procedure, set up qualifications and requirements, and selected a group of 110 potential astronauts.

By April, 1959, seven experienced jet aircraft pilots had been chosen to be the first astronauts (*Figure 1-8*). Their two-year group training programme included astronautical science, systems integration, spacecraft control, environmental familiarization, and egress and survival.

In designing the hardware for Project Mercury, the following four basic guidelines were established: (1) employ existing technology and equipment where possible; (2) adopt the simplest and most reliable approach to system design; (3) utilize an existing rocket launch vehicle for the orbital mission; and (4) conduct a progressive and logical test programme. Since man's capabilities in space were at that time unknown, automatic operation was provided for all critical systems of the Mercury spacecraft. Redundant systems were also included to give maximum reliability for safety and assurance of mission success.

The bell-shaped spacecraft that was developed was protected from re-entry heating by a heat shield covering the blunt end (*Figure 1-9*). In case of a launch vehicle failure on the pad, a

Figure 1-8. Astronauts taking flight training. Left to right: Malcolm Scott Carpenter, Leroy Gordon Cooper, Jr., John Herschel Glenn, Jr., Virgil Ivan Grissom, Walter Marty Schirra, Jr., Alan Bartless Shepard, Jr., and Donald Kent Slayton.

solid-propellant rocket mounted on top of the escape tower would propel the spacecraft clear of danger. In a normal mission, the escape tower was jettisoned after shut-down of the booster stage.

Inside the Mercury spacecraft, a form-fitting contour couch distributed the high-acceleration loads to which the astronaut was subjected. At all critical phases of the mission, the spacecraft was so oriented that the direction of acceleration forces was the same, from back to front. Thus, the spacecraft was pointed nose up at launch and heat shield forward upon re-entry.

The life support system protected the pilot from the hard vacuum of space as well as the extreme temperature variations associated with the orbital flight profile. It supplied oxygen, purified the air (by removing carbon dioxide and other foreign matter), and

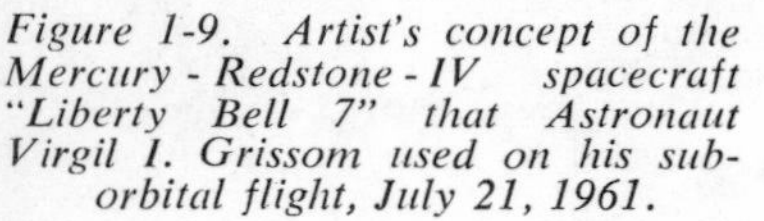
Figure 1-9. Artist's concept of the Mercury - Redstone - IV spacecraft "Liberty Bell 7" that Astronaut Virgil I. Grissom used on his sub-orbital flight, July 21, 1961.

Figure 1-10. Astronaut Alan Shepard, Jr., after the first Proje Mercury suborbital space flight.

controlled the temperature and air flow inside the astronaut's full pressure space suit. It also provided for food, drinking water, and waste management.

Systems development and production for the Mercury programme—from man-rating the rocket boosters to creation of the spacecraft subsystems—were characterized by extensive ground testing of each component, then of assembled components. All major aspects of the project were concurrently developed and tested. Thus, while the astronauts were being selected and trained, the world-wide tracking network was planned and built, and research and development of the flight hardware were accomplished. Only three years and four months elapsed between project initiation and the first manned orbital flight. The entire programme of six manned flights was completed in four years, seven months.

Of 25 major launches between August, 1959, and May, 1963, six were manned: two of them were sub-orbital ballistic profiles and the other four were Earth orbiting. In spite of launch failures and test anomalies, no astronaut injuries were suffered.

The first launches in the Mercury programme used the Redstone missile previously developed by the Army's Ballistic Missile Agency. The Redstone's thrust sufficed to place the Mercury spacecraft into a sub-orbital trajectory with an apex about equal to the planned orbital altitude, producing weightlessness for one-third of the flight. This was particularly important, since the biological effects of weightlessness upon man were yet unknown, save for limited data acquired through aircraft experiments where zero-g conditions could be simulated for only a few seconds at a time.

To test the spacecraft-booster combination, three Mercury-Redstone flights were accomplished: a systems test with an empty spacecraft; a second test with a chimpanzee as passenger; and a third flight to check improvements in booster accuracy. The fourth flight, on May 5, 1961, was the first United States manned space flight (*Figure 1-10*). During his 15-minute flight over a ballistic trajectory, Alan Shepard experienced about five minutes of weightlessness and reached a maximum altitude of 116 statute miles. A slightly modified spacecraft, with a window added, was piloted by Virgil Grissom on a similar flight in July, 1961.

For Project Mercury orbital flights, the Atlas intercontinental ballistic missile was used. In the "man-rating" process, pilot safety during countdown and launch was the primary consideration. An abort system was developed that shut down the engines and initiated spacecraft escape when trouble arose. From September, 1959, to May, 1963, 10 Mercury-Atlas launches were accomplished. Six unmanned flights tested the performance of booster and spacecraft before the first manned orbital flight was attempted.

On February 20, 1962, Astronaut John Glenn, Jr., piloted the first U.S. manned orbital flight for three orbits. No adverse effects were noted from weightlessness. Glenn controlled the attitude of the spacecraft manually for a large part of the mission and performed well under stress. Attainment of the basic Mercury programme objectives on this first orbital mission made it possible to extend the aims of the programme and conduct expanded space exploration and experimentation on subsequent missions. A second three-

Figure 1-11. Astronaut L. Gordon Cooper on the deck of the U.S.S. Kearsarge.

orbit flight was completed in May, 1962, with Scott Carpenter as pilot, followed in October by a six-orbit flight piloted by Walter Schirra, Jr. The Schirra flight provided information on extended exposure to the space environment, additional operational experience, and an opportunity to conduct a series of experiments and measurements in space.

The last Mercury flight was the 22-orbit, 34-hour and 20-minute flight by Astronaut Gordon Cooper, Jr., in May, 1963 (*Figure 1-11*). Effects of extended weightlessness were found not to be detrimental to the human body. Although Cooper had lost seven pounds due to temporary dehydration, he was in excellent condition upon recovery in the Pacific Ocean. During this same mission, several scientific experiments were conducted, including aeromedical studies, radiation measurements, photographic studies, and visibility and communications tests.

Mercury programme objectives had thus gradually expanded from the original goals of placing a man in orbit, recovering him safely and evaluating the data acquired, to a successful 34-hour orbital mission.

Many were the gains from Project Mercury. We learned to design, build and test manned spacecraft; to prepare launch vehicles for safe and reliable manned flight; and to operate a world-wide network of radio and radar tracking and communications with the spacecraft and pilot. We also learned to recover the spacecraft and pilot, select and train astronauts, and develop and operate life support and biomedical instrumentation systems. In addition, we acquired valuable experience in large-scale management and systems engineering; we gained important scientific knowledge about the space environment; and we made technological advances necessary for further progress in space exploration.

During Project Mercury, NASA's planning pointed to manned exploration of the Moon and the nearby planets as a goal for the indefinite future beyond 1970. In July, 1960, following a Congressional committee recommendation for a high priority manned lunar landing programme, NASA announced that the successor to Project Mercury would be Project Apollo. At that time, the goal of Project Apollo was to carry three astronauts in sustained Earth-orbital or circumlunar flight. Plans for an eventual manned lunar landing were to be studied.

Then, in May, 1961, President Kennedy recommended to Congress an expanded national space programme with the major accelerated goal of "landing a man on the Moon and returning him safely to Earth, during this decade". Congress subsequently endorsed the plan for expanding and accelerating Apollo including the development of spacecraft, large rocket boosters, and unmanned explorations to support the Apollo objectives.

Meanwhile, in December, 1961, the decision was made to extend the manned space flight effort beyond Mercury, to provide experience in space operations that would benefit the Apollo lunar landing programme. The Mercury follow-on programme, named Gemini, was designed to subject two men to long-duration Earth-orbital flights in order to obtain the experience and knowledge essential for trips to the Moon. Since the rendezvous and docking of two space vehicles comprise a key element of the Apollo lunar landing approach, one important objective of the Gemini project was to achieve a rendezvous and docking in Earth orbit between the spacecraft and another orbiting vehicle. Moreover, experiments were planned for the astronauts to perform mechanical and other tasks outside the spacecraft while in orbit. This extravehicular activity was another technique projected for use in more advanced missions, such as assembling structures and repairing equipment in space, and functioning outside the Apollo spacecraft on the lunar surface. Perfecting methods of controlled re-entry and landing at pre-selected sites was still another objective of the Gemini programme. This would be necessary for return from the lunar landing mission of Apollo. The information and experience gained from Gemini on the effects of weightlessness and the physiological reactions of crew members during long-duration missions would materially help in planning Apollo missions to the Moon.

There were several important changes and design advancements from Mercury to Gemini. Whereas the Mercury spacecraft had been designed for complete automatic control from the ground with a redundant capability for control by the single-astronaut crew, the Gemini spacecraft was designed to be controlled in flight by a two-astronaut crew, with ground control acting as the back-up. In Mercury, an impending launch vehicle failure was automatically sensed and the escape system was accordingly activated; in a Gemini launch, however, vehicle malfunction activated warning lights and gauges on the instrument panel, leaving it up to the astronauts to decide whether or not the situation was serious enough to abort the mission. In the Mercury spacecraft, almost all systems were stacked in layers in the pilot's cabin, which often made it necessary to disturb several systems in order to get at a particular problem area. In the Gemini spacecraft many of the systems were positioned outside the cabin area and arranged in modular packages, so that any system could be removed without disturbing others. Spare packages could be completely checked out and kept in reserve for replacement purposes.

When completed, the Gemini spacecraft comprised two major units: (1) the adaptor module, consisting of the retrograde and equipment sections; and (2) the re-entry module, consisting of the rendezvous and recovery section, the re-entry control system, and the cabin section. The heat shield was attached to the cabin section, in the manner of the Mercury capsule. The re-entry module was the only portion of the spacecraft recovered from orbit.

In the event of a mission abort, provisions for crew safety on Gemini were significantly different from those on Mercury and Apollo. Mercury had a tower atop the spacecraft with a tractor rocket to pull the entire spacecraft away from the danger area if needed; Apollo's design is similar. Gemini spacecraft, however, were equipped with crew ejection seats similar to those used in high performance aircraft, and parachutes for softening the return to earth. The use of ejection seats was possible because the Gemini launch vehicle fuel burned on contact with the oxidizer, minimizing the explosion hazard present in the Mercury launch sequence. An additional benefit of the ejection seat method was that it could be used during the re-entry phase, at low altitudes, in the event of trouble during the terminal portion of the mission.

The Gemini launch vehicle was 90 feet long by 10 feet in diameter, with a first stage engine that developed 430,000 pounds of thrust and a second stage engine that produced 100,000 pounds at the ignition altitude of approximately 200,000 feet (*Figure 1-12*). As noted above, the Titan II employed storable propellants which ignited on contact, providing the precision necessary for Gemini rendezvous and docking missions.

To attain the reliability and safety required for manned launches, a series of modifications were made in the basic missile. Redundant hydraulic systems and additional instrumentation were installed; and a malfunction detection system was added, which sensed critical problems in the rocket and warned the crew of them.

The target vehicle for Gemini rendezvous and docking missions was a specially modified Air Force Agena, with a docking collar and certain instrumentation peculiar to Gemini missions. An Atlas launch vehicle boosted the Agena to near orbital speed, at which point the Agena propulsion system took over to place the vehicle in orbit. Development and modification of the Gemini

Figure 1-12. The launching of Gemini/Titan I.

Figure 1-13. Astronaut Edward H. White II during the third orbit of the Gemini/Titan IV flight.

launch vehicle and the Atlas-Agena combination were Air Force responsibilities in response to NASA requirements.

Beginning with Gemini III in March, 1965, eight manned missions were flown in less than 16 months.

The programme objective of investigating long-duration flight was achieved in three missions of 4, 8, and 14 days' duration. These were the flights of Gemini IV, Gemini V and Gemini VII in June, August and December, 1965, respectively. The second of these, the Gemini V mission in August, established a new manned space flight record of 190 hours, 55 minutes, surpassing the previously held Russian record. Post-flight medical evaluations of the crews revealed that no adverse effects resulted from lengthy exposure to weightlessness; and a definite pattern of aircrew adaption to weightlessness was detected as blood pressure and heart rates followed a levelling trend after the first several days. This information was of great significance, since eight days was the length of time planned for the first Apollo lunar landing mission, while other flights—both lunar and Earth-orbital—would last for periods of 14 days and more.

In the first long-duration mission, Gemini IV, pilot Edward White, employing a 25-foot tether and attached oxygen umbilical hose, demonstrated the feasibility of conducting activity outside the spacecraft (*Figure 1-13*).

This was the first U.S. extravehicular activity, during which White remained outside the spacecraft for 23 minutes, manoeuvring himself about the spacecraft, taking pictures and making observations of equipment. During the early portion of the manoeuvre, he used a hand-held manoeuvring unit which provided propulsion by emitting jets of oxygen. A chest pack contained an emergency supply of oxygen and maintained pressure in his special protective suit.

In June, 1966, an expanded extravehicular mission was included in the Gemini IX flight, during which Astronaut Eugene Cernan worked outside the spacecraft for over two hours.

On the Gemini X flight in July, 1966, pilot Michael Collins performed two extravehicular assignments. During his first exposure to space outside the cabin, Collins stood up in the open hatch of the spacecraft for nearly 45 minutes and performed picture-taking and other assignments. Later in the flight, during rendezvous with the Agena used in the Gemini VIII mission, he egressed from the cabin and manoeuvred in space for 55 minutes with a hand-held manoeuvring unit. Using a 50-foot tether and umbilical hose, Collins manoeuvred to the Agena and retrieved a micro-meteoroid collection experiment.

The in-flight manoeuvring capability necessary for spacecraft rendezvous and docking was first demonstrated during Gemini III, when orbital attitude and orbital phase were changed. The Gemini VI mission in December, 1965, accomplished the first successful space rendezvous, as well as demonstrating the high degree of launch and preflight operational capability needed to carry out the rendezvous. The lift-off of the Gemini VI space vehicle occurred within 11 days of the Gemini VII launch from the same pad, and within one-tenth of a second of the scheduled lift-off time. Four orbits later, command pilot Walter Schirra manoeuvred Gemini VI to within 120 feet of Gemini VII to accomplish rendezvous of the two manned spacecraft. Later, the Gemini VI spacecraft was

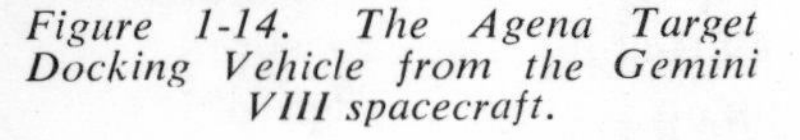

Figure 1-14. The Agena Target Docking Vehicle from the Gemini VIII spacecraft.

Figure 1-15. The Augmented Target Docking Adaptor (the Angry Alligator) as seen from Gemini IX.

manoeuvred to within less than one foot of Gemini VII, following station-keeping and fly-around manoeuvres.

But rendezvous and docking with an Agena Target Vehicle was actually achieved for the first time during the Gemini VIII flight in March, 1966. The mission was successfully terminated shortly after the docking, however, because of a spacecraft malfunction (*Figure 1-14*).

Three other types of rendezvous manoeuvres were performed during the Gemini IX flight. In place of the Agena Target Vehicle, an "augmented target docking adaptor" was used. Docking could not be performed, however, because the target vehicle shroud failed to jettison (*Figure 1-15*).

One of the most significant rendezvous manoeuvres carried out on the Gemini IX mission was a simulation of a rendezvous of a Lunar Module with an Apollo spacecraft in lunar orbit. The manoeuvre successfully performed by Gemini IX would be required during an Apollo lunar mission if a decision were made not to continue with a lunar landing after the Lunar Module had descended to the 50,000 foot level.

The third portion of the rendezvous and docking objective — manoeuvring while docked by using the Agena's primary propulsion system — was accomplished on the Gemini X flight. This was a dual rendezvous mission, with the first rendezvous made with an Agena X Target Vehicle launched one orbit earlier. After docking with the Agena X, the combined Gemini/Agena vehicle was manoeuvred to a higher orbit, with an apogee of 476 miles — the highest man had yet ventured in space.

This orbital change was the first of a series of orbital manoeuvres using the Agena X propulsion system, in preparation for rendezvous with another orbiting target. The second rendezvous target was the Agena vehicle which had remained in orbit since the Gemini VIII mission four months earlier. Terminal rendezvous of Gemini X with Agena VIII was successfully accomplished after the spacecraft had separated from Agena X.

Guided re-entry of the spacecraft to a particular target was included in all of the Gemini flights. With the aid of an on-board computer, the pilots were able to use the aerodynamic lift of the spacecraft as it re-entered the atmosphere to guide it toward a pre-selected landing area. The most accurate demonstrations of this capability were provided by Gemini IX and Gemini X, both of which "splashed down" within three miles of the intended target point.

Advances in Gemini operational techniques and equipment enabled a sizeable number of scientific and technological experiments to be conducted during Gemini flights. A total of 49 separate experiments were scheduled in the programme, many of which were repeated on several flights. These experiments ranged from specific physiological measurements of the crew to technological developments proving out new equipment and techniques.

One particularly interesting experiment concerned synoptic terrain photography, that is, successive colour pictures of geographic and geological points of interest from orbital altitude. Geologists, geographers and oceanographers obtained valuable new information from these colour photographs. For example, new geological knowledge was gained as a result of photographs taken during Gemini IV, which revealed new characteristics of a volcanic field in Mexico and a previously unknown fault in the lower California peninsula. Since the pictures were taken by men rather than pre-programmed equipment, cloud-free photographs were obtained.

Similarly, synoptic weather photography produced colour photographs of storm patterns and cloud cover from relatively low orbital altitudes. These pictures were helpful as additional data to assist in interpreting unmanned meteorological satellite results.

Another example of new knowledge gained from Gemini experiments were Professor E. P. Ney's zodiacal light photography experiments. Begun on Gordon Cooper's 34-hour Mercury flight in 1963, and continued into the Gemini programme, these experiments comprised photographs of the zodiacal light in the night sky — the visible manifestation of dust grains in orbit about the Sun. Taken from orbital altitudes according to directions provided by the experimenter, the photographs by Cooper confirmed theories about the zodiacal light which had not been subject to proof by earlier means.

On the Gemini V flight, in August, 1965, photographs of the gegenschein phenomenon were obtained. This first photographic evidence indicated that the gegenschein — a glow in a direction opposite the Sun from the Earth — is probably produced by back-scattering of sunlight through dust. Further astronomical advances are anticipated in the future through orbital photography and observations with telescopes in orbit.

Scientific achievements in the Mercury and Gemini programmes can be summarized under three interrelated categories. We have acquired a significant body of knowledge concerning both men and machines in relation to and in interaction with the space environment; we have gained invaluable experience in space operations; and we have demonstrated the value of manned spacecraft as vehicles for scientific experimentation and observation.

Since the early days of Project Mercury, when highly respected medical opinion warned against the unknown effects of weightlessness and rapid heart rates, great advances have been made. We now know that trained astronauts can function well for many days in space, supported by environmental control systems based on contemporary technology. We know, too, that these trained crewmen can also withstand the stresses of launch operations and re-entry into the Earth's atmosphere. We have learned that the radiation hazard in near-Earth orbit is acceptably small. We have developed and utilized biosensors to relay critical physiological data from spacecraft to Earth stations by means of telemetry. We have demonstrated that a crewman can perform useful work outside his spacecraft. We have gained experience in producing large rocket boosters with sufficient reliability of performance to be "man-rated". A world-wide tracking and communications network, tied into an advanced Mission Control Centre, enables us to execute positive mission control in real time. Crewmen have performed as experimenters, conducting a wide variety of scientific and technological investigations.

During the brief period of manned space flight, we have accumulated hundreds of man-hours of operating experience. The valuable knowledge gained from Mercury and Gemini included how to select and train crew members for space flight, and how to control and recover men and machines through missions of increasing length and complexity. Since our first hesitant steps into space in 1961, we have increased from five minutes to two weeks the time during which crewmen experienced weightlessness. We have developed and exercised techniques for manoeuvring in space and controlling re-entry flight paths to accurate landings in planned recovery areas on the Earth's surface. Our operations progressed from flight over a ballistic trajectory to dual and triple rendezvous manoeuvres with other spacecraft, manned and unmanned. We have joined spacecraft in space and exploited a rocket propulsion unit "stored" in orbit to propel the linked configuration through new manoeuvres, opening the door to future assembly, crew transfer, and re-supply operations in space.

Man's exploration of the space environment is accelerating at a rapid rate. The size and complexity of the spacecraft and launch vehicles have increased with each major programme and the missions to be accomplished become ever more complex. Mercury and Gemini established the basic knowledge of the space environment in combination with men and machines, and the operational techniques necessary for the next giant steps by Apollo astronauts in this great new age of exploration.

CHAPTER TWO

Development of the Saturn Launch Vehicle

By G. E. Mueller

INTRODUCTION

Even before the United States launched its first artificial satellite, weighing 30.8 pounds, in January, 1958, its scientists were making plans that envisioned launch vehicles having payloads of 6,000 to 12,000 pounds for Earth escape missions and 20,000 to 40,000 pounds for orbital missions.

By early 1957, designers had become convinced that it was possible to build a clustered-engine booster that would generate 1.5 million pounds of thrust and lift multi-ton payloads. In December of that year the von Braun group, then working for the Army, submitted to the Department of Defence a "Proposal for a National Integrated Programme" which called for the development of such a booster. Subsequent studies concluded that the development was feasible, and in August, 1958, the Army received formal approval to initiate a booster research and development project. In October of that year the programme objectives were expanded to include a multi-stage carrier vehicle capable of performing advanced space missions.

The vehicle was tentatively identified as Juno V. The initial objective of the research and development programme was to prove that large amounts of thrust could be produced by clustering several engines. To speed booster development, it was decided to use propellant tanks and components designed for the Army's Redstone and Jupiter programmes. Most of the hardware could be built with existing tools, using established fabrication and inspection procedures. It was also determined that the Thor-Jupiter engine could be modified and used for the booster. The approach was to be initially demonstrated by building and testing a single non-flight stage.

Early in 1959 the Juno V designation was changed to Saturn, a name suggested by the comparable positions of the Jupiter and Saturn programmes and the two planets in the solar system. Late that year, two policy decisions of far-reaching significance were made:

1. The Department of Defence decided that it had no immediate use for a large rocket and, in view of the emerging national space programme, turned the Saturn Project over to the newly formed National Aeronautics and Space Administration (NASA). The Marshall Space Flight Centre, in Huntsville, Alabama, was the NASA segment given responsibility for the development.

2. NASA formed the Saturn Vehicle Evaluation Committee, composed of NASA and Department of Defence officials. The committee decided that all upper stages of the Saturn should be powered by the high energy propellant combination of hydrogen and oxygen, and that a new hydrogen engine would be developed.

Based on the committee's decisions, NASA outlined a building block approach to launch vehicle development that would lead to a series of successively larger vehicles, beginning with the 1.5 million pound thrust capability.

An urgent need for an even more powerful vehicle was created in late 1961, when the late President Kennedy challenged the nation to place men on the Moon before the end of the decade. Plans were subsequently completed for the design, manufacture, and operation of the Saturn family of vehicles with three configurations (*Figure 2-1*): Saturn I would be used to place unmanned Apollo Command and Service Modules into Earth orbit; Saturn IB would launch these two modules plus the lunar landing craft — the Lunar Module — into Earth orbit for astronaut training and rendezvous practice; and the Saturn V would be used for the lunar landing. The IB configuration, a combination of Saturn I and Saturn V stages, was not in the original plans for the space programme; however, it was realized that development of an intermediate vehicle capable of carrying the final spacecraft configuration would make the goal much easier to meet. Manned Earth-orbital rendezvous flights began a year earlier without the expense of a completely new programme.

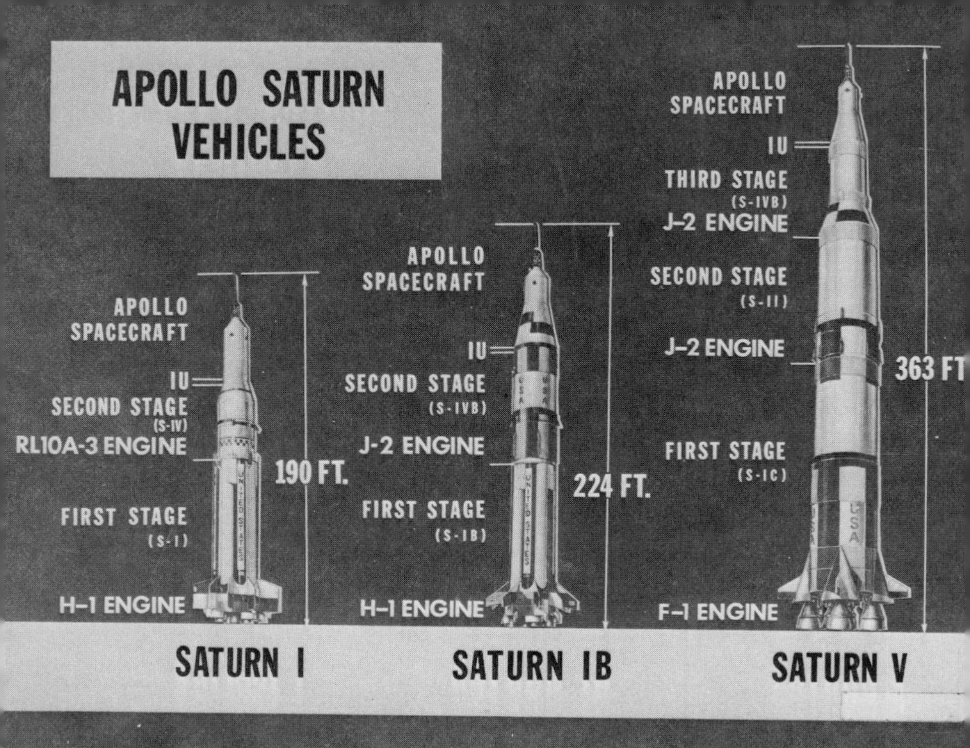

Figure 2-1. Apollo Saturn Vehicles.

SATURN I LAUNCH VEHICLE

Design studies showed that a booster built around a single high thrust engine was technically practical, but such an engine could not be developed and tested in time to meet flight schedules: development of a new engine would almost certainly be complicated by unpredictable technical problems. The approach selected was to secure the needed thrust by grouping eight engines of the Thor-Jupiter type. This engine, thoroughly tested and of proven reliability, was quickly simplified and uprated in thrust, and designated the H-1 *(Figure 2-2)*.

The other aspects of development of the Saturn I first stage, designated S-I, were solved in parallel. The S-I used the propellant tanks and components designed for the Redstone and Jupiter missiles, and was built using existing tools and established fabrication and inspection procedures. (*Figure 2-3*.)

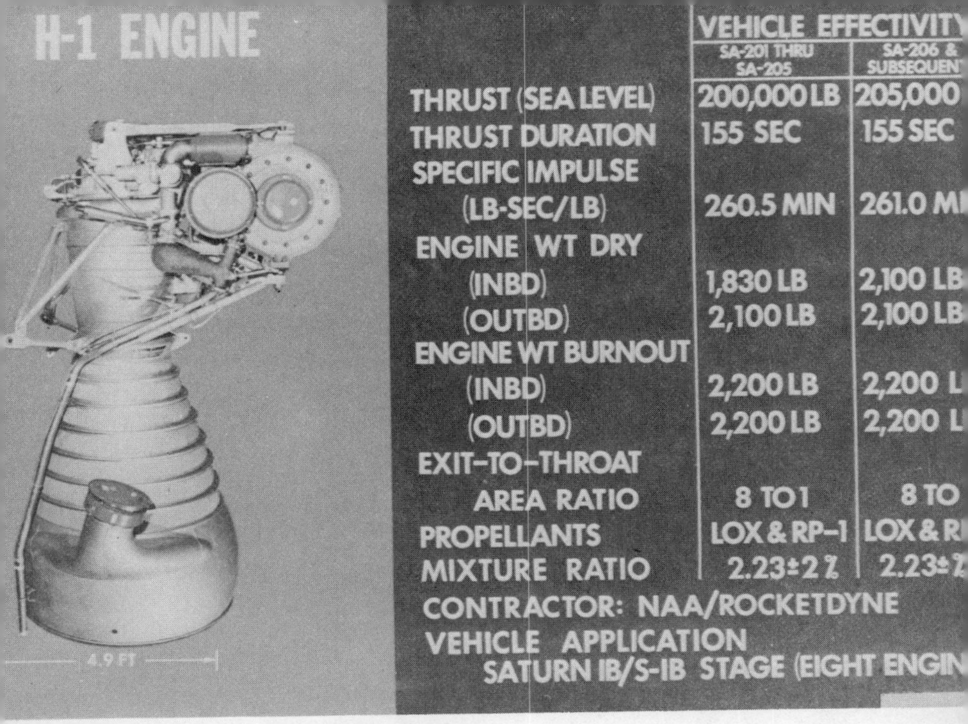

H-1 ENGINE

4.9 FT

	VEHICLE EFFECTIVITY	
	SA-201 THRU SA-205	SA-206 & SUBSEQUEN
THRUST (SEA LEVEL)	200,000 LB	205,000
THRUST DURATION	155 SEC	155 SEC
SPECIFIC IMPULSE (LB-SEC/LB)	260.5 MIN	261.0 MI
ENGINE WT DRY		
(INBD)	1,830 LB	2,100 LB
(OUTBD)	2,100 LB	2,100 LB
ENGINE WT BURNOUT		
(INBD)	2,200 LB	2,200 L
(OUTBD)	2,200 LB	2,200 L
EXIT-TO-THROAT AREA RATIO	8 TO 1	8 TO
PROPELLANTS	LOX & RP-1	LOX & R
MIXTURE RATIO	2.23±2%	2.23±%

CONTRACTOR: NAA/ROCKETDYNE

VEHICLE APPLICATION
SATURN IB/S-IB STAGE (EIGHT ENGIN

Figure 2-2. H-1 Engine.

Many design problems had to be solved in the development of a reliable booster. To use the maximum amount of propellants, a special propellant utilization system was devised. Special sliding joints were designed for the booster structure to compensate for shrinkage of the tanks when filled. To protect sensitive connections above the engines from the exhaust gases of 5,000°F, special insulation materials were developed. Tail shroud enclosures were designed for the rear of the booster to relieve in-flight aerodynamic pressures on the engines.

In the interest of advancing development as efficiently and rapidly as possible, NASA enlisted many private companies, universities, and other government agencies. From these in-house and out-of-house efforts came first designs and then hardware, with both being proved by thousands of tests. Several successful static firings of the S-I stage (*Figure 2-4*), conducted in the spring of 1960,

Figure 2-3. Saturn I Vehicle First Stage (S-1).

verified the clustered engine technique and established the basis for still larger vehicles.

During 1960, NASA awarded a contract for the second stage of the Saturn I (*Figure 2-5*), known as the S-IV, which was to be powered by six RL-10 engines (*Figure 2-6*). The S-IV stage would use the more potent and harder to handle combination of liquid hydrogen and liquid oxygen, which was relatively new fuel for propulsion: it had been applied only to the Atlas-Centaur vehicle, and in many areas the technology was still being defined. NASA working groups were formed with contractor participation; the design was reviewed and the proposed mating method defined to assure second stage systems' compatibility with the booster and to determine interface and interstage requirements.

The Saturn I was initially planned to be a three-stage vehicle;

Figure 2-4. S-I Stage Static Firing.

however, the added thrust obtained by use of LH_2 in the S-IV stage precluded the need, at that time, for a third stage. The two-stage vehicle with spacecraft was about 190 feet tall and weighed approximately 1,112,000 pounds at liftoff.

On October 27, 1961, the first Saturn I booster was flight-tested successfully from the Kennedy Space Centre in Florida (*Figure 2-7*). This vehicle, comprising the first stage booster with its dummy upper stage, was designated SA-1. Three successful flights followed, the SA-2, SA-3 and SA-4, each carrying dummy second stages.

The SA-5 vehicle, a combination of the S-I and S-IV stages, was successfully launched on January 29, 1964, with both stages live and functioning perfectly to place a 37,000-pound payload into Earth orbit (*Figure 2-8*). The SA-6 and the SA-7, launched in May and September, 1964, placed unmanned boilerplate configurations of the early Apollo spacecraft into Earth orbit. The SA-5,

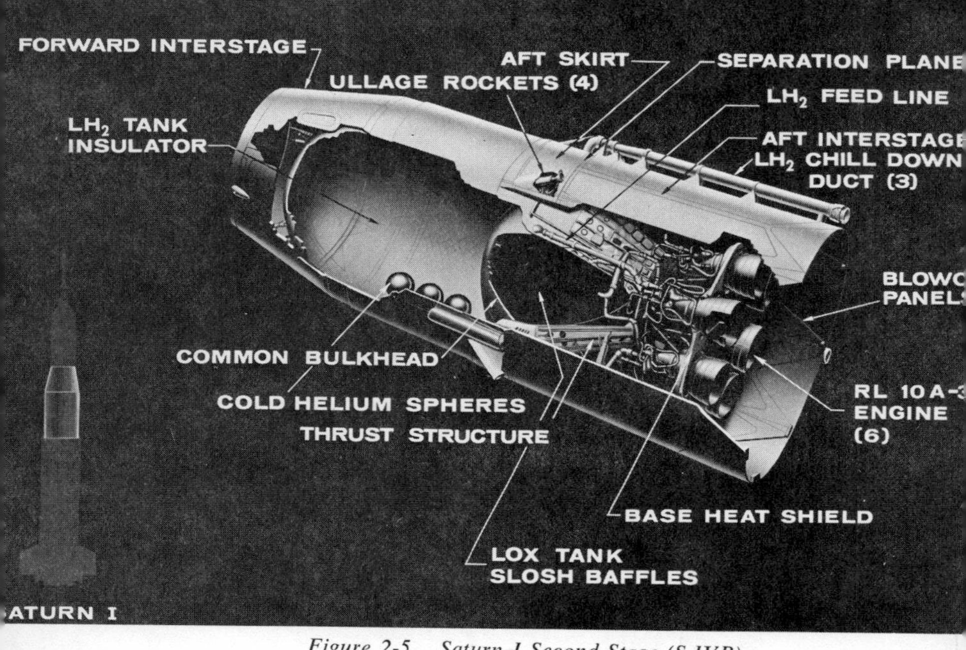

Figure 2-5. Saturn I Second Stage (S-IVB).

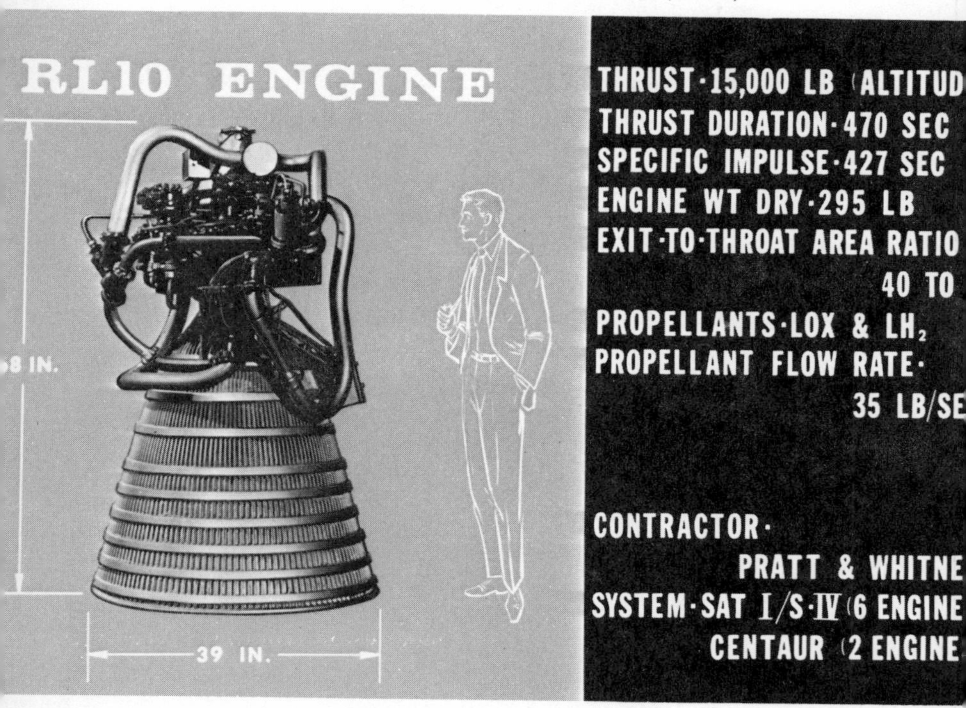

RL10 ENGINE
THRUST · 15,000 LB (ALTITUD
THRUST DURATION · 470 SEC
SPECIFIC IMPULSE · 427 SEC
ENGINE WT DRY · 295 LB
EXIT · TO · THROAT AREA RATIO
40 TO
PROPELLANTS · LOX & LH2
PROPELLANT FLOW RATE ·
35 LB/SE
CONTRACTOR ·
PRATT & WHITNE
SYSTEM · SAT I/S · IV (6 ENGINE
CENTAUR (2 ENGINE
8 IN.
39 IN.

Figure 2-6. RL-10 Engine.

ɪure 2-7. *First Saturn I Launch.*

Figure 2-8. Saturn I SA-5 Launch.

-6, and -7, were the first Saturn vehicles to fly an Instrument Unit, which constitutes the "brain" or "nerve centre" originating the commands for engine gimbaling, inflight operations of engine propulsion system, and staging operations. The components of these first units were available items not specifically designed for Saturn, and required pressurization and environmental control systems for their protection. The guidance computer was adapted from one developed for use in the Air Force's Titan missile.

The final three vehicles, SA-8, SA-9, and SA-10 were launched during 1965 with the assigned mission of orbiting Pegasus satellites (*Figure 2-9*). Each Pegasus, a 3,200-pound instrumented satellite, unfolded its wings to a total span of 96 feet after entering Earth orbit. This large exposed surface provided the means for gathering valuable meteoroid data, which the satellite's instruments transmitted back to Earth for evaluation of the meteoroid hazard in near Earth orbit. For these later Saturn Is an unpressurized version of the vehicle Instrument Unit had been developed, more compact in size and with a greatly improved inertial platform and control computer.

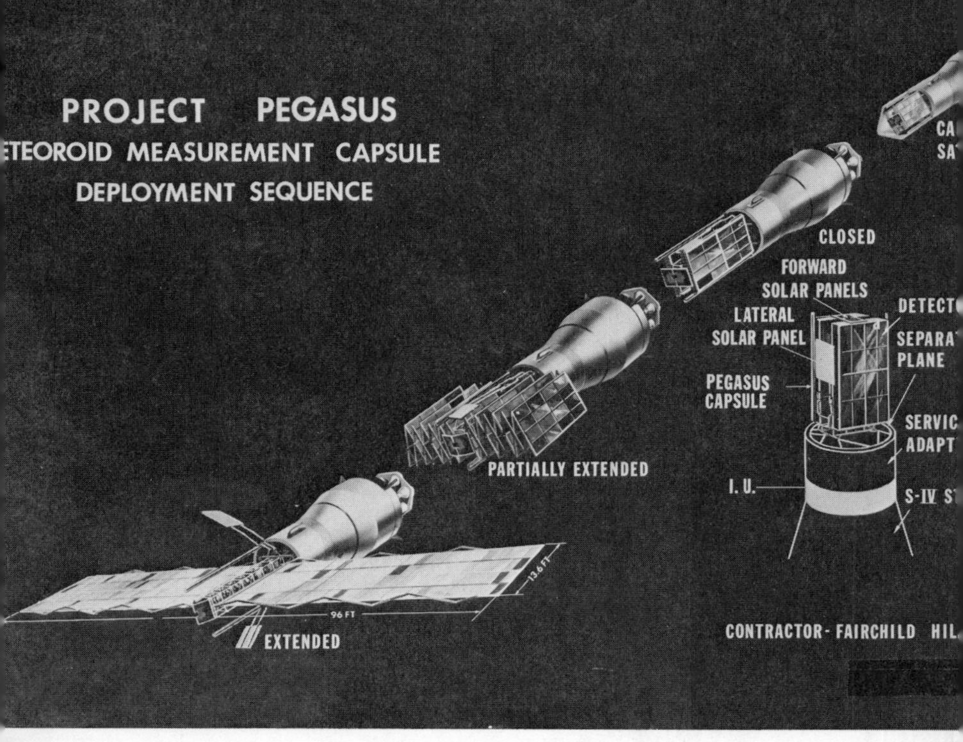

Figure 2-9. Project Pegasus.

The Saturn I test and launch programme, completed with the launch of SA-10 on July 30, 1965, had an unprecedented record of 100% success *(Figure 2-10),* and all agencies concerned were confident that equal reliability could be developed in larger, more powerful launch vehicles.

SATURN IB LAUNCH VEHICLE

The technological advances realized during the Saturn I programme made it possible to basically improve the design and efficiency of an already near-perfect vehicle. Original plans had not included any major vehicle development between Saturn I and Saturn V; however, the challenge of landing a man on the Moon by 1970 made it highly desirable to take immediate advantage of the added technology and insert a new vehicle configuration in the development schedule. In July, 1962, NASA announced that a new two-stage vehicle, to be known as the

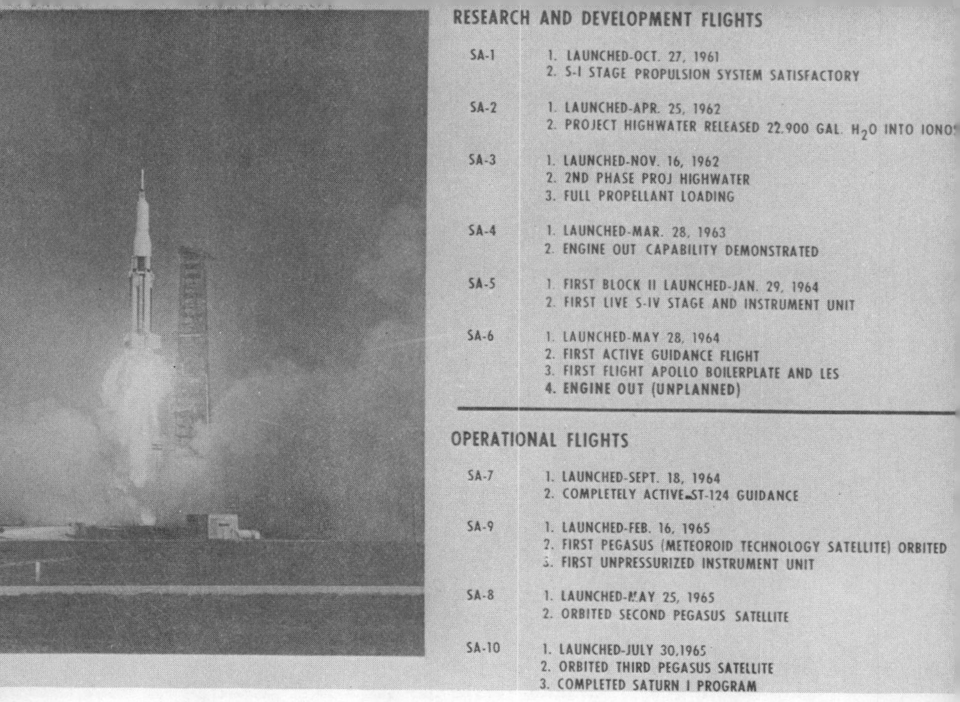

RESEARCH AND DEVELOPMENT FLIGHTS
SA-1
1. LAUNCHED-OCT. 27, 1961
2. S-I STAGE PROPULSION SYSTEM SATISFACTORY
SA-2
1. LAUNCHED-APR. 25, 1962
2. PROJECT HIGHWATER RELEASED 22,900 GAL. H_2O INTO IONOS
SA-3
1. LAUNCHED-NOV. 16, 1962
2. 2ND PHASE PROJ HIGHWATER
3. FULL PROPELLANT LOADING
SA-4
1. LAUNCHED-MAR. 28, 1963
2. ENGINE OUT CAPABILITY DEMONSTRATED
SA-5
1. FIRST BLOCK II LAUNCHED-JAN. 29, 1964
2. FIRST LIVE S-IV STAGE AND INSTRUMENT UNIT
SA-6
1. LAUNCHED-MAY 28, 1964
2. FIRST ACTIVE GUIDANCE FLIGHT
3. FIRST FLIGHT APOLLO BOILERPLATE AND LES
4. ENGINE OUT (UNPLANNED)
OPERATIONAL FLIGHTS
SA-7
1. LAUNCHED-SEPT. 18, 1964
2. COMPLETELY ACTIVE ST-124 GUIDANCE
SA-9
1. LAUNCHED-FEB. 16, 1965
2. FIRST PEGASUS (METEOROID TECHNOLOGY SATELLITE) ORBITED
3. FIRST UNPRESSURIZED INSTRUMENT UNIT
SA-8
1. LAUNCHED-MAY 25, 1965
2. ORBITED SECOND PEGASUS SATELLITE
SA-10
1. LAUNCHED-JULY 30,1965
2. ORBITED THIRD PEGASUS SATELLITE
3. COMPLETED SATURN I PROGRAM

Figure 2-10. Saturn I Launch Summary.

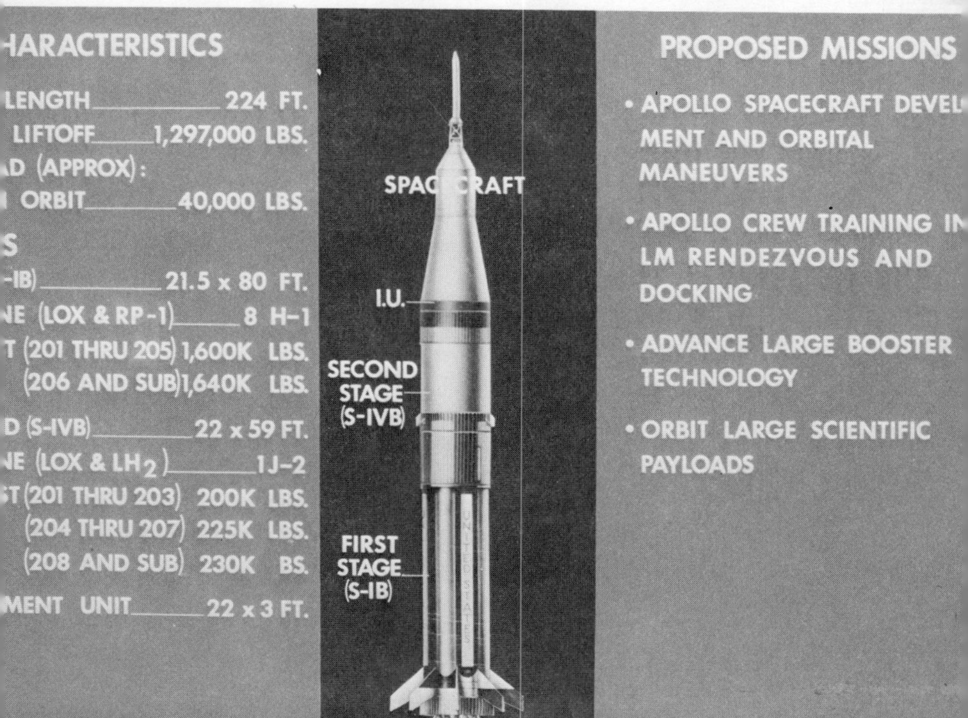

Figure 2-11. Saturn IB Launch Vehicle.

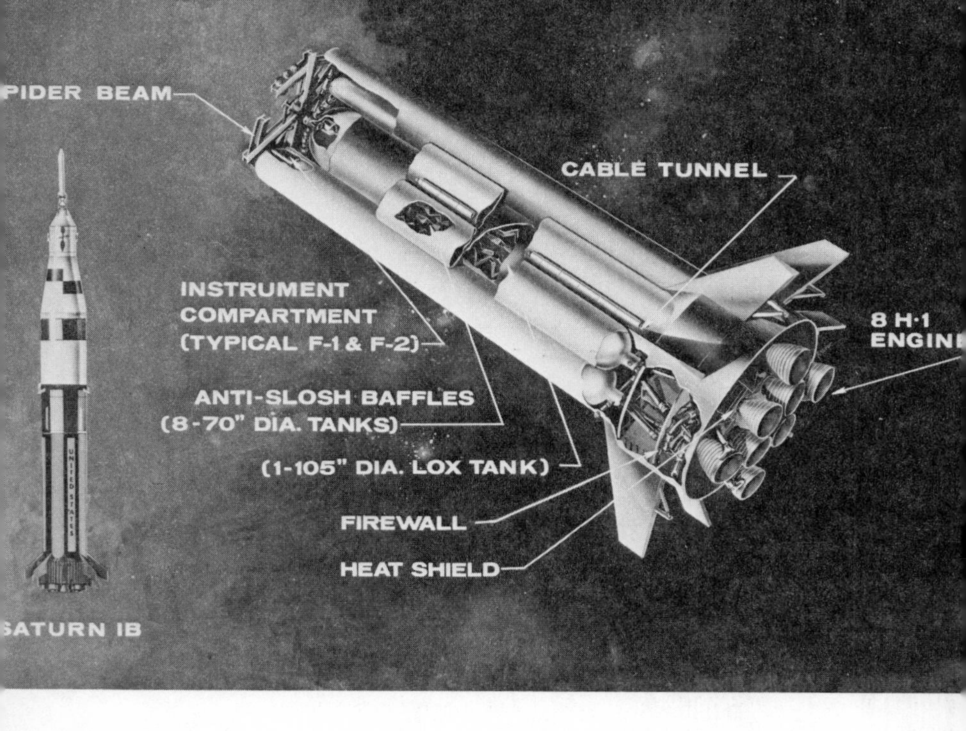

Figure 2-12. Saturn IB First Stage (S-IB).

Saturn IB *(Figure 2-11),* would be developed for manned Earth-orbital missions with full-scale Apollo spacecraft. With the Saturn IV, the Saturn-Apollo interface problems and re-entry heat shielding requirements could be studied in flight, thereby markedly shortening spacecraft development time by obtaining design information during actual flight and re-entry.

The Saturn IB consisted of the S-IB stage (*Figure 2-12*), which was a modified version of the S-I stage; and the S-IVB second stage. First stage weight was reduced by approximately 20,000 pounds to increase payload capacity. The reduction was accomplished by incorporating a new fin design, removing the hydrogen vent pipes and brackets unnecessary to the new design, resizing machined parts in the tail section assembly, redesigning the spider beam, and modifying the propellant tanks. The H-1 engine was uprated to 200,000 pounds of thrust, compared with 188,000

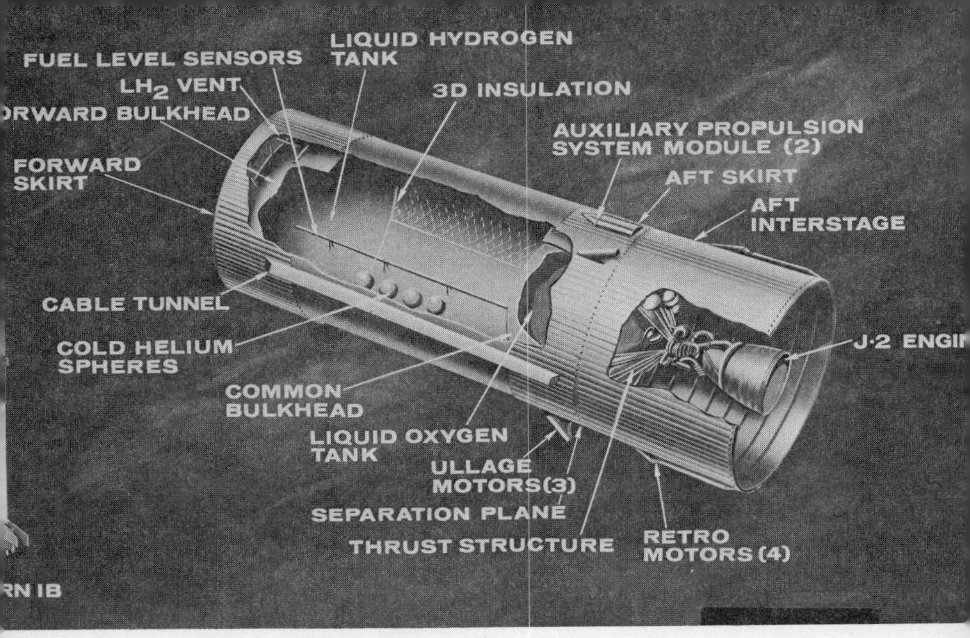

Figure 2-13. Saturn IB Second Stage (S-IVB).

Figure 2-14. J-2 Engine.

-2 ENGINE

6.8 FT

	VEHICLE EFFECTIVITY		
	SA-201 THRU SA-203	SA-204 THRU SA-207 & SA-501 THRU SA-503	SA-208 & SUBSEQUENT AND SA-504 & SUBSEQUENT
THRUST (ALTITUDE)	200,000LB	225,000LB	230,000L
THRUST DURATION	500 SEC	500 SEC	500 SEC
SPECIFIC IMPULSE (LB-SEC/LB)	418 MIN	419 MIN	421 MIN
ENGINE WEIGHT DRY	3,480 LB	3,480 LB	3,492 LB
ENGINE WEIGHT BURNOUT	3,609 LB	3,609 LB	3,621 LB
EXIT TO THROAT AREA RATIO	27.5 TO 1	27.5 TO 1	27.5 TO
PROPELLANTS	LOX & LH_2	LOX & LH_2	LOX & LH
MIXTURE RATIO	5.00±2%	5.50±2%	5.50±2

CONTRACTOR: NAA/ROCKETDYNE

VEHICLE APPLICATION:

SAT IB/S-IVB STAGE (ONE ENGINE)

SAT V/S-II STAGE (FIVE ENGINES)

SAT V/S-IVB STAGE (ONE ENGINE)

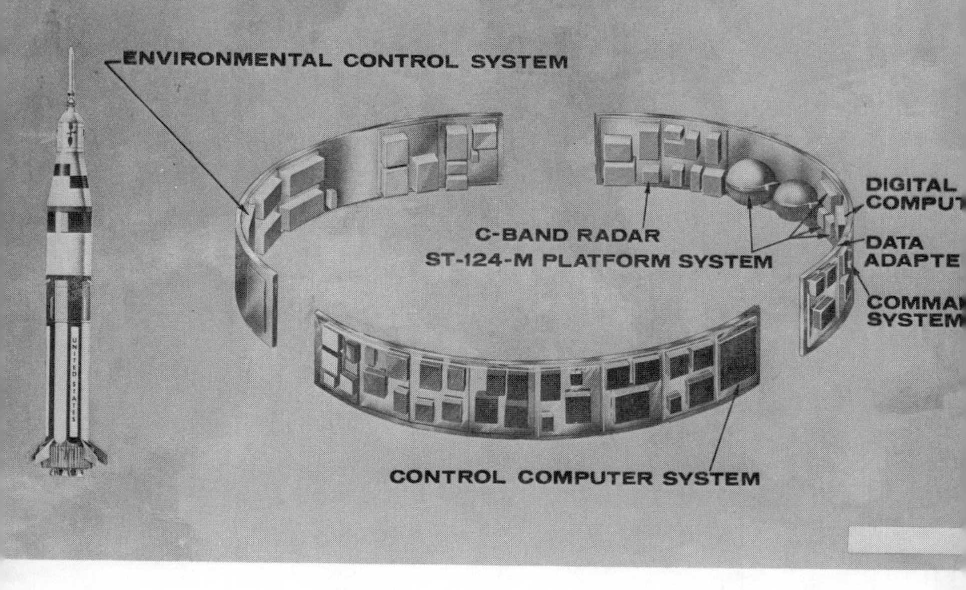

Figure 2-15. Saturn IB Instrument Unit.

pounds of thrust for each engine in the Saturn I. The S-IVB second stage *(Figure 2-13)* was originally designed for the future, larger launch vehicles. Its accelerated development was made possible by technology gained during the S-IV development. Power for the S-IVB stage was supplied by a single J-2 engine (*Figure 2-14*), a hydrogen-fueled type with a 200,000-pound thrust. Development of this engine had begun in 1960 and was near completion.

The design of the Saturn IB Instrument Unit (*Figure 2-15*) was based on that used in the Saturn I, but considerably modified, improved, and decreased in size. The early units used hardware that had been developed to meet military requirements, where the primary interest lay in automatic control systems for accurate delivery of inanimate payloads after a relatively short period of powered flight. The addition of man as an extremely important consideration meant that new systems had to be developed to respond to the longer duration flights, varied objectives, and an overriding concern for the safety of human passengers. The

Figure 2-16. First Saturn IB Launch.

Figure 2-17. Saturn IB Launch with LH_2 Experiment.

guidance computer used in the Saturn IB was of a completely new design that provided the increased flexibility necessary to carry out the programme's missions.

The Saturn IB, with the Apollo spacecraft, is approximately 224 feet tall and 22 feet in diameter. Empty, its weight is about 85 tons; when fully fuelled the liftoff weight is around 650 tons.

The first Saturn IB (*Figure 2-16*) was launched on February 26, 1966, and was the first "all up" or "live stages" launch of the new vehicle. The primary purpose was to flight test the launch vehicle and the Command and Service Modules of the spacecraft. In mid-1966 a second Saturn IB (AS-203) was launched with an LH_2 experiment as the prime objective (*Figure 2-17*). On-board television cameras recorded the behaviour and control of liquid hydrogen in the orbiting S-IVB stage. A third Saturn IB (AS-202) was launched in August, 1966, to further flight test the launch vehicle and develop the Command and Service Modules.

Following the tragic spacecraft incident in early 1967, Saturn vehicle launches were at a standstill for approximately a year. Flights were resumed in January, 1968, when an unmanned lunar module, a major stage of the Apollo spacecraft, was placed in orbit by the fourth Saturn IB for development flight testing. Nine months later a fifth vehicle was launched; its payload was the first manned Apollo spacecraft, and it was placed in Earth orbit.

In brief, the accomplishments of the five Saturn IB flights were:

1. Proved "all up" concept.
2. Verified engines and launch vehicle for manned flight.
3. Demonstrated mission support capability.
4. Demonstrated semi-automatic pre-launch checkout.
5. Provided flight experience with S-IVB stage, Instrument Unit, and Ground Support Equipment in support of larger Saturn vehicles.
6. Flight tested Lunar Module of Apollo spacecraft.
7. Placed first manned Apollo spacecraft into Earth orbit.

SATURN V LAUNCH VEHICLE

Introduction

The questions facing national space planners in 1961 and 1962 were complex. The United States had undertaken a manned lunar landing effort as the focal point for a broad new space programme, and there was no vehicle available that even approached the needed capability. Theoretically it was possible to use the Saturn I vehicle for a manned lunar landing, but it would have been extremely difficult. About six Saturn I launches would have been required to place the spacecraft components into Earth orbit, to be assembled there preparatory to the lunar trip. At that time no rendezvous and docking operations had taken place and the techniques still had to be perfected.

Following intensive concept studies, NASA announced in January, 1962, that a new rocket, much larger than any previously produced, would be developed to carry out the lunar landing. The new rocket, to be known as the Saturn V launch vehicle *(Figure 2-18),* would be composed of a small Instrument Unit containing guidance and control, and three propulsive stages.

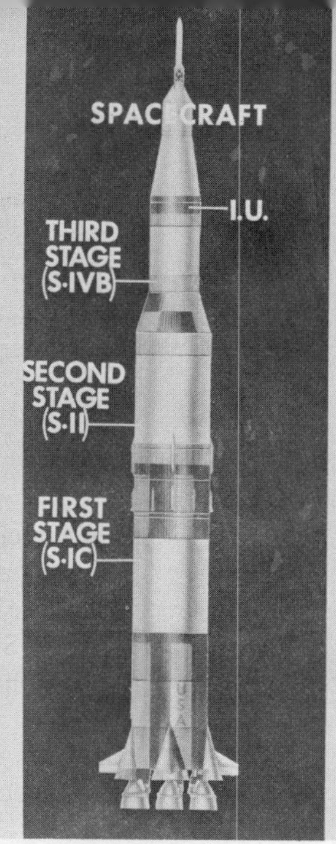

Figure 2-18. Saturn V Launch Vehicle.

The first stage, which would generate 4½ times the force of Earth gravity, would boost the entire vehicle to an altitude of about 38 miles and a speed of 6,000 miles per hour before separating from the upper stages. At a velocity of 15,300 miles per hour and an altitude of 115 miles, the second stage would shut down and be discarded. The third stage would ignite and burn briefly to boost the spacecraft to orbital velocity, about 17,500 miles per hour. The third stage would remain with the spacecraft, and its engine fired again to accelerate the spacecraft from its Earth-orbital speed of 17,500 miles per hour to about 24,500 miles per hour in a trajectory for a lunar trip. The spacecraft would then turn around, dock with the Lunar Module, pull the Lunar Module from the forward end of the third stage which is then abandoned. The launch vehicle's work would be completed at this point in the mission. Earth-orbital missions could be performed through the use of just the first two stages, but all three would be

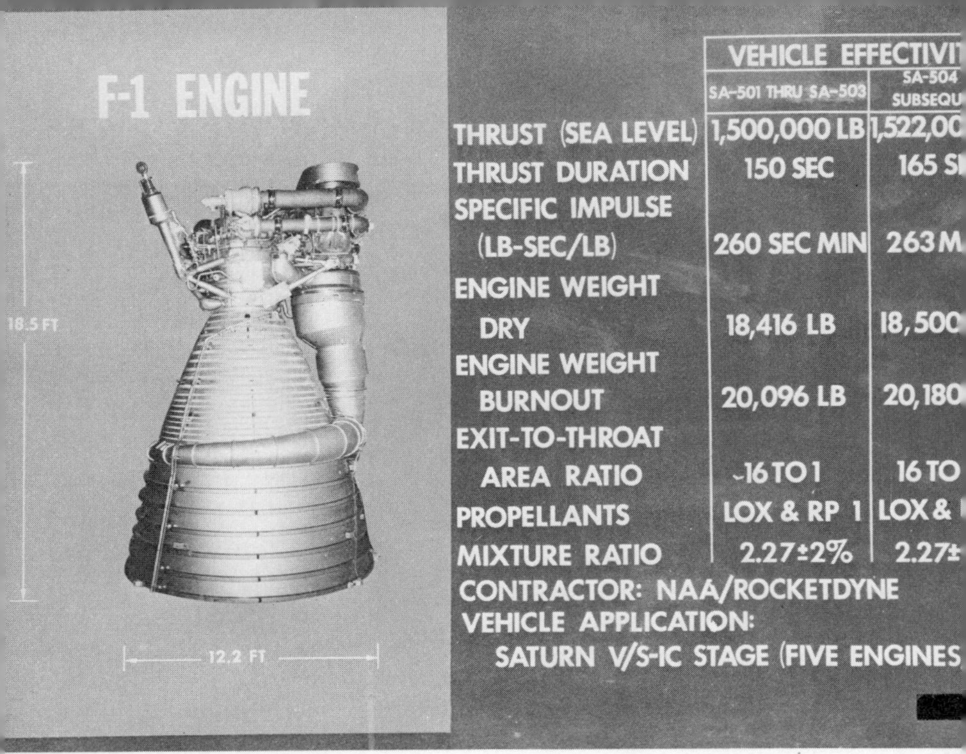

F-1 ENGINE
18.5 FT
12.2 FT
VEHICLE EFFECTIVI
SA-501 THRU SA-503
SA-504
SUBSEQU
THRUST (SEA LEVEL) 1,500,000 LB 1,522,00
THRUST DURATION 150 SEC 165 S
SPECIFIC IMPULSE (LB-SEC/LB) 260 SEC MIN 263 M
ENGINE WEIGHT DRY 18,416 LB 18,500
ENGINE WEIGHT BURNOUT 20,096 LB 20,180
EXIT-TO-THROAT AREA RATIO 16 TO 1 16 TO
PROPELLANTS LOX & RP 1 LOX &
MIXTURE RATIO 2.27±2% 2.27±
CONTRACTOR: NAA/ROCKETDYNE
VEHICLE APPLICATION:
SATURN V/S-IC STAGE (FIVE ENGINES

Figure 2-19. F-1 Engine.

required for lunar and planetary expeditions. The first stage would use the F-1 engine (*Figure 2-19*), which had been in development since 1958, and the second and third stages would use the hydrogen-fuelled J-2.

Saturn V Description

From these plans, aided by the early vehicle technology and given impetus by President Kennedy's challenge that "this nation take a clearly leading role in space achievement", the Saturn V evolved and is even today the United States' most powerful rocket. Completely assembled, with its three basic stages and the Apollo spacecraft, the Saturn V is 363 feet tall and weighs over 6 million pounds when fully loaded.

First Stage. The first stage of the Saturn V, designated S-IC (*Figure 2-20*), comprises six components in a vertical arrangement. At the bottom is a cluster of five F-1 engines. Upward from the engines, forming a cylindrical configuration, are the thrust structure,

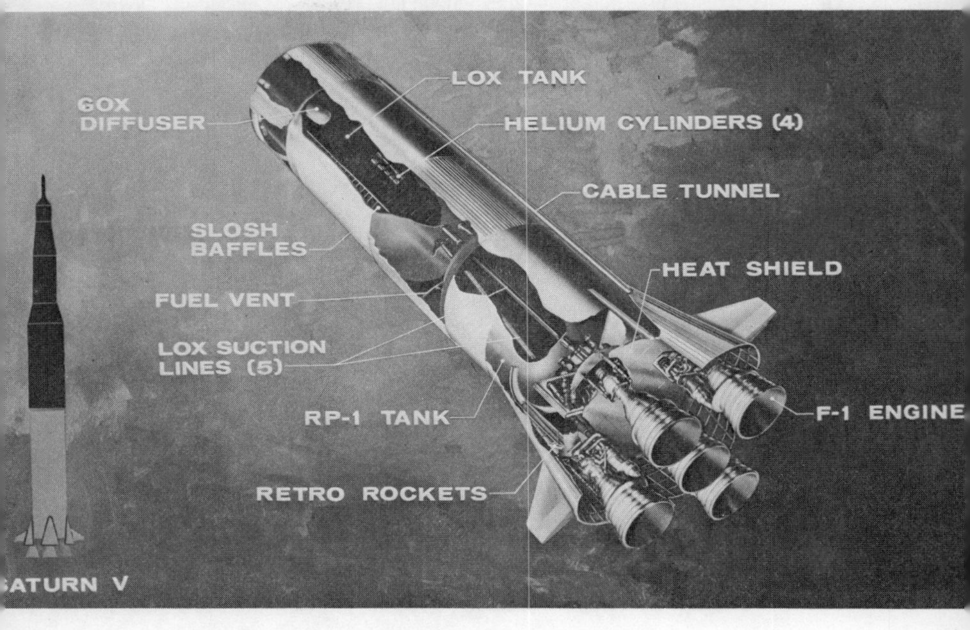

Figure 2-20. Saturn V First Stage (S-IC).

fuel tank, intertank structure, liquid oxygen (LOX) tank, and a forward skirt that connects the first and second stages. A 75-square-foot fin is mounted at each outboard engine position to provide stability during flight.

The S-IC is 33 feet in diameter (less the fins) and 138 feet long, and weighs 303,000 pounds without fuel and 4,881,000 pounds when loaded. In a firing, approximately 209,000 gallons of RP-1 (refined kerosene) and 334,500 gallons of liquid oxygen are consumed in about 2.5 minutes. (Propellant consumption varies with cut-off times tailored for different missions.) It contains its own instrumentation and safety systems, but receives guidance and control commands from the Instrument Unit.

Second Stage. The Saturn V second stage, the S-II *(Figure 2-21)*, measuring 81 feet in length and 33 feet in diameter, is powered by five J-2 engines, providing a total stage thrust of 1,000,000 pounds. It has a dry weight of 72,000 pounds and can carry

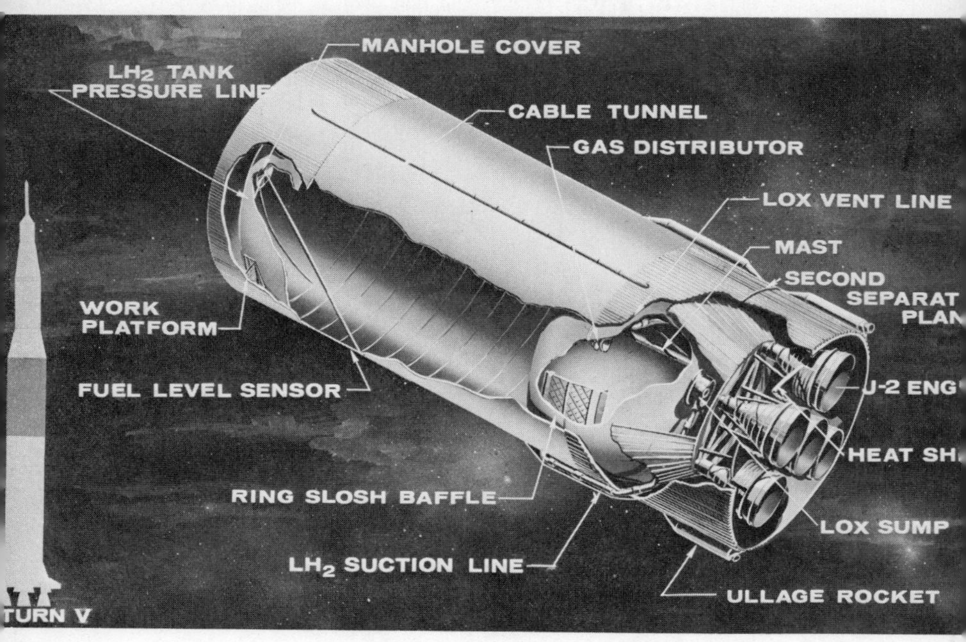

Figure 2-21. Saturn V Second Stage (S-II).

930,000 pounds of propellant. It burns approximately 275,000 gallons of LH_2 and 84,750 gallons of LOX during a typical six-minute flight. The major structures of the S-II are the forward skirt, liquid hydrogen tank, LOX tank, aft skirt, and thrust structure, and the interstage. It contains its own measuring, signal conditioning, telemetry, and electrical power systems; the engine servoactuators execute guidance command generated by the Instrument Unit.

Third Stage. Saturn V's third stage, the S-IVB (*Figure 2-22*), is approximately 22 feet in diameter and 59 feet long, and consists of one large tank with a bulkhead separating the fuel and oxidizer compartments. An interstage adaptor connects the larger diameter second stage to the smaller third stage. The power source is a single J-2 engine with a thrust of 230,000 pounds at altitude. Empty, the stage weighs 33,600 pounds; the fuelled weight is 265,600 pounds. Its LOX and LH_2 propellant capacity is 230,000

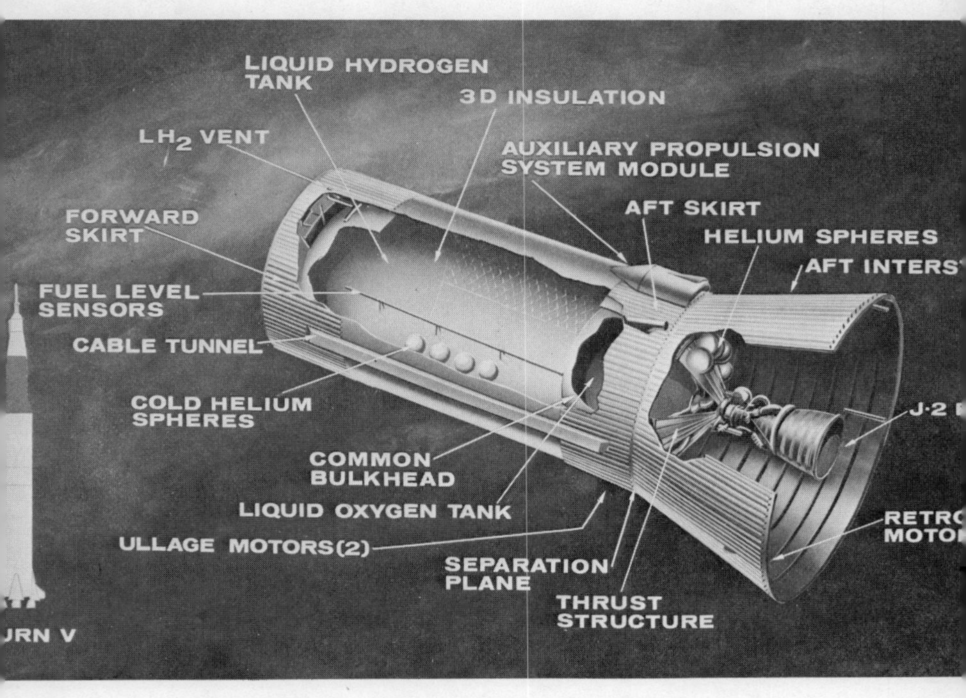

Figure 2-22. Saturn V Third Stage (S-IVB).

pounds. Measuring, signal conditioning, telemetry, and electrical power systems are internally contained.

The S-IVB stage is basically the same as that of the Saturn IB except that the Saturn V version has orbital restart capability required for the lunar mission. After the J-2 engine has placed the stage and spacecraft into an Earth orbit, the engine shuts down and all systems are checked out. When the spacecraft is properly oriented, the engine restarts and propels the spacecraft into a translunar trajectory. Typical burn times of the S-IVB are 2.5 minutes for the first burn to Earth orbit, and 5.5 minutes for the second burn to a translunar injection.

Instrument Unit. The Instrument Unit (*Figure 2-23*), located above the S-IVB stage, has a diameter of 21 feet 8 inches, a 3-foot height, and an average weight of 4,500 pounds. It is a highly flexible system that can provide guidance and control for a

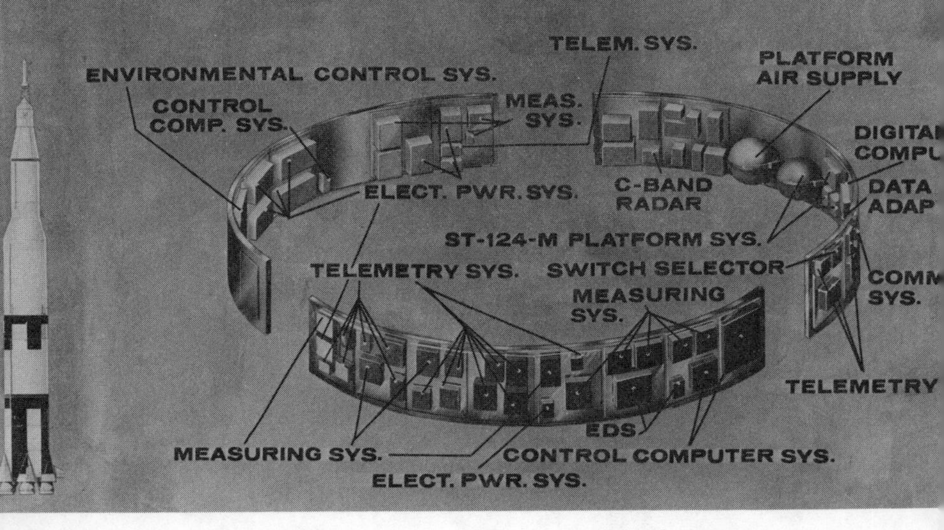

Figure 2-23. Saturn V Instrument Unit.

variety of vehicle configurations and flight paths. In addition to the basic functions, the Saturn V Instrument Unit has the following additional capabilities:

- Automatic checkout of the S-IVB and Instrument Unit systems prior to lunar injection.
- Guidance during injection of the S-IVB, Instrument Unit, and spacecraft into lunar transfer trajectory.
- Stabilization of the S-IVB, Instrument Unit, and Lunar Module during turn-around of the Command and Service Modules.
- Execution of manoeuvres to remove the S-IVB and Instrument Unit from the spacecraft orbital plane.

Lunar Landing Mode. As the project to accomplish the lunar landing goal was studied, it was determined that one huge rocket would not be built for a direct flight from the Earth and a soft landing of the entire spacecraft on the Moon. Instead, two orbital rendezvous approaches were considered:

- Bring together two Saturn V payloads in Earth orbit to form a Moonship, and then proceed to the Moon.

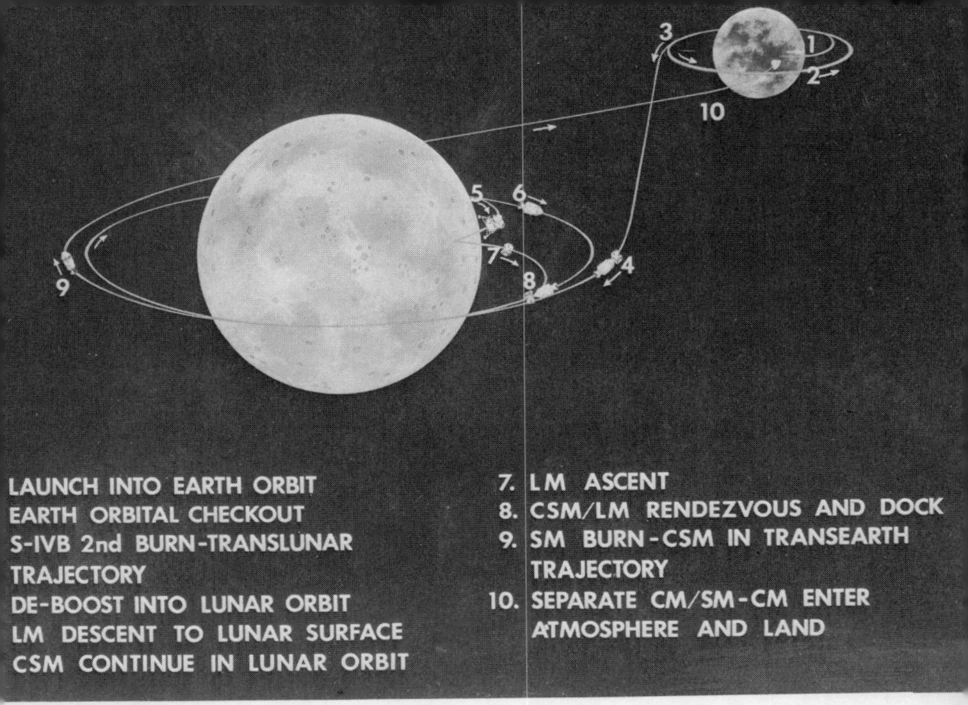

Figure 2-24. Apollo Lunar Landing Profile.

- Launch a single Saturn payload into lunar orbit, dispatch a small landing craft to the lunar surface, and later rendezvous the landing craft with the mother ship in lunar orbit for the return to Earth.

In July, 1962, it was announced that the lunar orbit rendezvous method *(Figure 2-24)* was favoured on the basis of cost, safety, and time. With all the experience to be gained from the Saturn I and IB programmes, it was clearly realized that the task ahead was still a momentous one.

Efforts in all aspects of the programme were broadened and intensified. Research continued with metals, insulations, processes and techniques. Design of advanced ground support equipment and launch support equipment extended proven design concepts to meet the increased needs of the larger more powerful vehicle. Tooling was adapted to the size and weight of the new, larger stages, and fabrication methods were improved and developed

Figure 2-25. Saturn V Facility Checkout Vehicle.

for handling new and sometimes exotic materials. New assembly methods were devised—assembly in the vertical position to prevent the stages from deforming through their own weight. Checkout procedures also had to be revised.

To assure that development progressed quickly and precisely, the facility design, construction and instrumentation were correlated with vehicle and ground support equipment programmes. As firm requirements were being set for static and dynamic test stands, techniques for handling, transporting and assembling the large stages at the launch site were also being determined. Because of their bulk, the new vehicles presented numerous problems: additional safety procedures had to be developed, telemetry stations had to be expanded, and improved methods of protecting and servicing the vehicle after assembly had to be devised.

By May 26, 1966, the first Saturn V vehicle, in the Facilities Checkout configuration, had been assembled and transported to

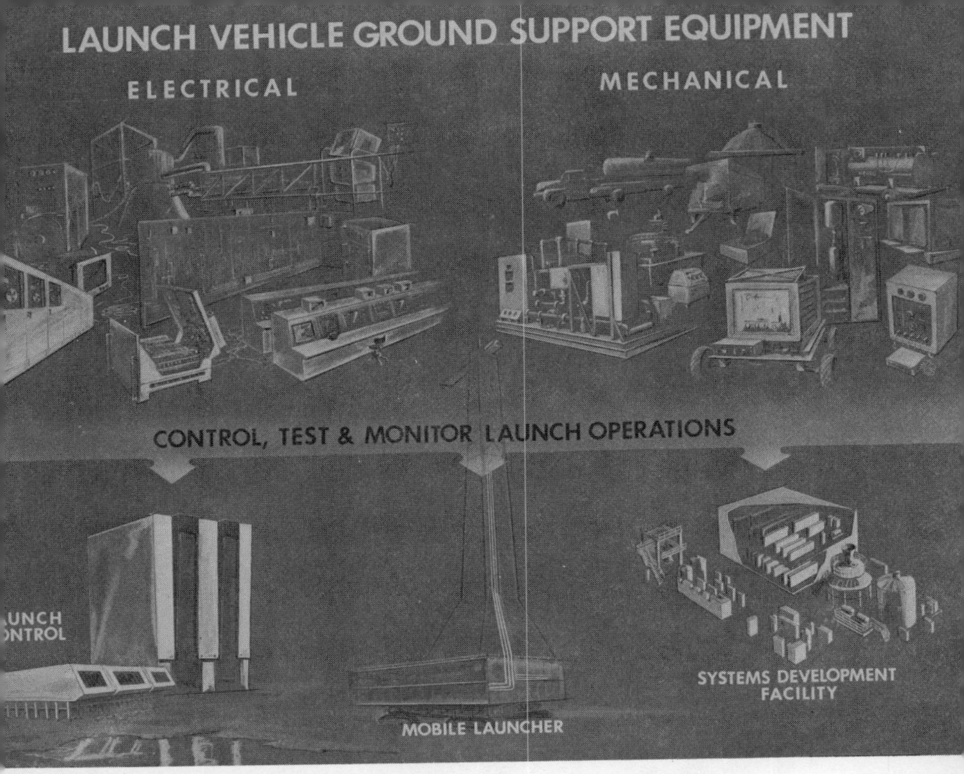

Figure 2-26. Launch Vehicle Ground Support Equipment.

the launch pad on a crawler (*Figure 2-25*). There it was used for a non-launch checkout of the launch facilities.

Launch Vehicle Ground Support Equipment

The third element of the system, following the vehicle and the Apollo spacecraft, is the Ground Support Equipment. The vehicle is totally dependent upon its Ground Support Equipment (GSE) —the electrical power sources, hydraulic pressure units, and the checkout equipment that probes the vehicle's well-being after assembly, and during countdown and launch. Development of the Ground Support Equipment *(Figure 2-26)* was intimately associated with development of the vehicle, and the schedules established for its design, fabrication and testing had to parallel those for the vehicle.

The complexity of the Ground Support Equipment programme can be illustrated by the more than 60,000 events monitored

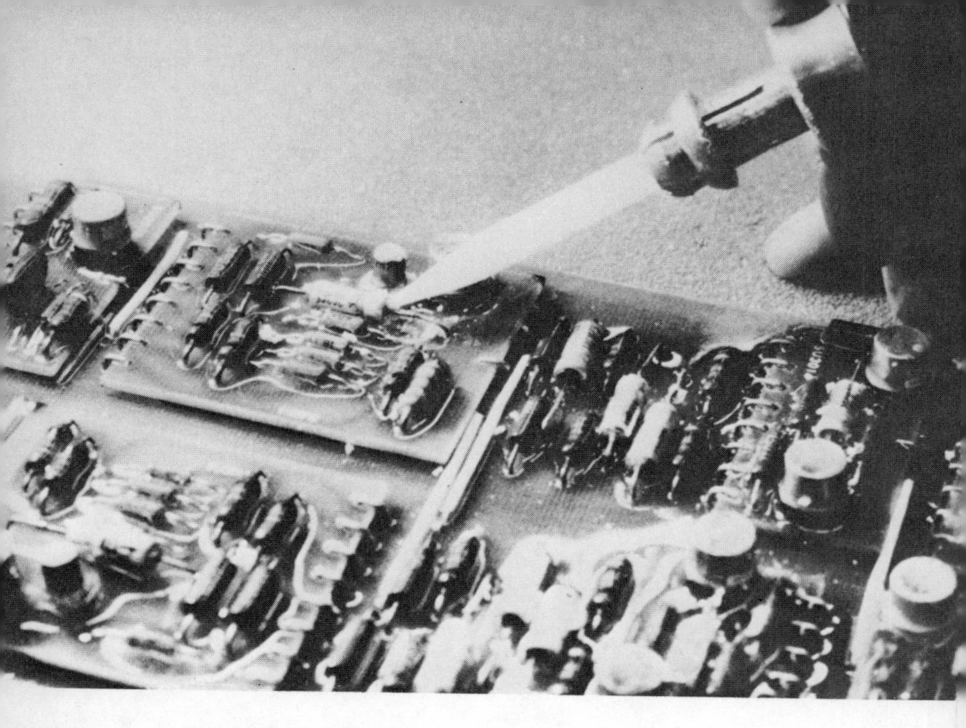

Figure 2-27. Cracked Solder Joints in Ground Control Computer.

during the development and manufacture of the mechanical and electrical Ground Support Equipment, as opposed to a total of 40,000 events for the Saturn V's three stages and Instrument Unit.

Problems were encountered in the GSE of the same major magnitude as those experienced in the development and operation of test and flight hardware. One typical example of GSE anomalies is the parity errors—discrepancies in computer input and output—which occurred in the ground control computers during prelaunch operations for the first Saturn IB flight. When finally isolated, the problem was determined to be cracks in circuit board solder joints *(Figure 2-27),* of which there are thousands. Thermal cycling of the conformal coated boards was established as the cause. A reliable fix was determined and the 64,000 boards involved were corrected with minimum impact to the programme.

A valuable accessory devised for the GSE development was the Systems Development Facility, more commonly referred to as

Figure 2-28. Saturn V Systems Development Facility.

the "Breadboard" (*Figure 2-28*). This facility is a simulation of the launch vehicle automatic checkout GSE at the launch pad, plus components simulating a completely assembled Saturn vehicle. Its primary purpose is two-fold: first, verification of vehicle and GSE compatibility and, secondly, development and verification of checkout tapes for actual flight missions. In addition, the facility permitted early identification and solving of problems that otherwise would have been encountered later at the launch pad and caused a delay in the launch.

LAUNCH VEHICLE TECHNOLOGY

Many problems were encountered early in the programme that required advancing the state-of-the-art in widely diverse areas. Some of the major difficulties, and the unique solutions that had to be devised, are described in the succeeding paragraphs.

Manufacturing/Welding

In manufacturing terms, the Saturn launch vehicle can be described as a large, lightweight, thin-skinned, cryogenic, high pressure vessel that requires extremely close manufacturing tolerances (±0.013 inch on 33 foot diameter). Because of the sheer physical size and the extremely lightweight structure, the manufacturing process demanded high-strength materials and miles of precision welding. The capability and reliability of the existing welding equipment were inadequate. The weld length in one pass for a 33-foot diameter tank was 100 feet, and the weld had to be perfect: one weak spot could have resulted in destruction of the entire vehicle.

The solution was an automatic weld machine that moved along a precision track, joining the aluminium sections of the tanks with a perfect or near perfect weld *(Figure 2-29).* Each weld was thoroughly inspected by X-ray, dye penetrant, and ultrasonic methods. When weld flaws were detected, repairs were made by hand-held equipment.

An interesting problem encountered was the tendency of the torch head to wander off the weld seam. It was discovered that the segments were too smoothly machined for the weld torch tracking system, which was based on detecting the discontinuity of induced eddy currents at the seam. The individual segments had been formed with such precision and size that the joint between the two segments offered no reasonable level of electrical discontinuity to the instrument. The solution was to scarf the segments and redesign the tracking mechanism for a much higher gain.

Insulation

S-IVB Stage. Early in the 1960s efficient insulation for large quantities of LH_2 was unknown. Even when properly insulated and filled to capacity, an S-IVB stage will lose some of the LH_2 by boil-off during the countdown procedure and must be replenished, at LH_2 flow rates up to 300 gpm. Numerous insulation materials for the LH_2 tanks were tested without success. Balsa wood was considered as a liner, but was not available in the size and quantity required for the 22-by-40-foot tanks. Another

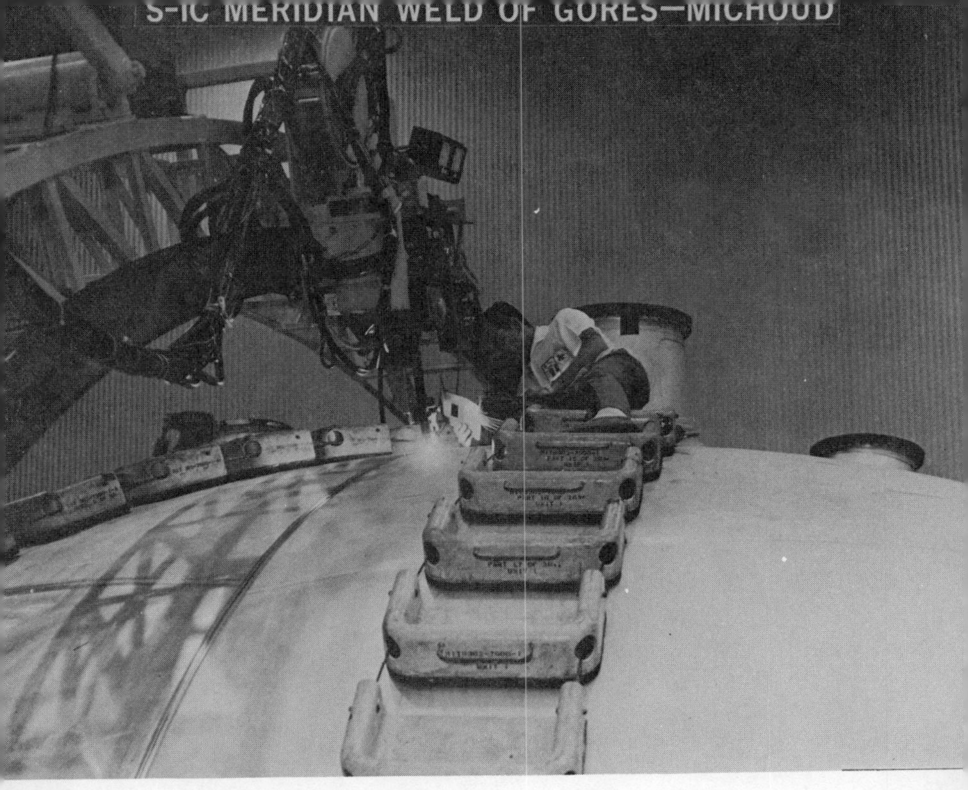

Figure 2-29. Automatic Weld Machine.

material considered was foam-filled fibreglass honeycomb, which proved unacceptable; the foam tended to shrink away from the sides of the honeycomb, allowing hydrogen to penetrate through to the wall. Attempts were made to force-press the fibreglass honeycomb a fraction of the distance through the foam, but this resulted in a shear plane located at the interface between the honeycomb and the foam, causing horizontal cracking.

The most successful material was a three-dimensional lattice of fibreglass threads filled with polyurethane foam. The threads were spaced approximately 3/16ths inch apart in all three planes to form a lattice work, and the foam was allowed to rise during the foaming process. Tests of this material were highly successful, but its fabrication presented a problem. Building a framework of threads in two dimensions (*X* and *Y*) is rather simple; however, ingenuity was required to design a machine that could weave the final thread in the Z plane.

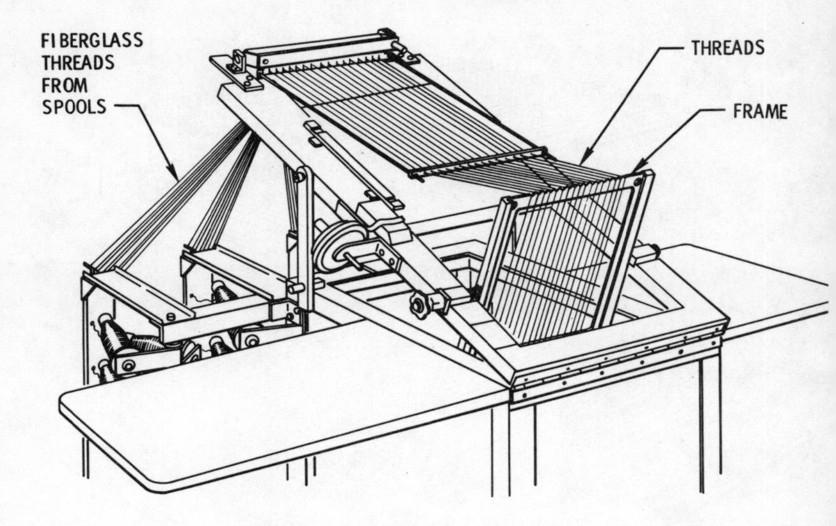

Figure 2-30. X-Y Thread Wrapping.

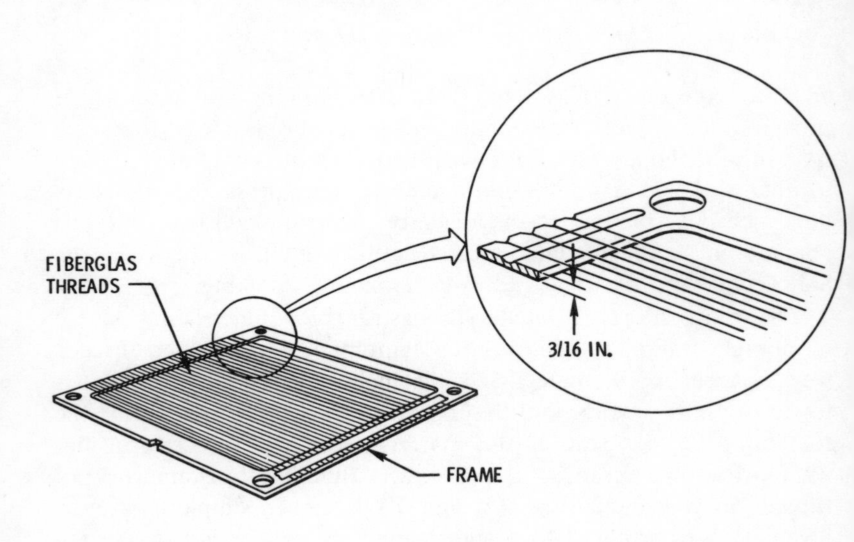

Figure 2-31. X-Y Frame.

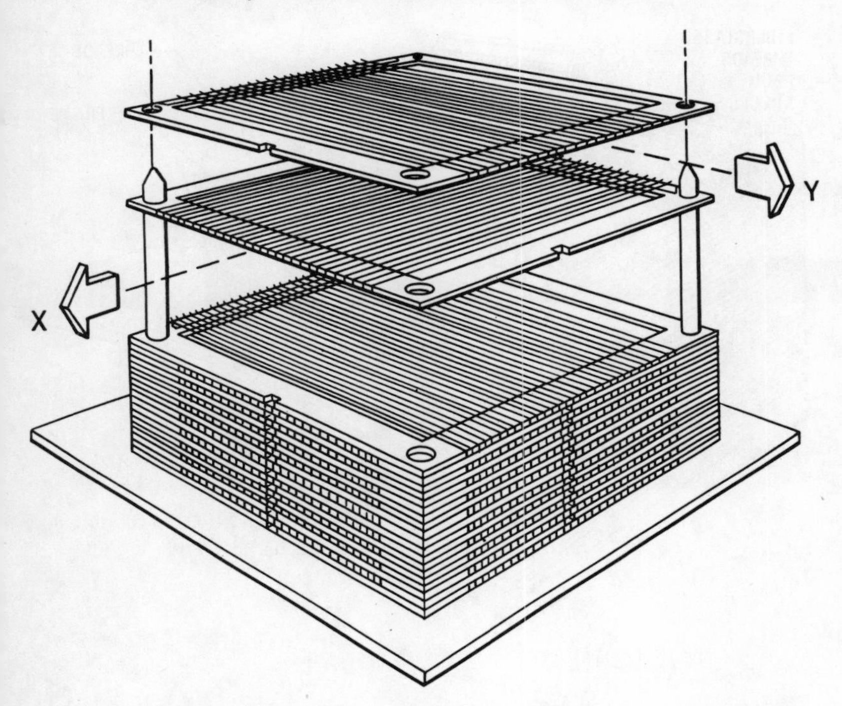

Figure 2-32. X-Y Frame Assembly.

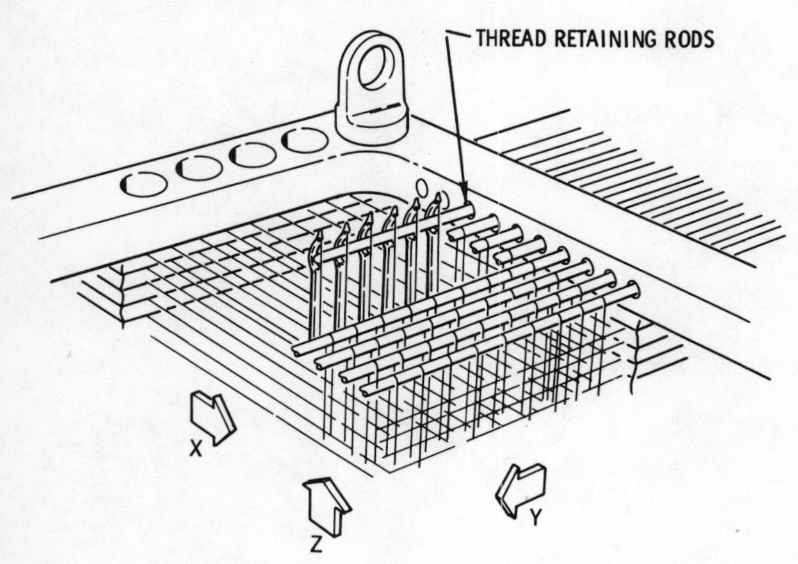

Figure 2-33. Vertical (Z) Thread Retaining Rods.

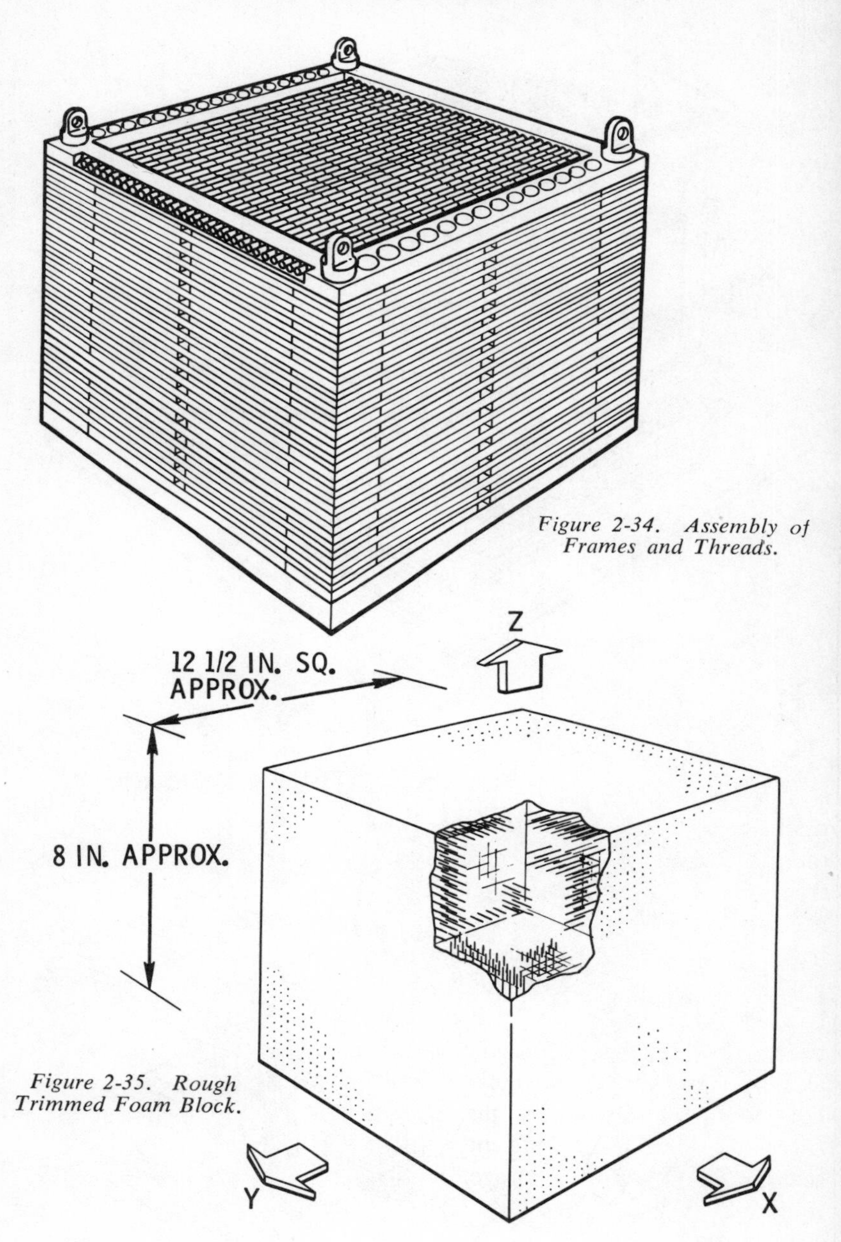

Figure 2-34. Assembly of Frames and Threads.

Figure 2-35. Rough Trimmed Foam Block.

Figure 2-36. Installation of Insulation tile in LH_2 Tank.

The *X-Y* machine on which threads are wound on special frames is shown in *Figure 2-30.* The frames are stacked alternately at right angles to each other *(Figures 2-31* and *2-32),* which arranges the threads in two of the three axes. The frames with *X* and *Y* threads are placed on the *Z* machine *(Figure 2-33),* where special needles weave the fibreglass as threads in the *Z* axis. After the frames are threaded they are placed in a mould (*Figure 2-34*) and polyurethane foam is poured over the threads. Once the foam is air-curved, the "cones" are removed *(Figure 2-35),* sawed into blocks, and machined into tiles with concave or convex contours as needed for lining the tank walls. This insulation is relatively lightweight and provides the necessary thermal characteristics *(Figure 2-36).* It is sufficiently free from maintenance problems, sensitivity to handling, storage, repeated thermal shock and transportation.

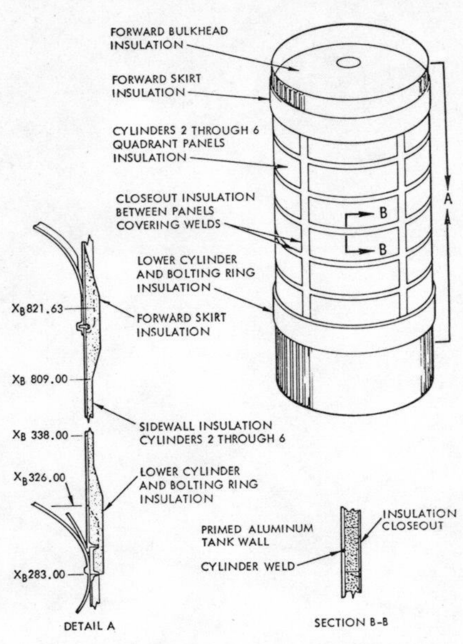

Figure 2-37. S-II Stage Insulation.

S-II Stage. For the Saturn V second stage, an external insulation was chosen (*Figures 2-37* and *2-38*) to take advantage of the gain in strength imparted to the aluminium alloy tank skin by the propellant at cryogenic temperature (–423°F.). To fulfil its purpose, the insulation had to limit the amount of heat leak to the LH_2 to meet the net positive suction head requirements of the feed pumps and to limit the ground hold boil-off of LH_2; remain structurally intact through all ground operations at the test site; and withstand flight environments of aerodynamic heating and shear. Certain environmental requirements were established, including the capability to remain thermally and structurally adequate when exposed to the natural environment for a minimum of three years, under minimum cold day conditions of 28°F. when tanked with LH_2 under flight conditions with temperatures up to 650°F., and when subjected to the sinusoidal and random vibration levels as experienced during static firing and launch conditions. The insulation also had to be flame retardant to reduce the fire

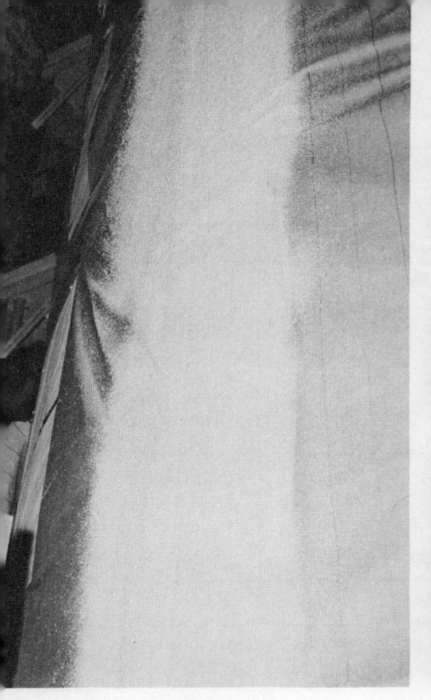

ure 2-38. *Spray Foam Insulation.*

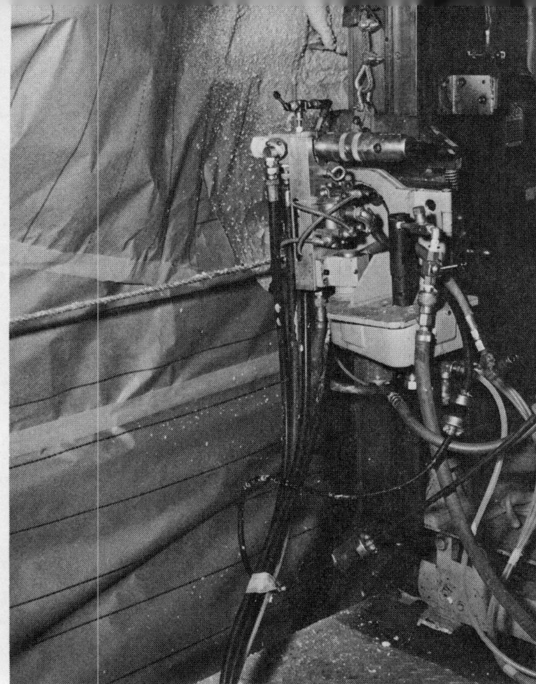

Figure 2-39. Application of Foam Insulation.

hazards associated with the use of LH_2 and *LOX*. Highly desirable was an insulation that could be applied by spray techniques, which would simplify application and reduce stage weight.

From the many considered and tested, the insulation selected was a foam-filled phenolic honeycomb core purged with helium *(Tables I and II)*. Helium provides an inert atmosphere and, to preclude ingestion of condensable gases, also allows for the practical fabrication of an insulation composite with less than a perfect external surface seal. The foam is sprayed onto the outside of the structure and allowed to cure. It is then machined to the required thickness and coated with a vinyl/polyurethane for protection against possible subsequent damage *(Figure 2-39)*.

Inflight Control of Liquid Hydrogen (LH_2)

The "super cold" LH_2 propellant selected for the early Saturn vehicles' upper stages and the Saturn V second and third stages was almost an unknown with respect to its behaviour under near weightless conditions and its effect on engine restart and vehicle

Property	Requirement	ASTM
Density, lb/ft^3	$2.0^{+0.2}_{-0.3}$	D1622
Closed cell, percent	89 minimum	D1940
Thermal conductivity, $Btu\text{-}in/hr\text{-}ft^2\text{-}°F$	0.2 maximum	C177 to -320F
Permeability at ambient conditions, Scc helium/sec	1×10^{-4} maximum	No ASTM
Coefficient of linear thermal expansion (RT to -150F, perpendicular to rise), in./in.-degree F	8.2×10^{-5} maximum	D696 below room temperature
Coefficient of linear thermal expansion (RT to -300F, perpendicular to rise), in./in. - degree F	6.9×10^{-5} maximum	D696 below room temperature
Flammability	Nonburning	D1692

Table I. Physical and Chemical Properties of Cured Foam.

Table II. Mechanical Properties of Cured Foam.

Property	Test Temperature degrees F,	Minimum Average Value psi (1)	Individual Value, psi, Minimum	Strain in./in. Minimum Individual	ASTM
Compressive strength (parallel to the rise)	70	25	20	0.021	D1621
	-423	45	40	0.021	D1621
Tensile strength (parallel to the rise)	70	48	38		In-house technique
	-423	50	38	0.012	
Tensile strength (perpendicular to the rise)	70	33	26		In-house technique
	-423	25	20	0.013	
Shear strength (perpendicular to the rise)	70	25	20		C273
	-423	20	15	0.022	C273

(1) A minimum of four specimens tested.

control. A single, carefully designed experiment, performed in conjunction with the flight of the Saturn IB vehicle, provided most of the knowledge necessary to achieve maximum and effective use of the propellant. The primary objective of the flight was to place the vehicle's second stage into a 100-nautical-mile circular orbit with 18,000 pounds of liquid hydrogen aboard. The four principal areas of investigation were the hydrogen venting system,

Figure 2-40. Saturn IB LH (Liquid Hydrogen) Experiment.

engine chilldown and recirculation system; tank fluid dynamics; and heat transfer into the liquid through the tank walls.

A television system, developed by NASA, enabled real-time observation of the liquid hydrogen behaviour throughout the flight. Mounted inside the tank of the S-IVB stage were closed-circuit TV cameras and lights, positioned so that the side and bottom of the tank as well as the hydrogen level could be clearly observed. Reference marks were painted on the tank walls to assist engineers in studying the action of the liquid during flight (*Figure 2-40*). The television picture was received and recorded by four ground receiving stations, one of which was located at Carnarvon, Australia.

At an altitude of approximately 40 nautical miles the first stage cut off as planned and separated from the second stage. The physical separation of the stages caused the liquid to slosh, and this reaction continued as the second stage ignited and accelerated.

Eventually the deflector ring and acceleration forces exerted a damping effect, and the fuel began to slow its movement and, finally, "settled".

At second stage cut-off, which occurs at orbital insertion, the liquid hydrogen began to rise toward the top and side of the tank because of a combination of physical phenomena, primarily zero gravity and amplification of the sloshing due to acceleration reduction. Heat transfer through the super-insulated tank wall to the liquid hydrogen (which boils at –423°F.) caused the density of the fluid near the wall to be reduced, resulting in a buoyancy effect with the liquid flowing upward along the tank wall and accumulating at the surface.

Since control of the fuel is essential to restarting the engine (for ejection from Earth orbit), a propulsive venting system to resettle the liquid propellant was included in the original design. The venting system makes use of hydrogen gas generated by heat transfer to provide a small thrust to overcome aerodynamic drag. In this test it was observed that the device operated as planned: when the propulsion venting system was operating, the liquid began to resettle to the bottom of the tank once the sloshing energy had been dissipated.

The other major factor in stage restart is chilling of the engine's propellant feed lines, turbopump, and thrust chamber to below the hydrogen boil-off temperature, to keep the liquid propellant from turning into gas. The recirculation system installed for this purpose, which keeps the liquid hydrogen flowing through the feed line for five minutes before restart, performed satisfactorily.

An additional experiment made on this flight was to learn how rapidly a propellant tank could be vented in orbit without losing some of the liquid itself. Vaporized hydrogen was observed moving towards the vent exits, behaving in a manner similar to that of a carbonated beverage that had been shaken before being uncapped. As the vapour moved forward it carried along globules of liquid hydrogen, ranging in size from one to six inches in diameter. This showed that at a high vent rate some liquid propellant would be lost, but that at lower venting rates the hydrogen vapours condensed into liquid and resettled to the tank bottom.

Figure 2-41. S-II Stage Structural Test Facility.

Figure 2-42. S-II Battleship Stage Test.

The liquid hydrogen experiments verified the adequacy of the propulsive venting system and the engine chill-down and recirculation system. In conjunction with the determination of tank fluid dynamics and heat transfer, they were major steps toward verifying engine restart and vehicle control capabilities. All of these data were of vital importance to the lunar landing programme.

Saturn Test Philosophy

The first concern in the development testing of any flight hardware is determining how many of the components and systems can be tested on the ground as opposed to those that must be flight-tested. In the early phases of rocket propulsion and launch vehicle technology, nearly all of the tests were performed by actual flights: not enough was known of space environment effects to permit simulation; it was feasible because no lives were involved, and the systems were relatively simple and inexpensive and could be expended. With the advent of more complex launch vehicles and manned spacecraft, this picture changed. It was necessary to place the maximum emphasis on ground testing, and the advances in technology as well as the increasing knowledge of the space environment made it possible. Most of the ground tests for the Saturn launch vehicles were nondestructive, and in practically all cases the resulting design decisions were verified in the subsequent flight tests.

Structural Tests. To verify their structural integrity, all Saturn stages were subjected to simulated flight loads (*Figure 2-41*) and the components were tested to optimize and prove the design load-carrying capability and to establish a margin of safety beyond the maximum expected operational environment.

Battleship Tests. "Battleship" tests (*Figure 2-42*) were conducted on propulsive stages to investigate overall propulsion system compatibilities and to establish system limits. The battleship configuration duplicates the flight stage in all respects except for the propellant containers, which are of heavier thickness than the flight article. During these tests, the engine was repeatedly fired to evaluate the engine and stage performance, propellant feed system operation, and compatibility of all stage systems with engine systems.

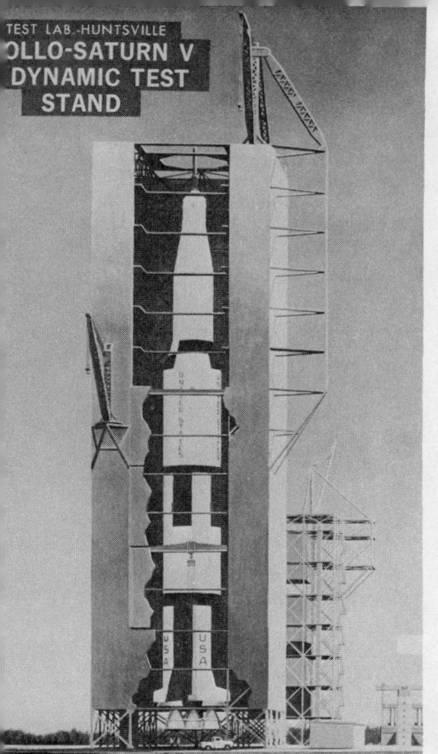

Figure 2-43. Apollo-Saturn V Dynamic Test Stand.

Figure 2-44. S-IVB Facility Checkout Stage.

Dynamic Tests. In the dynamic tests (*Figure 2-43*) the response of the complete launch vehicle was monitored under all flight conditions such as launch, maximum aerodynamic loading, and stage separation. These tests confirmed vehicle flight control system design and verified vehicle structure dynamics analysis. The tests included evaluation of bending modes, frequencies, interaction between engine gimballing motion and vehicle structures, damping characteristics, and local modes and frequencies.

Facility Checkout Tests. Facility checkout tests (*Figure 2-44*) were conducted at each stage test stand and launch complex to verify facility/vehicle compatibility. Some of the test objectives were to prove Ground Support Equipment capability for handling and transporting the vehicle and providing environmental control and propellant servicing, to develop test procedures and flow sequences, to train operating personnel in the various servicing and launch operations, and to develop safety methods and procedures.

Figure 2-45. Prestatic Checkout of S-IB Stage.

Launch Vehicle Checkout. In addition to the development ground testing, a comprehensive programme of launch vehicle checkout (*Figure 2-45*) was developed to determine the readiness of the vehicle for launch. The programme included the individual stage checkout at the factory, pre- and post-static testing at the test sites, and the prelaunch checkout of the integrated launch vehicle at the launch site. Automatic checkout instrumentation was installed at factories, test sites, and launch sites (*Figure 2-46*). As the stages and the completely assembled launch vehicles have processed through the automatic checkout stations, test engineers have analyzed and evaluated the data. From this analysis and evaluation, the engineers have been able to detect and isolate malfunctioning hardware and perform the maintenance actions that have returned the flight hardware to a state of operational readiness.

The exhaustive levels of many ground tests were the solid foundation of the Saturn flight test programme. The test plan

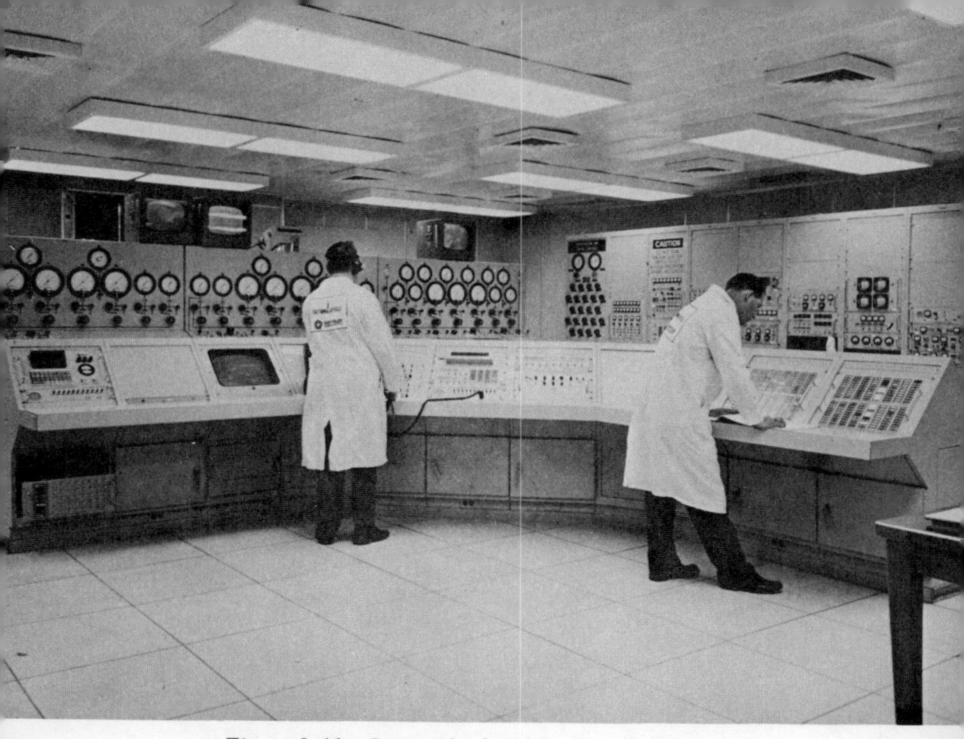

Figure 2-46. Stage Checkout Station.

philosophy was to fly as early as ground tests permitted, but not until the vehicle was proven to be completely ready and reliable.

The Saturn ground test programme has been characterized by a heavy reliance on applicable portions of test history from each predecessor vehicle. Relatively few test specimens were required in the programme considering the size and complexity of the Saturn family of launch vehicles. Some spectacular and well-publicized failures have been scattered throughout the ground test programme; however, these failures have been fully exploited for their learning potential, and the hardware remaining after each has been repeatedly used.

Flight tests were conducted only on those parameters of stages and systems that could not be proven by ground tests. These tests demonstrated operational capability, provided for a high degree of crew safety, and involved maximum use of the capabilities of each type of launch vehicle. More specific objectives included validation of all vehicle and spacecraft sub-systems, re-entry and landing

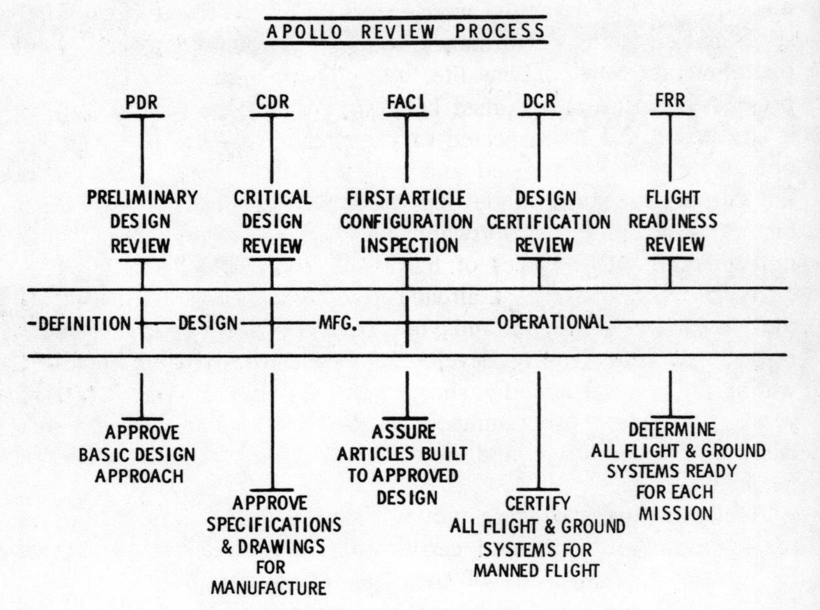

Figure 2-47. Apollo Review Process.

performance, testing of abort procedures, development of full operational capabilities of the propulsion system of the launch vehicles and spacecraft, perfection of navigation and guidance capabilities, rendezvous and docking exercises, and development of ground control capabilities.

Programme Review Process

The size and complexity of the Saturn programme dictated the need for an organized process whereby management could keep a finger on the pulse of the overall programme. One very effective innovation in this area was the systematic review procedure (*Figure 2-47*) implemented to track the pace and progress of the launch vehicle from definition, to design, to manufacture, and through the operational phase.

During the definition phase alternate approaches were studied and trade-offs between approaches were made to select the best. Once the approach had been determined, detailed performance and design requirements were generated. These requirements stipulated system performance during checkout, launch, and operations throughout the life span of the vehicle. They also prescribed features that must be designed into the vehicle so that it is reliable, can be inspected to assure quality, is safe to use and operate, and can be tested and maintained. Considered early in the design phase and made a part of the requirements were such factors as spare parts, software, facilities, and the handling and movement of large pieces of hardware from location to location.

By taking advantage of already developed missile hardware for the first configuration, and by devising special management procedures, the Saturn development schedule was compressed within an unprecedentedly short time. Within a period of five years, the Saturn programme proceeded from definition, through basic and final design and manufacturing, to the completion of the first Saturn V.

As the Saturn hardware moved through these phases, a series of design, acceptance, and certification reviews were conducted. For example, the results of the Preliminary Design and Critical Design reviews were the basis for approval of the basic design approach and the specifications and drawings required for manufacture. Any changes or open questions that arose during the reviews became action items, which were assigned for resolution and monitored by a strict schedule and follow-up system. Changes that had to be made were fed back into the design specifications. As the programme continued into the next phase, the first completely assembled stages were subjected to a strenuous configuration inspection (First Article Configuration Inspection) to ensure that all details were built according to the approved design.

The Design Certification Review was primarily a verification that all flight and ground systems could support manned flight. This portion of the process included the vendors, subcontractors, prime contractors, and project and programme managers, each of whom in turn had to certify that his hardware was operational and sufficiently reliable for manned flight.

Two to four weeks prior to the actual flight of each vehicle a Flight Readiness Review was conducted by key management staff members to determine that all flight and ground systems were ready for each of the missions assigned to the particular flight.

Each of these reviews provides a common meeting ground for all the technical and management disciplines that impinge on the programme. It is at this apex that representatives of all disciplines — reliability and quality, safety, test engineering, operations — participate as a team to evaluate technical progress from preliminary design through flight readiness.

Saturn Launch Vehicle Payload Growth

Adjustments in launch vehicle payload capability and spacecraft or payload weight are dictated by such factors as hardware changes, individual stage propellant loads, specific mission profile requirements, and specific engine performance as demonstrated by static firing. Strict controls were maintained to cope with the frequently changing picture of weight versus capability. This demanded the continuous attention of both managers and engineers to ensure that the payload weight and the launch vehicle capability remained compatible and the lunar landing could be achieved within the established time frame.

Over the past several years, as the Apollo spacecraft matured and became operational, its weight grew significantly, necessitating a corresponding increase in the launch vehicle's performance. The first three Saturn Vs were initially committed to an 85,000-pound Earth-escape capability; however, a payload capability of 100,000 pounds became necessary to accomplish the initial lunar landing. The succeeding flights required further increases in launch vehicle performance to carry additional scientific equipment and experiments to the lunar surface. As a result, the launch vehicle Earth-escape capability has grown to its present 106,500 pounds.

To obtain the approximately 21,500 pound increase, the vehicle's engines were uprated, stage weights were reduced, and optimum utilization was made of the system performance during the operational phase. For example, structural changes to the Y-ring tank supports in the Saturn V first stage reduced its weight by about 7,000 pounds; propellant loads on the first and second stages were increased by 293,000 pounds and 40,000 pounds, respectively, for longer burns of the engines; the thrust was increased

Figure 2-48. Internal view of Instrument Unit at Stabilized Platform Location.

by 15,000 pounds for the F-1 and by more than 5,000 pounds for the J-2; and the capability of specific vehicles has been matched with specific flight mission requirements. These improvements realized a corresponding increase in the Saturn V's Earth-orbital payload capability, from 265,000 to 285,000 pounds, which is sufficient to launch even heavier payloads than the currently planned Skylab and space stations.

Examples of Launch Vehicle Flight Anomalies

Despite all the reliability considerations in the design, manufacture, and test of the Saturn vehicles, anomalies of varying consequences have occurred during flight tests. Four typical instances and their solutions are described below.

Low Frequency Vibration (30-50 Hz) in the ST-124M3 Stabilized Platform. The ST-124M3 Stabilized Platform, located in the Instrument Unit (*Figure 2-48*), is a critical item in the Saturn vehicle inertial guidance system. The guidance system

Figure 2-49. Saturn V Instrument Unit Vibration Test Article.

consists of an inertial subsystem (ST-124M3) and a computer subsystem that provides guidance by (1) determining instantaneous vehicle acceleration, velocity, and position in a geometric inertial co-ordinate frame, and (2) determining the required thrust direction and thrust termination time to reach the trajectory or satisfy attitude requirements (steering).

On a Saturn IB flight, a low frequency vibration, encountered during the first five seconds of flight, forced the platform accelerometer pickups to drive against mechanical stops causing erroneous velocity pulses to be accumulated. This problem was also observed in the ground tests of the ST-124M3 Stabilized Platform which were being used to qualify the system at the higher vibration levels of the Saturn V. Since the S-IVB stage and Instrument Unit are common to the Saturn IB and Saturn V, it was possible to combine and check out these two indications by utilizing data developed from the following: (1) The Saturn V Instrument Unit vibration test article (*Figure 2-49*), which was

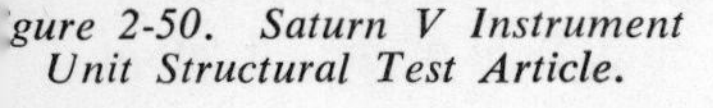

gure 2-50. Saturn V Instrument Unit Structural Test Article.

Figure 2-51. Saturn V Dynamic Test Article.

used to dynamically qualify the Instrument Unit structures, supplied the resonant frequencies, the corresponding mode shapes, and the driving point impedance; (2) the Saturn V Instrument Unit structural test article (*Figure 2-50*), used to statically qualify the same subsystem, supplied verification of driving point impedances obtained from the vibration article; (3) the Saturn V dynamic test article (*Figure 2-51*), which was used to dynamically qualify and evaluate the entire launch vehicle, provided verification of the stress and eflection analysis.

The analysis indicated a coupling of the rigid body structural modes with resonant frequencies of the accelerometer servo loop. The resonant frequency of the accelerometer servo loop was 30 to 50 hertz, and approximately coincides with the natural resonant frequency of the Instrument Unit honeycomb structure.

The alternatives for alleviating the problem were to redesign the accelerometer servo loop, which was prohibitive due to schedule requirements and cost; or to uncouple and/or reduce the amplitude

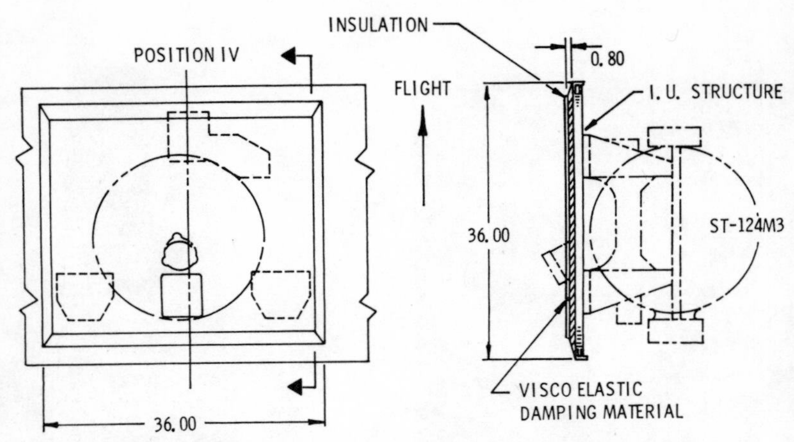

Figure 2-52. Instrument Unit Mass Damping Application.

of vibration to a level that could be tolerated. Attempts to uncouple the rigid body structural modes from the resonant frequencies of the accelerometer servo loop proved unsuccessful. However, it was possible to attenuate the amplitude of the vibration to an acceptable level by mass damping. The mass damping consists of bonding an 0.80-inch layer of damping compound to the outer Instrument Unit skin at the ST-124M3 mounting location (*Figure 2-52*). A minor modification was also made in the ST-124M3 servo loop, increasing the mechanical stop range from $\pm 3°$ to $\pm 6°$.

Longitudinal Oscillations (POGO). Liquid propellant boosters have frequently exhibited longitudinal oscillation in varying degrees during their development phase. The phenomenon occurred on the Atlas, Thor and Titan vehicles. It is caused by engine thrust oscillations resulting from coupling of vibrations in the structure with vibrations in the propellant feed system. These vibrations can involve the entire vehicle structure or just local structures. The first such occurrence of any significance in the Saturn programme was with the second and unmanned Saturn V flight, when these oscillations were noted during the first stage (S-IC) portion of powered flight. The problem was resolved by lowering the resonance of the liquid oxygen lines to uncouple it from that

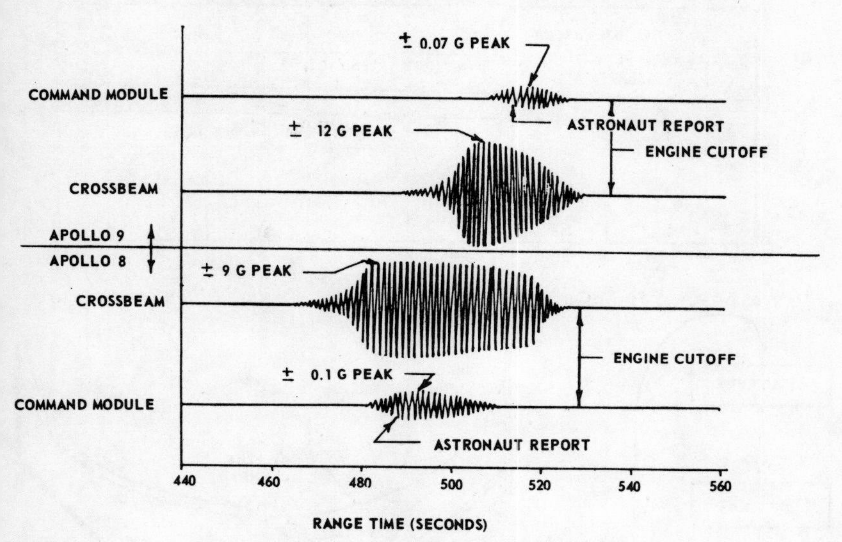

Figure 2-53. S-II Oscillation Responses.

of the structure. A helium gas accumulator system was added to the four outboard liquid oxygen (LOX) lines. The solution was verified on subsequent flights of the first stage.

During the third and fourth Saturn V missions the same type of condition arose with the second stage. The astronauts reported low frequency (18 Hz) longitudinal oscillations (vibration), commonly referred to as POGO, late in the stage burn. *Figure 2-53* shows the peak g's experienced in the second stage centre engine support structure (crossbeam) of the two vehicles, along with the Command Module peak g's during the two flights.

Although the structure and flight subsystems in the S-II second stage aft section were qualified to sustain loads of this magnitude, the variations in amplitude sensitivity among vehicles was of sufficient concern to require correction on the later vehicles. The investigation that followed indicated that the centre engine crossbeam support was sensitive to vibrations in the 18-Hz frequency, and that the centre engine produced this frequency after about 455 seconds of vehicle flight (*Figure 2-54*). The

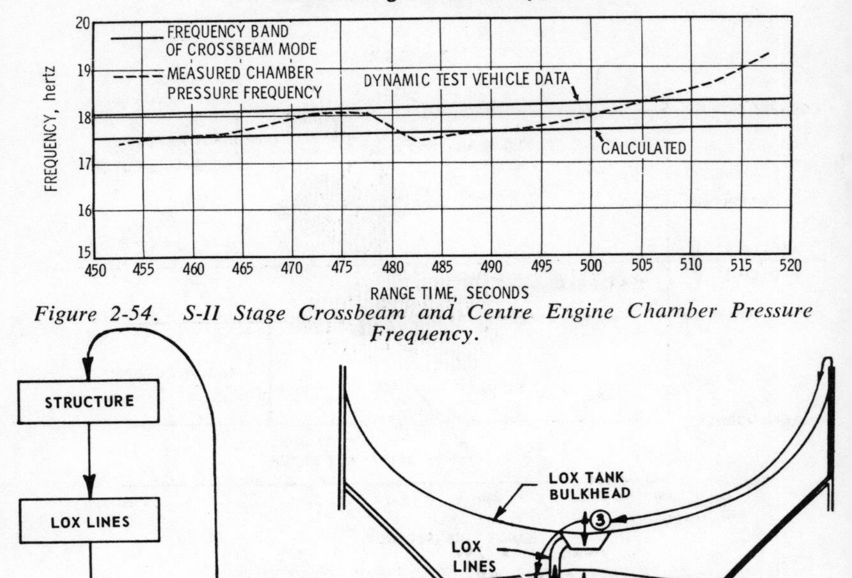

Figure 2-54. S-II Stage Crossbeam and Centre Engine Chamber Pressure Frequency.

Figure 2-55. Engine-Structure-Feedline, Closed-Loop Oscillation.

fourth flight was simulated by computer, using the simplified block diagram of the closed loop system shown in *Figure 2-55*, and the flight data and computer simulation correlation was verified (*Figure 2-56*).

Review of flight data and static test data from S-II stage firings revealed a correlation that was of great value in prescribing and verifying a resolution. Static firings exhibited the same kind of motion and frequency at the same vehicle LOX levels as was experienced in the flight vehicle, although at a much lower amplitude due to test stand constraints. Both sets of results supported the conclusion that the 18-Hz POGO was primarily due to regenerative feedback through the centre engine which occurred when the LOX had been consumed down to a particular level.

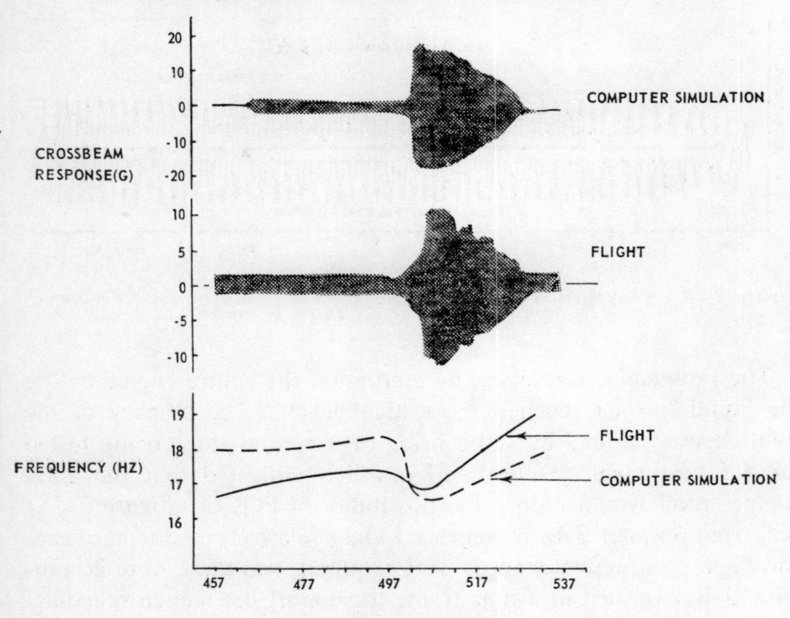

Figure 2-56. Flight Data/Computer Simulation Comparison.

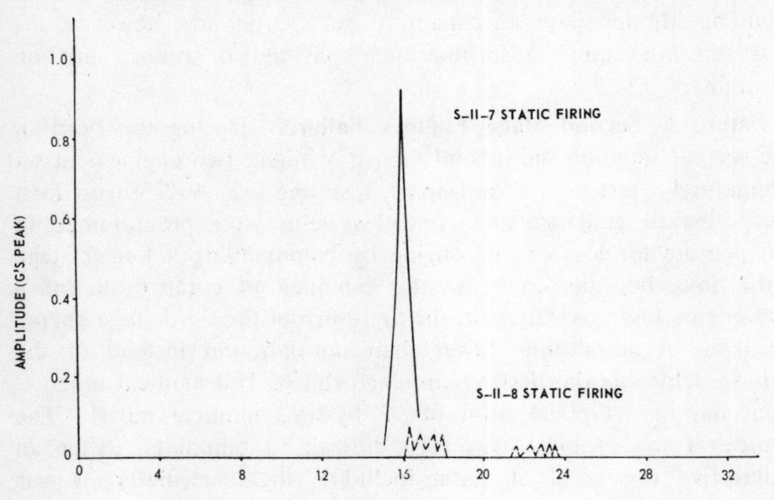

Figure 2-57. Centre Engine Thrust Pad Peak Response at Critical Liquid Level.

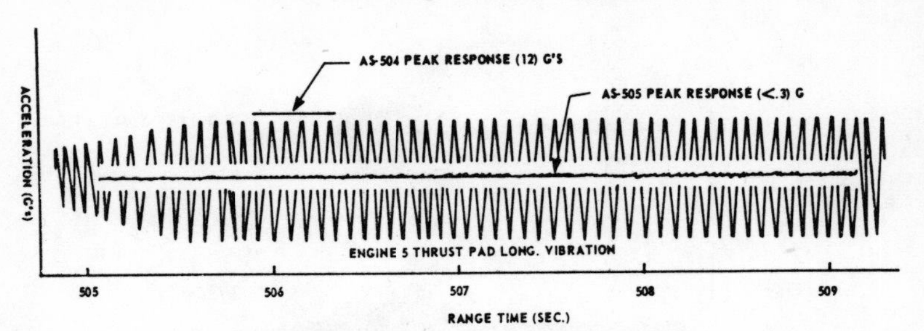

Figure 2-58. Comparison of Fourth and Fifth S-II Stages Low Frequency Vibration Response.

The problem was resolved by cutting off the centre engine before the liquid oxygen reached the critical level. The efficacy of the solution was verified by static firing of a second stage being tested for a future flight (*Figure 2-57*), which indicated that the early engine cutoff would reduce the possibility of POGO (*Figure 2-58*) occurring on later Saturn vehicles. Only relatively minor hardware and flight programme changes were required, and these were accomplished and verified in a time frame to support the launch schedule. This change in engine cutoff time decreased the engine burn time by approximately 80 seconds, and reduced the Saturn V Lunar Landing Mission payload capability by 500 pounds; however, the loss did not compromise the lunar payload or reduce mission flexibility.

Saturn V Second Stage Engines Failure. During the burn of the second stage on the second Saturn V flight, two engines cut off prematurely, just one second apart, resulting in a 40% thrust loss. The onboard guidance and control systems were programmed to compensate for loss of one engine by commanding a longer stage burn time, but this could not be extended to compensate for a two-engine loss. As a result, the first burn of the third stage engine occurred at an altitude lower than normal, and instead of the vehicle achieving the desired circular orbit of 100 nautical miles, it went into an elliptical orbit of 92 by 192 nautical miles. The source of the problem was very difficult to pinpoint. After an exhaustive review of all data, including flight telemetry, it was established that the second engine, which had been operating perfectly, was cut off by the failing first engine because of a

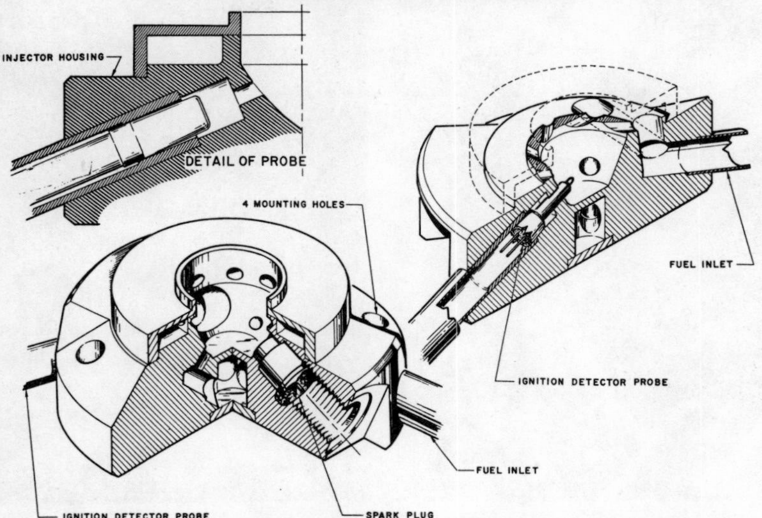

Figure 2-59. Engine Injector and ASI Assembly.

complicated wiring error. The error had not been detected during static and other tests due to an odd combination of several human errors.

S-IVB Stage Engine Failure. One prime mission objective of the second Saturn V flight, to restart the third stage engine to take the spacecraft out of orbit and into a simulated translunar trajectory, was not accomplished because the third stage engine failed to reignite. Here again, locating the source of the trouble involved a long and exhaustive search on the part of NASA and its contractors. A wealth of information was obtained from the flight data, but the issue was confused by the presentation of many secondary effects. One of these was a slight drop in performance about two-thirds of the way through the first burn of the third stage. Also, environmental temperatures differed from previous flights. In the ensuing investigation, employing intensive flight data analysis, computer simulation, and ground testing, all of these discrepancies proved to be due to the same defect.

The J-2 engine injector, its LOX dome, and the Augmented Spark Igniter (ASI) are shown in *Figure 2-59*. The ASI is

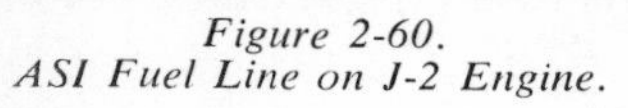

Figure 2-60.
ASI Fuel Line on J-2 Engine.

Figure 2-61.
First Saturn V Launch.

basically a redundant spark plug arrangement augmented by injection of liquid oxygen and liquid hydrogen to sustain the pilot flame for main chamber ignition. To prevent system complexity, the flow of oxidizer and fuel is maintained during engine operation.

Figure 2-60 shows the ASI fuel line with braided sections that protect the single-ply flexible metal bellows in the line. Ground testing revealed that these bellows failed in the vacuum environment of the AS-502 flight. Failures did not occur during ground tests at sea level conditions since liquid air formed on the bellows surface, which then reduced the vibration loads induced by the critical flow rate of liquid hydrogen through the line. However, failures were experienced during tests under vacuum conditions because no liquid air formation occurs in the vacuum environment. The line failure started with a leak and was followed by complete breakage.

Deterioration in the line structure caused a shift in the ASI mixture ratio, which in turn caused ASI assembly erosion, damage to the main chamber injector, a drop in engine performance and eventually engine cutoff. The failure was repeatedly duplicated by ground tests, which confirmed the flight data analysis and computer simulations. It was obvious then that the third stage failed to restart in orbit because insufficient fuel flow to the ASI assembly prevented proper ignition mixture rates. A redesign of the assembly was initiated immediately. The modification that proved effective, based on extensive component and engine system testing, was to substitute hard line sections for the flexible bellows.

Although Saturn vehicle flights have shown no serious design problems, the data from all flights are subjected to an extensive analysis, and all potential problem areas are kept under close surveillance.

SATURN RELIABILITY

Reliability is the probability of successful operation of a component or assembly over a given time period under a specified operating environment. In major technological undertakings, empirical studies and tests were once an accepted philosophy: the V-2 development required several hundred research and development flights; about 50 Redstone missiles were launched as test flights; prior to military deployment of the Jupiter missile, approximately 30 were test flown. Ten R&D flights were initially planned for the far more complicated Saturn I, but after seven flights, all of them successful, the vehicle was declared operational.

The dimensions, complexity, and cost of the later Saturn vehicles precluded the use of the trial and error methods, so that test flights prior to a manned flight were severely restricted. These limitations stressed more than ever before the critical need for intensive engineering, compulsive meticulousness in manufacture and assembly, quality control, and extensive testing on the ground.

NASA's reliability programme is based on the fundamental belief that a high mission success can be achieved only through a continuous sequence of closely controlled actions and events that begin with the design phase of individual parts, components, and systems. The approach to hardware design has been conservative, with emphasis on simplicity and reliance on proven building blocks and techniques. Reliability is designed into hardware, then proven in test.

Reliability considerations have been emphasized so strongly in the Saturn programme that they have, in effect, constituted a notable design parameter. Throughout the in-house, prime contractor, subcontractor, and vendor activities, reliability has been planned and conducted as a special and controlled effort. For example, a 100% in-process inspection is recommended for all critical components. Secondly, when final design is initiated, parts used in that design must be selected from a NASA Preferred Parts List. This standardization of parts across the entire launch vehicle reduced the qualification testing required, and gives assurance that only parts with known reliability histories are used.

The specific approach, beginning during the design phase, has been to reduce the system concept to its simplest functional unit, and to analyse that unit to determine all the possible ways it can fail; the effect of that failure on the subsystem, stage, vehicle, and mission; and how the failure compares in severity with other possible failures. The individual analyses are then combined to reflect failure effects on the subsystem, stage, and mission.

When a component fails to meet its goal, it is redesigned until the goal is met. In cases where the predicted reliability of a system was lower than that established, redundancies were considered. When prediction figures proved adequate, the system was developed using the most reliable parts, and the system was subsequently tested to prove that the established reliability goals were achieved. Additional changes and redundancies were again considered when inadequacies were revealed. Redundant components or subsystems had to be equated against weight constraints and the possibility of decreased reliability, plus any added complexity in certain modes of operation.

The Instrument Unit provides a good example of redundant systems. This unit, the "brains" of the entire launch vehicle, contains guidance, control, measuring, telemetry, power, tracking, sequencing, and emergency detection equipment. Its guidance and control system represents one of the most extensive applications of redundancy in flight systems. The various forms of redundance applied to critical components and subsystems include: duplex, triple-modular, prime reference standby, multiple-parallel-element, and quad redundancy.

Flight experience has proved that maximum reliability and quality assurance can be obtained by devoting equal and concurrent attention to the most minute part as well as the total system.

The final proof of success, of course, depended on the live launches. Since the Saturn V launches, their individual missions and results, are covered in detail in a separate chapter, only a brief summary of the events is included here.

On November 9, 1967, approximately five years after Saturn V development was officially authorized, the first Saturn V was launched; all stages performed as programmed (*Figure 2-61*). In addition to flight testing the vehicle for the first time, the launch included a simulated flight out to the Moon and back with the spacecraft Command and Service Modules. This provided the first test of the spacecraft heat shield at lunar re-entry velocities, as well as the first rehearsal of all the recovery forces, including the aircraft, ships, and tracking network. Although this was the first time that all elements of the vehicle and spacecraft had to work together in a launch environment, the only problems that arose were minor and had no effect on the mission.

Approximately five months later the second Saturn V was launched, with liftoff again occurring exactly on time. More significant anomalies were encountered this time requiring corrective action. After carefully studying the results of the first two flights, NASA announced on May 1, 1968, that the next launching would be manned.

The third Saturn V, carrying the Apollo 8 spacecraft, was launched on December 21, 1968, with Astronauts Borman, Lovell and Anders aboard. Apollo 8 completed 10 revolutions of the Moon during the 20 hours and 11 minutes spent in lunar orbit. Engineering evaluation of the launch vehicle confirmed that all test and mission objectives were met. The mission of the fourth Saturn V vehicle, launched on March 3, 1969, proceeded just as smoothly. Among the primary objectives successfully accomplished were astronaut extravehicular activity outside the spacecraft, and separation and subsequent rendezvous of the Lunar Module and

Figure 2-62. Rendezvous of Lunar Module and Command Module.

Command Module (*Figure 2-62*). The fifth successful launch of a Saturn V vehicle took place on May 18, 1969, and was the first mission in which the complete Apollo spacecraft was orbited around the Moon. The manned Lunar Module, previously flight tested in Earth orbit only, descended to within eight nautical miles of the lunar surface before docking with the lunar orbiting Command Module.

In its ultimate test, the manned lunar landing, the vehicle's performance was again flawless. The sixth Saturn V, launched on July 16, 1969, carried the first men to set foot on the Moon's surface. Of 15 million parts in the Apollo-Saturn vehicle, only one part failed and it was not mission critical. This is a demonstrated reliability of .999,999,996. The second lunar landing mission—the seventh Saturn V vehicle—was successfully launched on November 14, 1969, with all stages performing as planned.

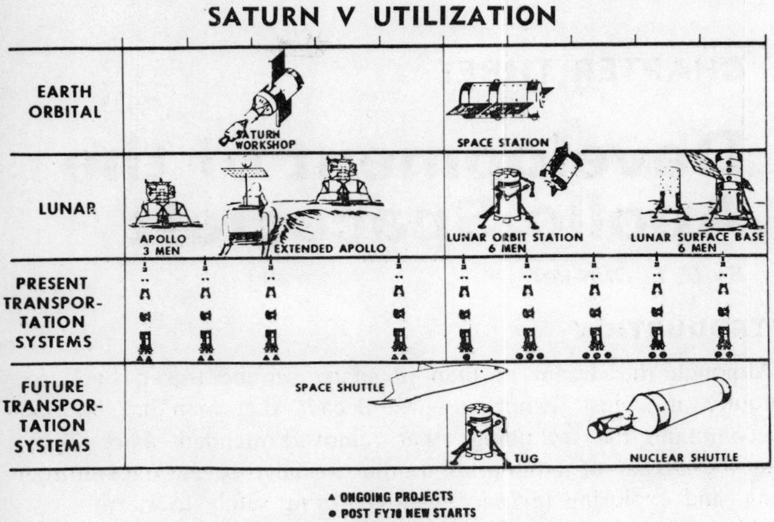

SATURN V UTILIZATION

▲ ONGOING PROJECTS
● POST FY70 NEW STARTS

Figure 2-63. Saturn V Utilization.

SATURN LAUNCH VEHICLES IN THE POST-APOLLO PERIOD

The early demonstration of Saturn V operational readiness resulted in a shift of Apollo missions from the Saturn IB to the Saturn V (*Figure 2-63*). The remaining seven Saturn IB vehicles will be utilized for ferrying services in the Skylab programme, which is discussed in a later chapter.

After the first successful lunar landing, the Apollo launch rate was reduced from six to two flights per year to permit careful scientific analysis of the findings and use of the resulting data in plans for future missions. Of the 15 Saturn Vs procured, seven have been launched. One is scheduled for launch in April, 1970, one in late 1970, two in 1971 and one in early 1972. One Saturn V, previously scheduled for an Apollo flight to the Moon, will launch the first experimental space station (Skylab) into Earth orbit in 1972. Subsequent to the Skylab launch, a two-year hiatus is planned to allow the scientific community to develop plans for additional lunar missions using the two remaining vehicles.

The production of Saturn V vehicles for the Apollo programme is near completion, and procurement of additional vehicles has been indefinitely suspended.

CHAPTER THREE

Development of the Apollo Spacecraft

By G. E. Mueller

INTRODUCTION

Although the dream of man in space can be traced back to antiquity, it is just within the past decade that man has had at his command the technology that removed manned space flight from the dream of visionaries to the actuality of man travelling, living, and exploring in space, and returning safely to Earth.

The primary block to manned space flight through the centuries was lack of a propelling force powerful enough to achieve the velocity needed to boost man into space. Not until the late 1950s had the requisite power been developed in the form of large liquid propellant rockets. By that time scientists were already convinced that man could survive in space, and plans were set in motion for practical application of the powerful engines.

The next hurdle was to build a craft adequate to carry man into space. This was an even more ambitious undertaking, requiring development of technology for which very little basis existed, and involving practically every scientific discipline. For the first time, man would go beyond the protective layer of the Earth's atmosphere and face the unknown hazards of airlessness, weightlessness, temperature extremes, and radiation. The craft to be developed, then, not only had to provide a safe, habitable environment and protection against the physiological stresses and incidental hazards of space flight, but also had to be sufficiently lightweight to be lifted with the rocket power available, sufficiently strong to sustain the forces of launch, flight and re-entry, and capable of withstanding heat, cold, vibration, and radiation.

As soon as the first steps into space, suborbital and Earth orbital flights in the single man Mercury spacecraft, approached realization more ambitious goals began taking shape. The plans for a programme named Apollo were first announced in July, 1960, and at that time had only one broad objective, to provide the capability for manned exploration of space. As envisioned, this advanced spacecraft programme would allow man to perform useful functions in space. The spacecraft would be capable of manned circumlunar flight as a logical intermediate step toward future goals of landing man on the Moon and other planets. The design would be flexible enough to permit its use in conducting scientific experiments in space.

During the next several months, NASA continued studies of manned spacecraft, particularly for long range, long duration flights. Then in May, 1961, the late President John F. Kennedy, in a message to Congress, committed the United States to the goal of landing men on the Moon and returning them safely to Earth before 1970. The Apollo programme had the new specific objective of lunar landing. At this time the total manned space flight time of the United States was represented by a 15-minute suborbital flight by Astronaut Alan Shepard in a Mercury spacecraft. This meant that much of the technology needed to sustain man in Earth orbital flight, to say nothing of a lunar landing, had yet to be developed. A new, intermediate programme named the Gemini was immediately inserted in the schedule and designed to fill the most critical gaps in spacecraft technology.

This "building block" philosophy, using the knowledge gained in each of the increasingly rigorous and sophisticated programmes as the basis for subsequent programmes, was one of the most important concepts in achieving, within an exceedingly short time, the ultimate goal, the development of a spacecraft capable of carrying men to the Moon and back. The rapidity of the achievement also depended heavily on the Nation's industrial and technical potential and capability, and on devising new management concepts for regulating and co-ordinating the numerous and diverse tasks involved.

Because the building block philosophy is integral to United States manned spacecraft development, a study of the Apollo spacecraft should include a review of the two programmes that preceded it, Mercury and Gemini, and their contributions.

THE MERCURY SPACECRAFT

Introduction

Several years of intensive, largely theoretical, investigation directed by the United States Air Force and the National Advisory Committee for Aeronautics (forerunner of NASA) culminated in 1958 with the conviction that manned space flight was possible. In September of that year, a committee comprising representatives from the Department of Defence and the National Advisory Committee for Aeronautics recommended a single-man spacecraft programme, designated "Project Mercury", and assigned the programme to the newly organized National Aeronautics and Space Administration. Objectives of the Mercury programme were: (a) to place a manned spacecraft in orbital flight around the Earth; (b) to investigate man's performance capabilities and his ability to function in the environment of space; and (c) to safely recover both the man and the spacecraft.

Design and Development

Providing a habitable environment for humans in space presented unprecedented challenges. On Earth, man functions in a gaseous environment consisting of about 20% oxygen and 80% nitrogen at a pressure of 14.7 pounds per square inch at sea level. Although man is quite an adaptable mechanism and can function in other atmospheres, his range of adaptability does not accommodate survival in the vacuum of space; he must take his atmosphere with him. He must also be protected from the physiological effects of environmental factors incidental to space flight—noise, vibration, acceleration and deceleration, impact, weightlessness, isolation, confinement and altered diurnal cycles. Weightlessness was one of the most critical unknowns and potentially one of the greatest limitations to manned space flight, since a weightless condition could be realistically simulated on Earth only for very brief periods and little was known of its effects on man.

To offer some assurance of reasonably rapid development, and in effect to counterbalance all these unknowns, the basic guidelines agreed upon were to follow the simplest and most reliable approach

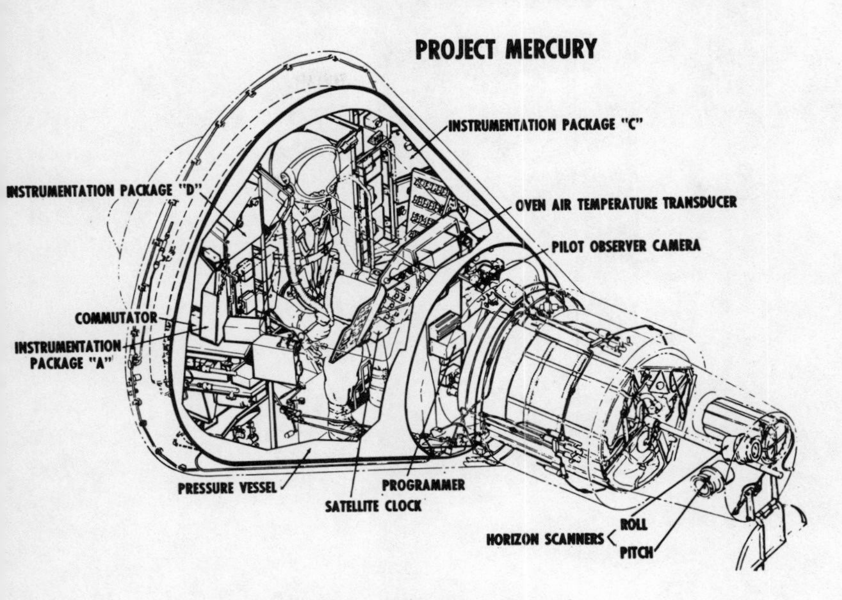

Figure 3-1. Mercury Spacecraft.

to systems design; to make maximum use of existing technology and off-the-shelf equipment, including use of an existing launch vehicle to place the spacecraft into orbit; and to conduct a logical and totally effective test programme aimed at both reliability and safety for human space flight.

One of the first design questions was that of shape. There were no precedents for spacecraft, and only the generally accepted feeling that streamlining was essential. However, streamlining was not relevant beyond Earth's atmosphere, and might aggravate the problem of re-entry from orbital speed. It was believed that a blunt end on the craft pointed in the direction of flight would slow the craft on the return trip by utilizing atmospheric drag. In addition it would form a shock wave that would dissipate some of the tremendous heat of re-entry. Wind tunnel tests supported this theory and the spacecraft configuration evolved as a somewhat bell-shaped blunt body, protected from re-entry heating by a heat shield covering the curved surface of the blunt end (*Figure 3-1*).

Figure 3-1.
Mercury Spacecraft.

The main body was 9 feet 7 inches long from the heat shield to the tip of the nose, 6 feet 2 inches wide at its widest point, and weighed about 3,000 pounds. The tapered portion was of double construction. The outer wall was formed of overlapping shingles of Rene, a thin refractory metal, corrugated for strength. The shingles allowed for thermodynamic expansion and contracting. The inner cabin, which was a pressure vessel to maintain the astronaut's life support environment, was constructed of two layers of thin-gauge titanium. A solid propellant rocket was mounted on top of the escape tower to lift the spacecraft clear of danger in case of a launch vehicle failure on the pad (*Figure 3-2*). In a normal mission the escape tower would be jettisoned after the booster stage had shut down.

Interior design of the Mercury spacecraft provided a form-fitting contour couch designed to distribute the high acceleration loads imposed on the man during launch and re-entry. At all critical phases of the mission the spacecraft was oriented so that the direction of acceleration force was the same, from back to front. Thus, the spacecraft was pointed nose up at launch, and heat shield forward on re-entry.

The life support system inside the pressurized crew compartment supplied breathing oxygen, purified the air by removing carbon dioxide and other foreign matter, and controlled the temperature and air flow inside the astronaut's full pressure suit. Thus the pilot was protected from the hard vacuum of space as well as the extreme temperature variations associated with the orbital flight profile. This system also included provisions for food and drinking water, as well as waste management within the weightless environment of space.

Since man's capabilities in the space environment were unknown, all critical systems of the spacecraft were operated automatically. Redundant systems were included to give maximum reliability for safety and assurance of mission success.

Contributions

When the Mercury programme was ended in May, 1963, its 25 major launches, including six manned launches and four Earth orbital missions, had established a base of technology and confidence for the follow-on Gemini and Apollo programmes. The most important contribution was proving the feasibility of the basic approaches selected to resolve the problems of placing man in orbit and returning him safely to Earth. It verified the selection of the blunt shaped body configuration and that man could exist, observe, and navigate in a space environment; it proved spacesuit and couch design, life support and biomedical instrumentation systems, guidance and control systems, communication and tracking techniques, and the Earth landing system. It set the pattern for quality assurance and reliability programmes and furnished valuable experience in systems engineering and large scale programme management.

The successful conclusion of this programme was not reached without some difficulties, but fortunately no lives were sacrificed. After the completely smooth ride of Alan Shepard, the second flight brought some anxious moments for its pilot Gus Grissom: the explosive mechanism on the escape hatch activated accidentally and blew the hatch open, but Grissom was able to swim free. During John Glenn's orbital flight, the first by an American astronaut, trouble developed with control of the capsule and he was forced to take over manual control of the small rockets that automatically control altitude.

THE GEMINI SPACECRAFT

Introduction

In December, 1961, the United States decided to extend manned space flight beyond the limits inherent in the first spacecraft, primarily to support the lunar landing programme. The main objectives of the Gemini, as the new programme was named, were

Figure 3-3. Gemini Spacecraft.

to develop a two-man spacecraft, improve and advance the manned space flight technology developed with the Mercury, and provide experience in space operations, such as rendezvous and docking of two orbiting vehicles. Information would also be accumulated on the effects of weightlessness and the physiological reactions of crew members during long duration missions.

Design and Development

Gemini design evolved logically from the Mercury spacecraft. It consisted of two major modules (*Figure 3-3*): the adaptor module, consisting of the retrograde section and the equipment section; the re-entry module containing the rendezvous and recovery section, the re-entry control system, and the cabin section. The configuration of the spacecraft was a blunt body similar to that used in Mercury. The heat shield was attached to the cabin section in the manner of the Mercury capsule. The re-entry module was the only portion of the spacecraft recovered from orbit: the adaptor module was allowed to burn up in the Earth's atmosphere. The astronauts' capsule, or re-entry module (*Figure 3-4*), was 11 feet

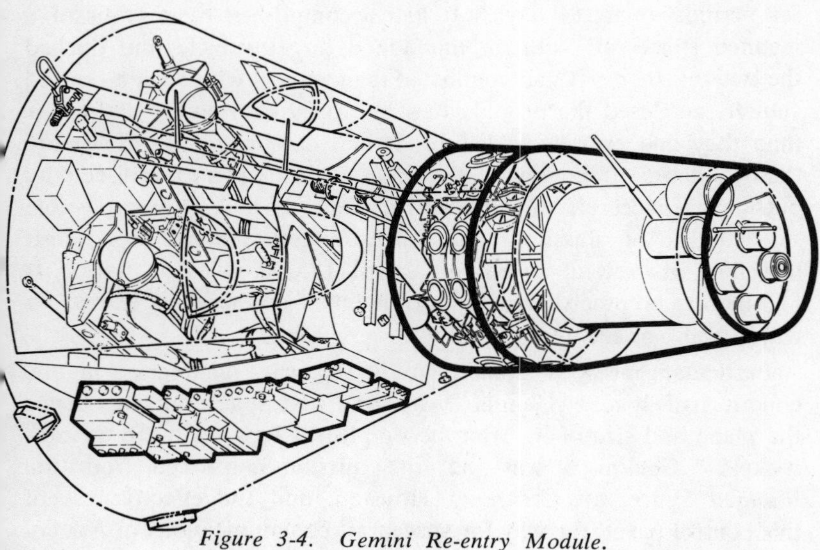

Figure 3-4. Gemini Re-entry Module.

high and 7½ feet wide at the base. Overall dimensions were 10 by 19 feet, and the spacecraft weighed approximately 7,000 pounds.

Unlike the Mercury, the Gemini placed emphasis on man by designing the spacecraft so it could be controlled by the astronauts with ground control as backup, and placed the systems and subsystems outside the cabin area in modular form so any system could be repaired or replaced without disturbing the others. Almost all the Mercury systems were in the pilot's cabin and stacked in layers, making it necessary to disturb a number of other systems to gain access to a problem area. This modular design concept was to prevent much loss of time in the Gemini and Apollo programmes.

Contributions

With the twelfth and final Gemini flight, the manned space flight programme had logged almost 2,000 hours in space. The Gemini programme had accomplished all objectives, contributing heavily to the knowledge and technology vital to the development of the Apollo spacecraft programme. Specifically it had demonstrated that man can live and work effectively in weightless space flight for periods up to 14 days. It had accomplished rendezvous of a manned spacecraft with an unmanned target vehicle and docked the two together. It had conducted manoeuvres with such a docked vehicle, and used the propulsion stage to fly men higher and faster than they had ever flown before. It had demonstrated, after more than 12 hours of experience, that man could perform useful activity outside a spacecraft in a protective suit. It had proved possible the control of missions and operation of manned spacecraft travelling in orbit at speeds of almost 18,000 miles an hour. It had performed precision landings of manned spacecraft (*Figure 3-5*) within sight of the recovery ships.

Particular missions stand out as historic landmarks in the conquest of space. Gemini 3 was the first spacecraft to change the plane and size of its orbit, accomplished by a set of 16 thruster rockets. Gemini 4 was the first mission controlled from the Manned Spacecraft Centre at Houston, and the effectiveness of this control paved the way for spacecraft communications in Apollo.

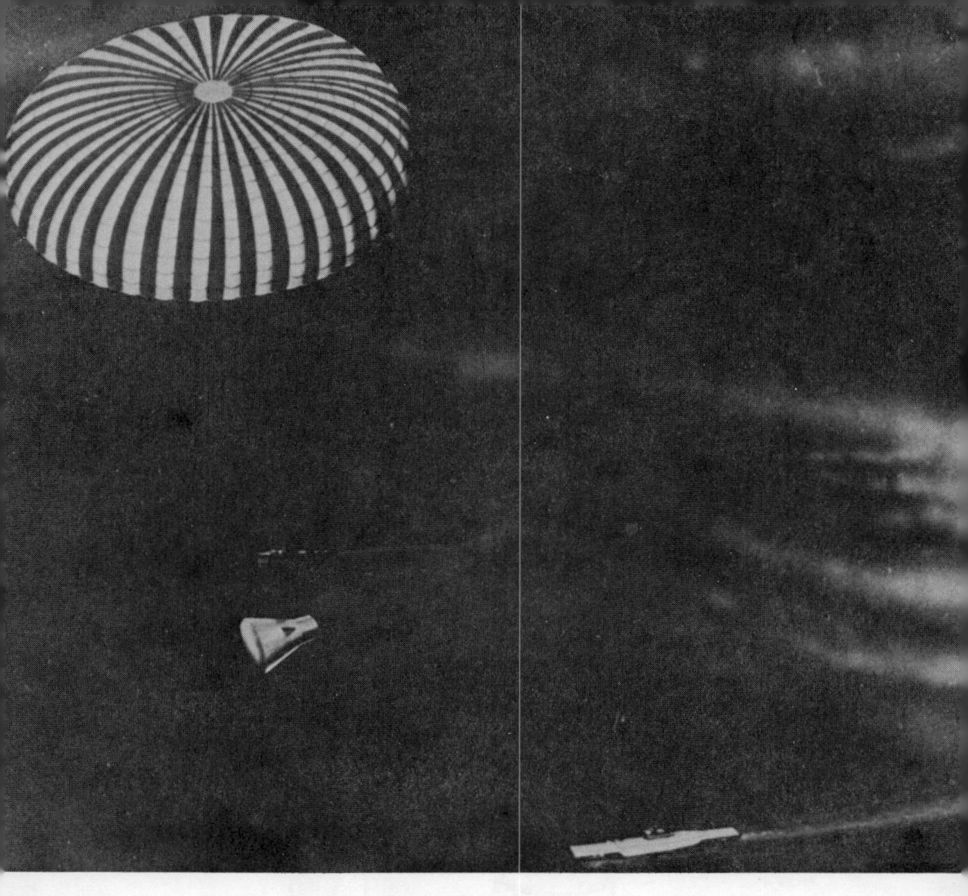

Figure 3-5. Gemini Spacecraft Landing.

Another highlight was the 100,000-mile chase in space, when Walter Shirra steered Gemini 6 at a speed of more than 17,000 miles an hour to catch Gemini 7, and the two craft circled the Earth for several hours, sometimes within a foot of each other.

THE APOLLO SPACECRAFT

Introduction

In May, 1961, when the Apollo programme was committed to a lunar landing, very little of the necessary technology was available. Many assumptions made in the early phases of the Apollo spacecraft were dependent on successful completion of the Mercury and Gemini programmes. The preliminary studies and research were obviously well performed: none of the major or critical assumptions made in the early development phases had to be changed. For

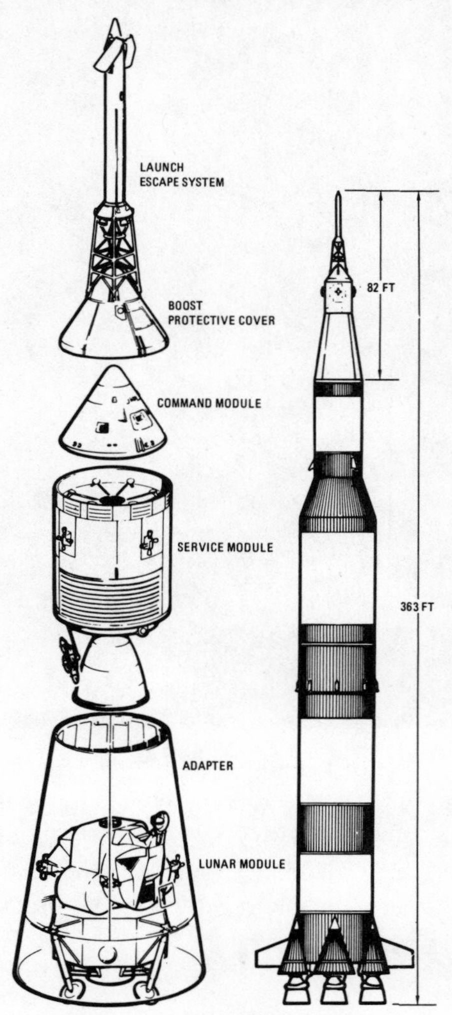

Figue 3-6. Apollo Spacecraft.

example, it was well known that the Apollo spacecraft could not tolerate the weight and size of batteries with the necessary life to power it for a lunar mission. Therefore, the design engineers selected fuel cells as the power source for the craft, and demonstrated the feasibility of their use during the Gemini programme.

In basic terms, the design requirements of the Apollo spacecraft were the same as those for the Mercury and Gemini, that is, a structure adequate to withstand the stresses of launch and re-entry and protected against space hazards, provisioned for a life supporting environment, and systems for navigation, guidance, and control. However, the Apollo objectives added numerous complicating features and greatly compounded the design and construction problems. While Mercury and Gemini operated in orbits close to Earth, the Apollo spacecraft had to carry its occupants approximately 240,000 miles to the Moon and back. The long-distance flight, plus the larger crew, called for greater supply of expendables such as oxygen, food, fuel and water. The functions of navigation, guidance and control were far more complex, and advanced systems of communications were needed. The environment of deep space required a superior structure and imposed new considerations for protection of the crew. The much higher speed of re-entry dictated an entirely new approach to descent and landing. Everything added weight and mass, increasing the need for propulsive energy. There was one constantly recurring theme: everything must be more reliable than any previous aerospace equipment, because the vehicle would become, in effect, a world in miniature, operating with minimal assistance from Earth.

The three-man spacecraft designed and developed to meet these requirements was divided into three major modules (*Figure 3-6*): the Command Module (CM), to house the astronauts and their equipment; the Service Module (SM), to provide support to the Command Module in terms of power, environmental control, etc.; and the Lunar Module (LM), to provide the means for actual descent to the lunar surface, liftoff, and return to the Moon-orbiting Command and Service Modules (CSM). The primary purpose of this three-module design was to reduce weight to a minimum particularly at such critical points as re-entry. For example, returning only the Command Module to Earth reduced the amount of spacecraft area to be covered with the relatively heavy heat shield. The division of the Lunar Module into two stages (*Figure 3-7*), with only the ascent stage returning from the lunar surface, permits a reduction in the size of the rocket engine and amount

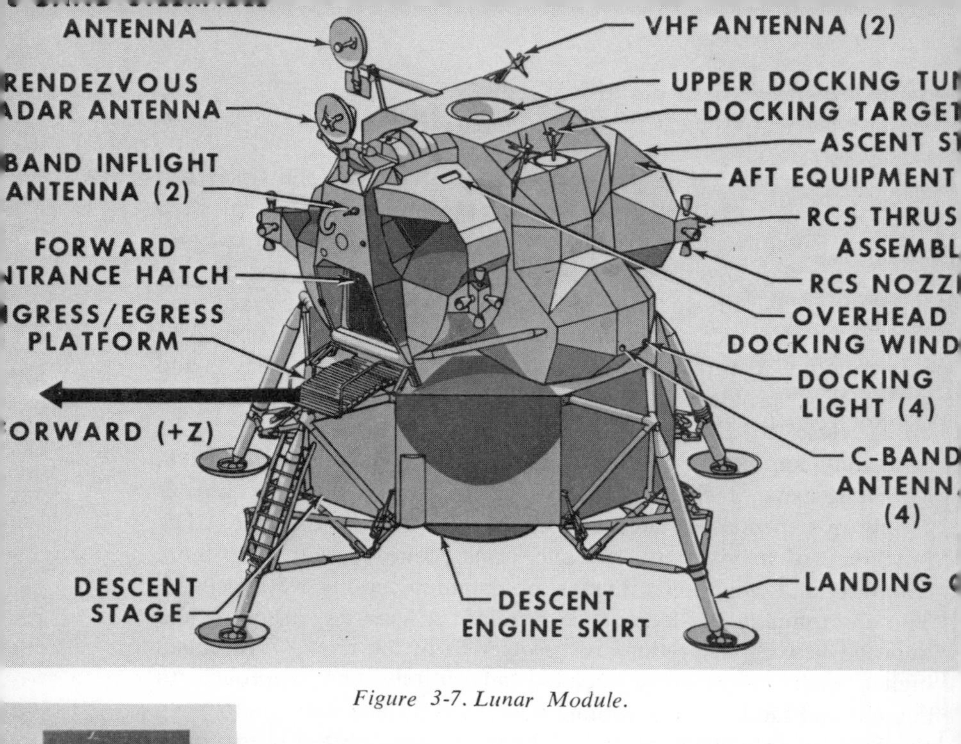
ANTENNA
VHF ANTENNA (2)
RENDEZVOUS
ADAR ANTENNA
UPPER DOCKING TU
DOCKING TARGET
ASCENT S
BAND INFLIGHT
ANTENNA (2)
AFT EQUIPMENT
RCS THRUS
ASSEMBL
FORWARD
NTRANCE HATCH
RCS NOZZ
GRESS/EGRESS
PLATFORM
OVERHEAD
DOCKING WIND
DOCKING
LIGHT (4)
ORWARD (+Z)
C-BAND
ANTENN
(4)
DESCENT
STAGE
DESCENT
ENGINE SKIRT
LANDING

Figure 3-7. Lunar Module.

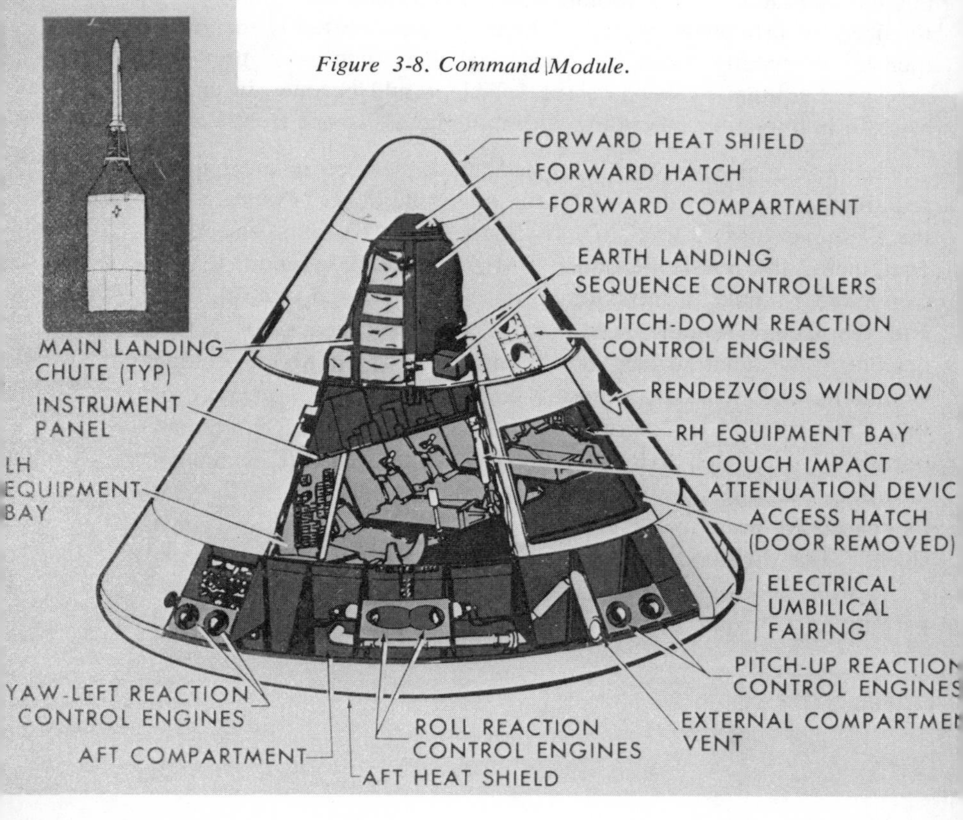
FORWARD HEAT SHIELD
FORWARD HATCH
FORWARD COMPARTMENT
EARTH LANDING
SEQUENCE CONTROLLERS
PITCH-DOWN REACTION
CONTROL ENGINES
MAIN LANDING
CHUTE (TYP)
RENDEZVOUS WINDOW
INSTRUMENT
PANEL
RH EQUIPMENT BAY
LH
EQUIPMENT
BAY
COUCH IMPACT
ATTENUATION DEVIC
ACCESS HATCH
(DOOR REMOVED)
ELECTRICAL
UMBILICAL
FAIRING
PITCH-UP REACTIO
CONTROL ENGINES
YAW-LEFT REACTION
CONTROL ENGINES
EXTERNAL COMPARTME
VENT
ROLL REACTION
CONTROL ENGINES
AFT COMPARTMENT
AFT HEAT SHIELD

Figure 3-8. Command Module.

of propellants needed. Other Apollo spacecraft components include the Launch Escape System (LES), the Spacecraft Lunar Module Adaptor (SLA), and the spacesuit.

Command Module

The Command Module houses the flight crew, the equipment necessary to control and monitor the spacecraft systems, and equipment and supplies required for the comfort and safety of the crew (*Figure 3-8*).

The primary structure of the Command Module (the crew compartment or pressure vessel) is encompassed by a metal structure carrying three heat shields and forming the conical shaped exterior. These heat shields are coated with ablative material, which protects the spacecraft from aerodynamic heating caused by friction when the Command Module re-enters the Earth's atmosphere at a velocity of approximately 36,000 ft./sec. The mechanisms by which the ablative heat shields protect the structure are:

1. Heat is absorbed during the thermal decomposition of the outer layer of the ablative material, which is then carried away by the air stream, exposing new and cold material.
2. The gases formed by the decomposition are injected into the aerodynamic boundary layer to reduce the heat input.
3. The heat charred outer layer reduces the heat input by narrowing the temperature difference between the boundary layer air and the surface.
4. The charred layer also radiates heat away from the surface.
5. The undecomposed virgin material below the surface acts as an insulator to absorb heat and slow down the rate at which heat is transferred to the underlying structure (*Figure 3-9*).

The conical blunt configuration of the Command Module was adapted from the successful Mercury and Gemini designs, which verified its superiority in space over the streamlined type. When a streamlined re-entry body enters the sensible atmosphere from outer space, half of the heat generated by the resulting friction is absorbed by the body. However, a blunt body colliding with stratospheric pressures at re-entry speeds will produce a strong bow shock wave in front of and detached from the body. This

Figure 3-9. Command Module Heat Shield after Re-entry.

shock wave, the air itself, absorbs much of the kinetic energy transformed into heat as the object enters the atmosphere. With less heat affecting the spacecraft, less heat shield material is needed to protect the spacecraft and its crew. This in turn decreased weight, a necessary goal in every space vehicle programme.

The Command Module is divided into three sections — forward compartment, crew compartment, and aft compartment. The forward compartment is an area between the apex of the forward heat shield and the upper side of the forward bulkhead. Its centre portion is occupied by a forward tunnel which permits crew members to transfer to the Lunar Module and return to the crew compartment during the performance of lunar mission tasks. The perimeter is divided into four 90° segments containing the recovery equipment, two reaction control motors, and the mechanism to jettison the heat shield. The major portion of this area houses the active components of the Earth Landing System (ELS), consisting of three main parachutes, three pilot parachutes, two drogue parachutes, and drogue and pilot parachute mortars. Four thruster-ejectors are installed in the forward compartment to eject the heat shield during landing operations. The thrusters operate in conjunction with the heat shield release mechanism to produce a rapid, positive release of the heat shield and prevent parachute damage.

The crew compartment is a pressurized three-man cabin, with pressurization maintained by the environmental control system. It contains spacecraft controls and displays, including guidance and navigation equipment, electrical and electronic equipment, observation windows, access hatches, food, water, sanitation and survival equipment. The compartment incorporates windows and equipment bays as part of the primary structure.

The aft compartment is encompassed by the lower portion of the crew compartment heat shield, aft heat shield and lower portion of the primary structure. It contains 10 reaction control motors; an impact attenuation structure consisting of four crushable corrugated aluminium ribs and eight struts that connect the crew couches to the Command Module structure; instrumentation; electrical power; and storage tanks for water, fuel, oxidizer and gaseous helium.

Service Module

The Service Module, which contains the main spacecraft propulsion system and supplies most of the spacecraft's consumables (oxygen, water, propellants, hydrogen, etc.) (*Figure 3-10*), is attached to the Command Module until just before re-entry into the Earth's atmosphere, when it is jettisoned. The structure is a cylinder formed by six panels of one-inch aluminium honeycomb, with its interior unsymmetrically divided into six sectors by radial beams or webs fabricated of milled aluminium alloy plate. The equipment contained within these sectors is accessible through maintenance doors located around the exterior surface of the module.

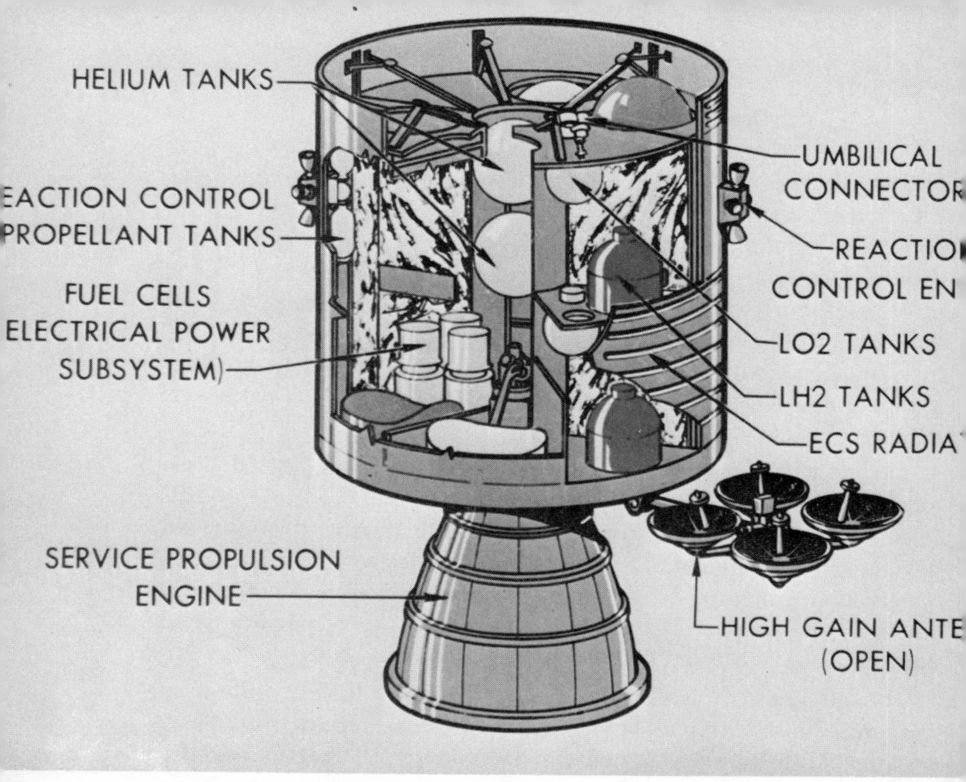

Figure 3-10. Service Module.

An area between the Service and Command Modules contains the system for separating the two modules, which consists of an explosive charge attached to each mechanical link The entire separation system is enclosed by a fairing 26 inches high and 13 feet in diameter.

The major functions of the Service Module propulsion unit, which consists of a 20,500-pound thrust engine using nitrogen tetroxide and hydrazine compound as propellants, are to:

1. Permit midcourse corrections to the lunar trajectory both going to and returning from the Moon.
2. Reduce the spacecraft's velocity so lunar gravity can pull it into a lunar orbit.
3. Increase spacecraft velocity so it can escape lunar gravity and return to Earth.
4. Provide the velocity changes necessary to change lunar orbits.

5. Provide power for abort manoeuvres after the emergency escape system has been jettisoned.

In performing these functions, the system may have to fire as many as 15 times during a single launch mission, and is the only means the astronauts have for breaking out of the lunar gravity. If the engine should fail and could not be repaired by the crew, the spacecraft would remain in lunar orbit indefinitely.

Every possible means of insuring simplicity and reliability have been incorporated into the Service Module engine. As an example, the propellants are fed from the tanks to the engine by helium pressure, in contrast to the more complicated pumping systems used for the launch vehicle propulsion system.

The Service Module also carried reaction control systems for attitude control and manoeuvring, the fuel cells that are the prime energy source for the electrical power subsystem and other systems, and two antenna systems for long-range communications between the Earth and the spacecraft. One system is a cluster of four 2,000 microhertz high gain antenna dishes; the other system is two 2,000 microhertz VHF omni-antennae. The major loads transmitted are the voice link and telemetry data.

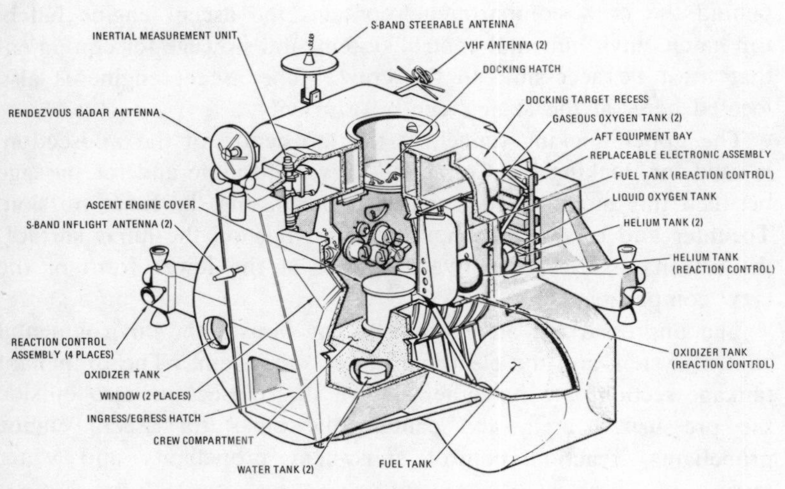

Figure 3-11. Ascent Stage.

Lunar Module

The descent and ascent stages of the Lunar Module (*Figure 3-7*) mate to form a self-sustaining structure, with provision for separating the stages and the umbilicals at lunar launch, or for abort at any time during the lunar landing phase of the mission.

The manned ascent stage (*Figure 3-11*) contains the crew compartment, the midsection, the aft equipment bay, and the tankage section. Both the crew compartment and the midsection can be pressurized at 5 psi, with a 100% oxygen atmosphere, and a temperature maintained at about 75°. The astronauts are housed in the crew compartment, and from there control the flight, lunar landing, lunar launch and rendezvous and docking with the Command and Service Modules. The compartment is also used as the operations centre for the astronauts during the lunar stay. In addition to the controls and indicators, and items necessary for crew comfort and support, this section also contains the forward hatch and tunnel.

Two triangular windows in the front face of the crew compartment provide visibility during all phases of the mission. Both windows are canted down and to the side to permit adequate lateral and downward visibility.

The midsection, which is a smaller, cylindrical section directly behind the crew compartment, contains the ascent engine hatch, top hatch, environmental control system, and stowage for equipment that must be accessible to the crew. The ascent engine is also located here, at the stage's centre of gravity.

The upper docking tunnel, at the top centre of the midsection, is used for docking the Command Service Module and for passage between this module and Lunar Module during the lunar mission. To enter and leave the Lunar Module while on the lunar surface, the astronauts use the forward tunnel, at the lower front of the crew compartment.

The unpressurized aft equipment bay houses the environmental control system and the electrical power subsystem. The propellant tankage sections are on either side of the midsection, also outside the pressurized area, and contain the tanks for ascent engine propellants, reaction control subsystem propellants and water tanks.

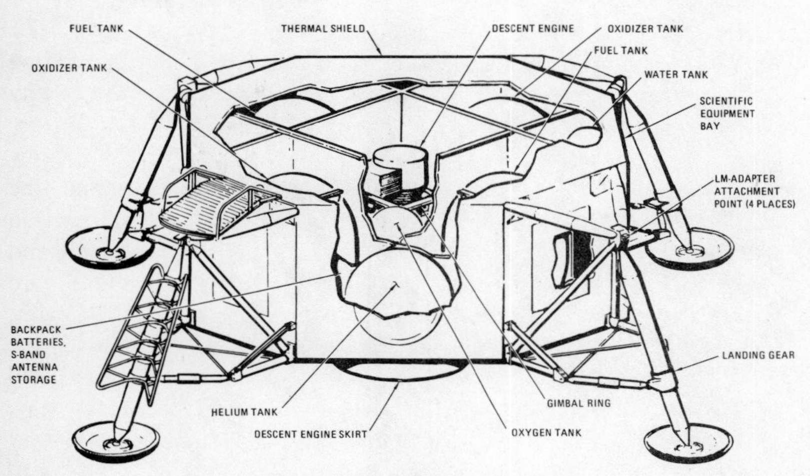

Figure 3-12. Descent Stage.

The descent stage (*Figure 3-12*) is the unmanned portion of the Lunar Module. It holds the equipment needed to land on the lunar surface, and serves as a platform for launching the ascent stage after completion of the lunar stay. In addition to the descent engine and related components, this stage houses the descent control instrumentation, which includes scientific equipment and the tanks for water, oxygen and hydrogen used by the environmental control system and the electrical power subsystem. The landing gear is attached externally.

In a centre compartment formed by the structural beams is the descent engine, a throttlable, gimballed type capable of 1,050 to 9,870 pounds of thrust, which provides the power for the complex manoeuvres required to take the Lunar Module from orbit down to a soft landing on the Moon. Four main propellant tanks surround the engine. Scientific equipment, helium, hydrogen, oxygen and water tanks are adjacent to the propellant tanks. The descent water tank has a capacity of 333 pounds at 0.75 fill ratio, and supplies most of the water required until staging occurs: after staging water is supplied by the two ascent stage tanks.

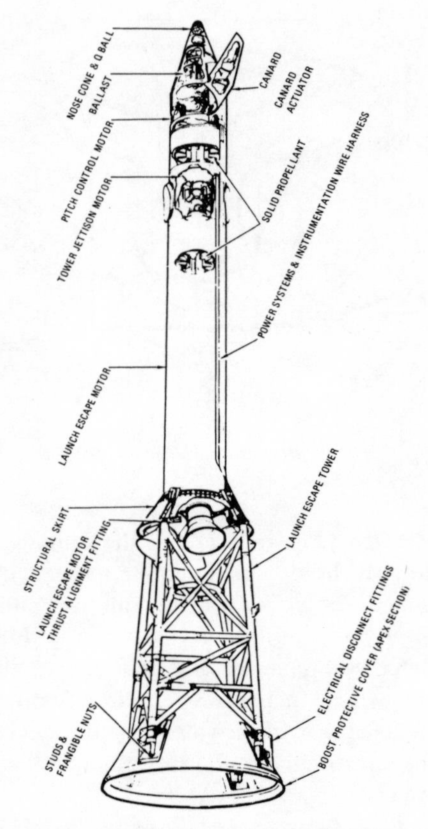

Figure 3-13. Launch Escape Tower.

The cantilever-type landing gear consists of four equally spaced legs or struts, connected to outriggers that extend from the ends of the descent stage structural beams. Each landing gear leg consists of a primary strut and footpad, a drive-out mechanism, two secondary struts, two downlock mechanisms and a truss. All struts have crushable attenuator inserts to absorb the impact loads from landing on the Moon.

At launch, the landing gear is stowed in a retracted position and remains retracted until shortly after Lunar Module separation from the third stage. Next, landing gear locks are pryrotechnically released and springs in the drive-out mechanism extend the landing gear. The landing gear is then locked in place by the downlock mechanism. With the landing gear in position, the Lunar Module is ready for touchdown on the lunar surface.

The Lunar Module is the first spacecraft to be designed for pure space travel, since it operates only near or on the Moon. For example, because it never encounters atmospheric pressure, the Lunar Module's outer skin is about the thickness of heavy-duty aluminium foil and the ladder used to reach the lunar surface is so light that, if the astronauts used it on Earth instead of the low gravity of the Moon, the ladder would collapse under their weight.

Launch Escape System

The Launch Escape System (LES) provides a method for emergency removal of the Command Module (CM) from the rest of the space vehicle should there be a pad abort or suborbital flight abort (*Figure 3-13*). Such a system is mandatory from a safety standpoint; if the launch vehicle were to fail structurally, the propellants would combine and form an explosive mixture, and the resulting explosion could destroy the spacecraft and its crew.

The system consists of a nose cone with an angle of attack meter (Q-ball), a ballast compartment, canard system, three rocket motors enclosed within an Inconel housing, a structural skirt, an open-frame tower, and a boost protective cover. The structural skirt is secured to the launch escape tower, which transmits stress loads between the launch escape motor and the Command Module. The boost protective cover, which protects the Command Module exterior during the launch and boost, is fastened to the lower end of the tower. Four explosive bolts, one in each tower leg well, secure the tower to the Command Module structure. After a successful launch, or during abort mode initiation, explosive squibs fracture the bolts and free the tower together with the boost-protective cover. The rocket motors, canard, and explosive squibs are activated by electronic sequencing

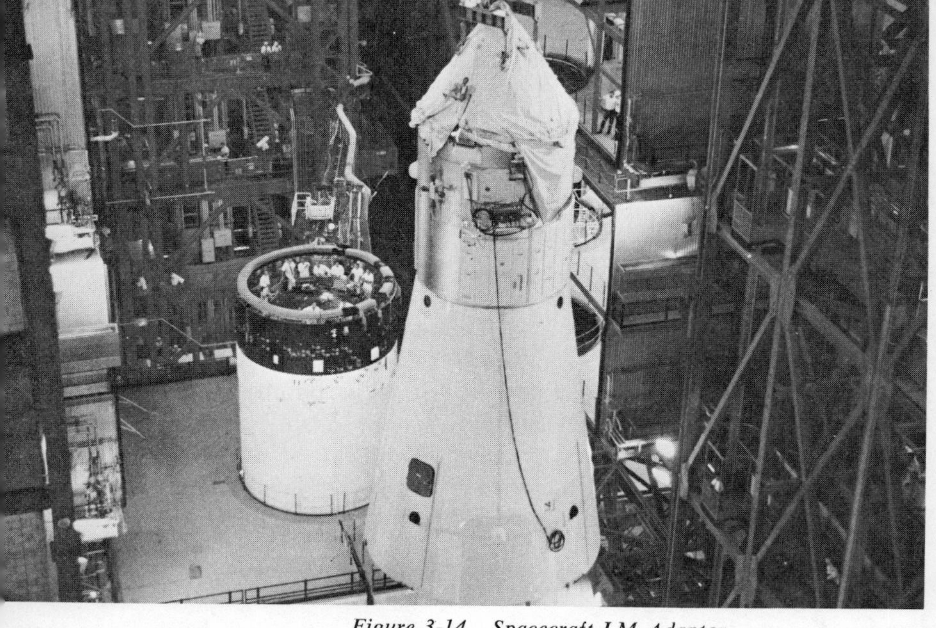

Figure 3-14. Spacecraft LM Adaptor.

devices. The Launch Escape System is jettisoned after first stage burnout when the vehicle is above the sensible atmosphere and the danger greatly reduced.

Spacecraft LM Adaptor

The Spacecraft LM Adaptor (SLA), which is the structural interstage between the launch vehicle and the spacecraft, and formed of four aluminium panels, houses the service propulsion engine expansion nozzle and the Lunar Module (*Figure 3-14*). At the time of SM/SLA separation, the linear-shaped charges installed at panel junctions are fired, the explosive force cuts through the SLA structure, and the panels fold back to expose the Lunar Module.

The Spacesuit

By design and function, the spacesuit may be considered a part of the spacecraft. The earliest designs were based mainly on experience and criteria established for high-performance jet aircraft pilots and were originally intended only to maintain a pressurized environment for the astronaut in the event of a failure of the spacecraft amospheric pressure sysem. This initial design objective was later expanded to develop a suit sufficiently complex to permit astronauts to breathe and walk and work on the Moon's surface.

Mercury Spacesuit. In Project Mercury flights, spacesuits were worn primarily to satisfy the requirement for capsule decompression protection. The spacesuit was considered a backup system for emergency use only. In normal flight it served primarily as a flight suit and ventilation garment, although it also provided protection for the critical launch phase in the event the capsule pressure system failed to stabilize. The spacesuits for Mercury were also designed to provide sufficient pressurized mobility to permit the astronauts to continue manual flight control of the spacecraft even after cabin depressurization. Prior to the manned Mercury flights, tests and training programmes had established a high degree of confidence in the astronaut's ability to perform all required functions in a decompressed capsule.

Gemini Spacesuit. The Gemini spacesuits also performed the function of decompression protection during launch and orbital phases of the flight, but with the objective of extravehicular excursions they became a prime life-support system (*Figure 3-15*). Redundancy had to be added to insure reliability consistent with mission objectives; the suit had to permit maximum mobility;

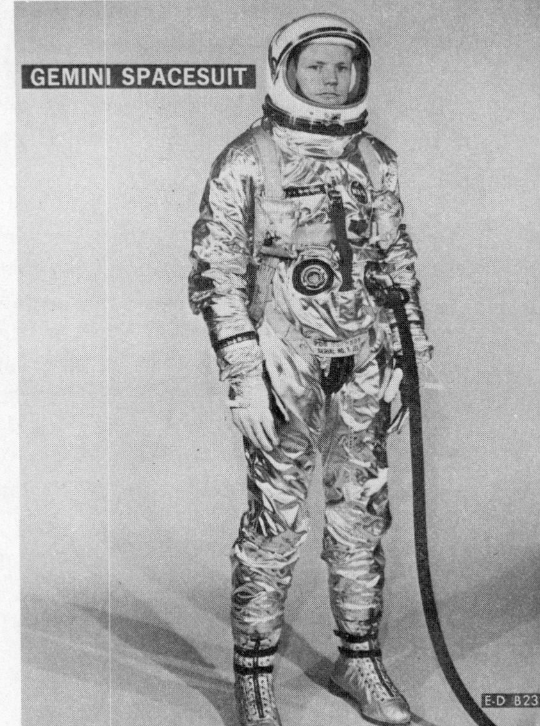

Figure 3-15. Gemini Spacesuit.

micrometeoroid, thermal, and visual protection systems had to be incorporated into suit design. Qualification and test programmes were made more comprehensive and more rigid. Since the size of the Gemini spacecraft did not permit the astronauts to don or doff the spacesuit completely in flight, one of the prime suit design requirements was long-term comfort.

All Gemini spacesuit design criteria were met. The first powered extravehicular manoeuvre in history was performed by a space-suited astronaut, Edward White, whose life support system consisted of a small chest pack called the Ventilation Control Module, with oxygen supplied through a 25-foot umbilical hose assembly. He wore an extra cover layer for micrometeorite and thermal protection, and a special sun visor to protect his vision while "walking in space".

Apollo Spacesuit. The major design objective of the Apollo spacesuit programme was to provide an overall system, or "Extra-vehicular Mobility Unit", that would permit one or more of the Apollo astronauts to explore the lunar surface. A secondary objective was to provide a spacesuit compatible with the Command Module under both pressurized and unpressurized operations. In addition to providing maximum mobility, with thermal, micro-meteoroid, and visual protection systems, it had to provide the capability of operating on the lunar surface for extended periods. Safety, redundancy, reliability, and quality control were of prime concern.

In the basic Apollo spacesuit, a suit, helmet, and a pair of pressure gloves form the environmental retaining envelope (*Figure 3-16*). A pressurization and ventilation system supplies the astronauts with a habitable atmosphere within the spacesuit. Ducts from the ventilation system distribute the pressurization and ventilation gas flow from either of the two inlet gas connectors to the helmet and torso. The gas then passes to either of two exhaust gas connectors and is conveyed through hoses to the Environmental Control System of the spacecraft, or during lunar explorations to the Portable Life Support System carried on the astronaut's back.

For lunar landings, the astronauts don an Integrated Thermal Micrometeoroid Protection Garment (ITMG) for protection against

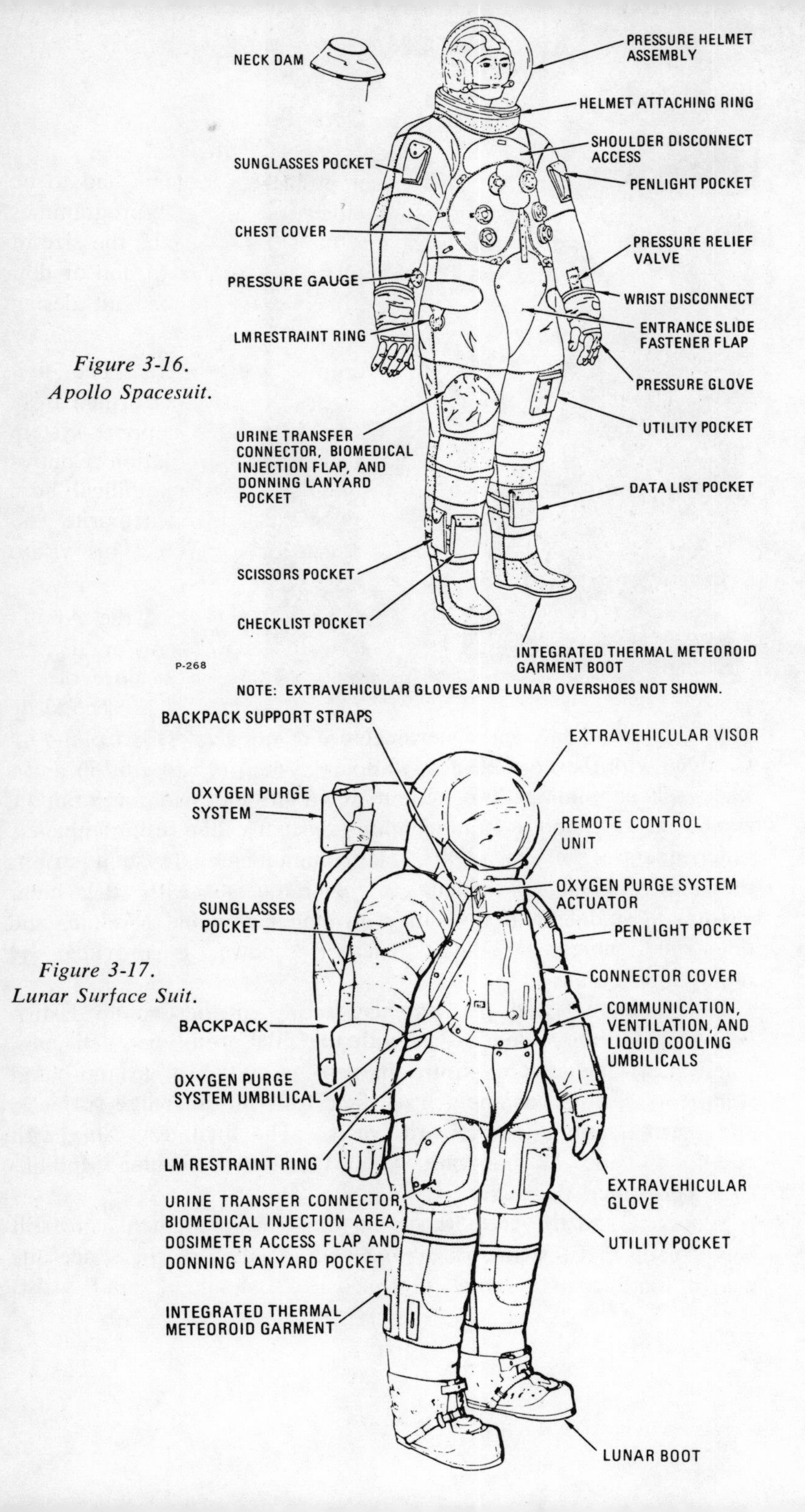
NECK DAM
PRESSURE HELMET ASSEMBLY
HELMET ATTACHING RING
SHOULDER DISCONNECT ACCESS
SUNGLASSES POCKET
PENLIGHT POCKET
CHEST COVER
PRESSURE RELIEF VALVE
PRESSURE GAUGE
WRIST DISCONNECT
LM RESTRAINT RING
ENTRANCE SLIDE FASTENER FLAP
PRESSURE GLOVE
UTILITY POCKET
URINE TRANSFER CONNECTOR, BIOMEDICAL INJECTION FLAP, AND DONNING LANYARD POCKET
DATA LIST POCKET
SCISSORS POCKET
CHECKLIST POCKET
INTEGRATED THERMAL METEOROID GARMENT BOOT
P-268
NOTE: EXTRAVEHICULAR GLOVES AND LUNAR OVERSHOES NOT SHOWN.
BACKPACK SUPPORT STRAPS
EXTRAVEHICULAR VISOR
OXYGEN PURGE SYSTEM
REMOTE CONTROL UNIT
OXYGEN PURGE SYSTEM ACTUATOR
SUNGLASSES POCKET
PENLIGHT POCKET
CONNECTOR COVER
COMMUNICATION, VENTILATION, AND LIQUID COOLING UMBILICALS
BACKPACK
OXYGEN PURGE SYSTEM UMBILICAL
LM RESTRAINT RING
EXTRAVEHICULAR GLOVE
URINE TRANSFER CONNECTOR, BIOMEDICAL INJECTION AREA, DOSIMETER ACCESS FLAP AND DONNING LANYARD POCKET
UTILITY POCKET
INTEGRATED THERMAL METEOROID GARMENT
LUNAR BOOT

Figure 3-16. Apollo Spacesuit.

Figure 3-17. Lunar Surface Suit.

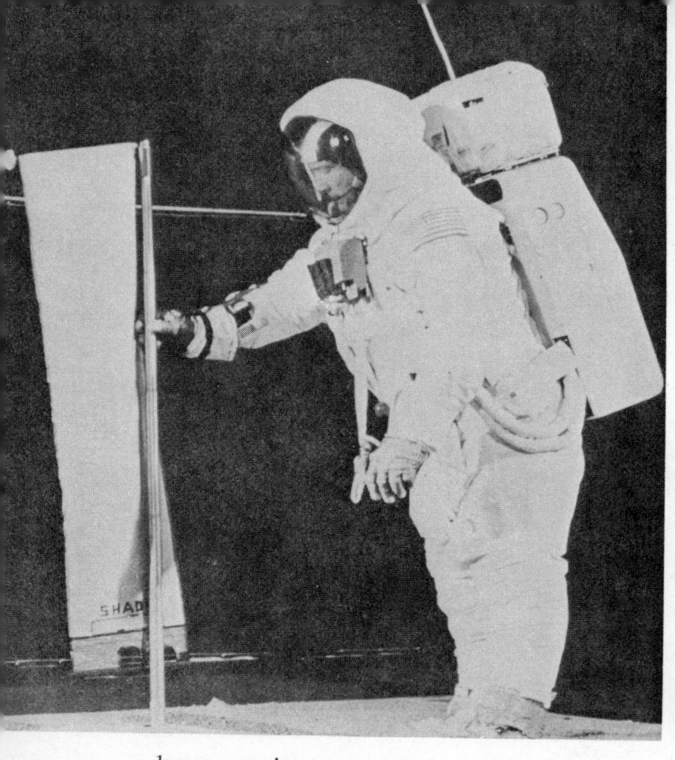

Figure 3-18. Portable Life Support System and Backpack Communications System.

lunar environment and micrometeoroid impacts (*Figure 3-17*). Coupled with the Portable Life Support System (*Figure 3-18*) and a backpack communications system, the ITMG provides ventilation, pressurization, and communications systems that are completely independent of the spacecraft. The communications system permits the extravehicular astronaut to communicate with the lunar excursion module and the lunar orbiting Command Module, and telemeters biomedical data. Electrical power is provided by replaceable batteries.

Developmental and performance testing of the Apollo Extravehicular Mobility Unit began with the first prototypes delivered. These tests included exposure in zero g and $\frac{1}{6}$ g ground-based simulators, altitude chamber tests, field tests on lunar-like surfaces, and thermal and micrometeorite tests. The final test came with Apollo 11 and 12 missions, when the Extravehicular Mobility Unit performed flawlessly.

Spacesuit and life support system design improvements are still being made and tested. Beginning with Apollo 16, the spacesuits will be modified to provide improved neck, shoulder, hand, waist, and leg mobility, durability of the materials, easier donning and doffing, and greater comfort and visibility. The Portable Life Support System will also be modified by several improvements, and a Secondary Life Support System will be provided.

Spacesuit use for future missions such as long-term space stations, extended lunar base operation, and interplanetary travel will require even more advanced approaches to design and operation. Spacesuits cannot be used for long-term intravehicular operation: the crewmen must be provided a true shirt-sleeve environment for normal operation. Sections of future spacecraft will be designed so that they can be sealed off like a compartment within a submarine, and lightweight emergency-type pressure suits will be sufficient to permit the crew to make internal repairs on a section of the spacecraft which is decompressed.

Applications of Spacesuit Technology. The need for fireproof materials for Apollo spacecraft demanded a complete testing of the flammability characteristics of hundreds of materials. These results have been computerized and are available to all of industry for use in making safer draperies, upholstery fabrics, mattresses and clothing. Fireproof Beta cloth developed in these programmes is already being used for firefighter suits in municipal departments as well as on board aircraft carriers at sea.

The spacesuit itself proved very valuable in a medical emergency: A pressure suit obtained from NASA's Ames Research Centre was used by a California hospital to stop severe and near-fatal abdominal haemorrhaging in a young woman patient, after nine operative procedures had failed to halt the bleeding. We can expect even more such benefits from future developments in spacesuits.

Apollo Communications

In each Apollo mission, the world-wide Manned Space Flight Network (MSFN) provides continuous reliable, and instantaneous communications with the astronauts, the launch vehicle, and the spacecraft, from liftoff to splashdown (*Figure 3-19*). Following the flight, the network continues to support the link between Earth and the Apollo experiments left on the lunar surface by the crew.

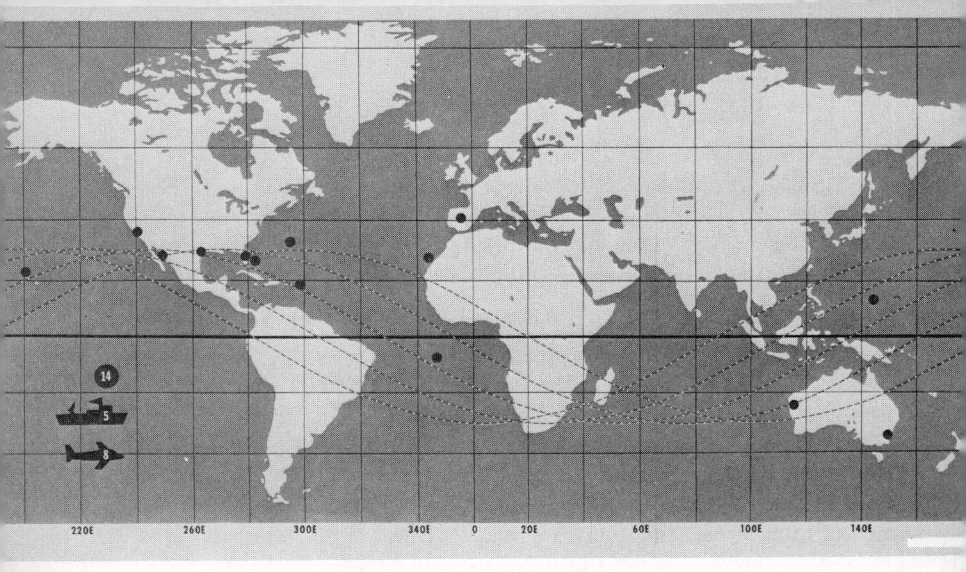

Figure 3-19. Manned Space Flight Network.

The centre of the Apollo Manned Space Flight network is at Mission Control in Houston, Texas (*Figure 3-20*). With this globe-spanning network and data refinement assistance from the Goddard Space Flight Centre, located in Maryland, this control centre maintains constant voice and tracking contact with the spacecraft. If for any reason the Mission Control in Texas becomes seriously impaired for an extended time, the Goddard Centre can act as a mission control centre and maintain voice and tracking contact with the spacecraft on an emergency basis.

The communications between the ground and the spaceship can be divided into three general categories. The first category involves monitoring the condition of launch vehicle, the spacecraft, and the crew, and requires voice communication between the astronauts and the ground. Since the astronauts cannot be expected to personally take and report the thousands of data readings needed to monitor spacecraft performance, these measurements must be made and relayed to the ground automatically. This process, which amounts to remote metering, is called "telemetry" and uses its own equipment and transmission channels.

The second category covers trajectory measurements for both the launch vehicle and the spacecraft, obtained from a network

Figure 3-20. MSC Mission Control Room.

of stations by a "tracking" process. The network is linked together by the NASA Communications Network (NASCOM), and all tracking information flows to and from Mission Control (Houston) and the Apollo spacecraft over this communications system. The NASCOM consists of almost three million circuit miles of diversely routed communications channels. It uses satellites, submarine cables, land lines, microwave systems, and high frequency radio facilities for access links. NASCOM control centre is located at Goddard, with regional communication switching centres in London, Madrid, Canberra, Honolulu and Guam.

The third category includes all other information transmitted from the ground to the spacecraft. Since the transmissions in most cases involve remote control commands, this channel is called the "command uplink". It allows ground controllers to do with the spacecraft what the crew cannot, either because they have not the time, the capability, the equipment, or, in extreme cases, are incapacitated. The command uplink system is comparable in coverage to both the voice and the telemetry systems.

Figure 3-21. KSC Radar Antenna.

The second category, tracking, is the most complicated communications area. Three basic systems are used for tracking: one is based on optical equipment and two on electronic equipment, with the choice at any moment depending on the space vehicle distance. During the early phases of boost-to-orbit, while the space vehicle is visible from ground, high-precision optical instruments track and determine the vehicle's position and velocity in space. Around the launch pad at Cape Kennedy are several dozen engineering cameras, including an intercept ground optical recorder. This is a heavy 18-inch-aperture reflecting telescope with a 35- or 70-mm. camera attached to its ocular. It is manually controlled in azimuth and elevation, has automatic focus and exposure control, and variable focal length up to 500 inches.

Distances beyond visual range require electronic gear, which basically use either pulsed radio signals or continuous radio waves to track the spacecraft (*Figure 3-21*). The pulse radar emits short radio pulses which are reflected back either from the vehicle's metal skin (called "skin tracking") or by a transponder, located inside the vehicle and connected through an antenna to its outside, that strengthens signal intensity ("beacon tracking"). Receiving the returning pulses, the ground station can determine not only the vehicle range, but also its azimuth and elevation above the horizon. The continuous wave radar system determines the vehicle velocity and its position in space by measuring the phase shift of the radio beam caused by the Doppler effect. This is possible since a well-stabilized frequency signal reaching to and returning from a moving vehicle appears to shift in frequency.

The conventional radar distance-measuring technique, successfully employed for orbital distances, will not suffice over the lunar distance. Therefore, NASA developed a new ranging system based on pseudorandom codes. Instead of using pulses, this ranging system transmits a long code which is repeated by the vehicle, allowing accurate range measurement. Because the system requires a wide bandwidth, the S-Band was selected both for ranging systems and communication systems. This Unified S-Band (USB) system handles all communications between the ground and the spacecraft, requiring only one amplifier and one antenna on board the Apollo spacecraft with identical antennae and support equipment at various ground stations, for the entire mission, from liftoff to landing. The simplicity of this single-band system contributes to reliability; multifrequency communications would require more radio systems and more antennae, each one increasing the opportunity for failure.

Most of the Unified S-Band ground stations are clustered around the North Atlantic Ocean, where the launch and insertion into parking orbit occurs. The stations with 30-foot antennae are used for distances up to about 12,000 nautical miles. To provide tracking and communication at lunar distances, NASA added a Deep Space Network, consisting of three stations with 85-foot dish S-Band antennae. These stations are located 120° apart in longitude: one is located at Canberra, Australia; one at Goldstone, California; and the third is at Madrid, Spain (*Figure 3-22*). At least one of these three stations will always keep the spacecraft in its antenna beam, which at the Moon covers an area about 1,400 miles in diameter. The land stations are supplemented by an ocean-going ship and a fleet of converted jet cargo aircraft (C-135A) to compensate for any gap that may occur in land-based coverage due to shifts in the spacecraft's Earth orbital angle to the equator.

All of the data to and from this complex network of stations around the world are brought into the Communications Centre at Goddard Space Flight Centre, where it is sorted, somewhat condensed, and relayed to the Mission Control Centre at Houston. Outbound communications follow the reverse route.

Facilities

Although Apollo facilities are scattered throughout the world, the management agency for the spacecraft programme is located at the Manned Spacecraft Centre in Houston (*Figure 3-23*). Here are the management offices, test facilities, and Mission Control,

Figure 3-22. Deep Space Network Antenna.

Figure 3-23. Manned Spacecraft Centre.

the heart of Apollo during mission periods. One of the largest computer complexes in the world supports this operation. Major research, development and manufacturing facilities for the Command and Service Modules are at Downey, California. Major research, development and manufacturing facilities for the Lunar Module are located at Bethpage, New York.

The numerous test facilities include, for example, one of the largest centrifuges in the world for g-level testing and two vacuum chambers capable of duplicating every condition of outer space except zero g. These facilities are used to test spacecraft, spacesuits and simulators, and can duplicate, under controlled conditions, any event or emergency that could occur during an actual flight mission. The test facilities at White Sands, New Mexico, include engine test stands. The Kennedy Space Centre has a spacecraft simulator for the final astronaut mission training.

Systems Engineering and Development

The concept of "systems engineering" for complex research and development programmes evolved in the United States over the last two decades. Although the term has been applied to many variants of technical management, systems engineering is basically an approach in which the interaction between each unit or subunit is considered in the design of the whole. The systems engineering role in the manned space flight programme was to provide programme-wide technical analysis for management to ensure that functional and performance requirements placed on all elements of a system were within the present or projected state of the art and could be developed within the scope of the project.

In the Apollo programme, the task being undertaken was so complex that the system had to be designed to maximize the probability of success and safety. This meant continuing to examine all possible ways of performing a particular function, and selecting the way that provided the highest degree in functionally matching the systems in the three major modules of the spacecraft.

Specific examples of application of the systems engineering approach were the following:

(a) To assure the astronauts' safety during the first seconds of flight, the design engineers had to develop a system to detect malfunctions, determine if these endangered the astronauts, and activate the launch escape system that would pull the spacecraft clear of the problem. The design of this sensing system provided the design engineers with many evenings of interesting work.

(b) Myriad considerations and problems were involved in integrating the interfaces between the space vehicle (spacecraft plus launch vehicle) and the launch facilities. For example:

1. The high pressure gas lines, the fuel loading lines, and the electrical lines carried by the swing arms on the service tower had to fit exactly the corresponding umbilical connectors on the vehicle and be able to pull clear at the first motion of the space vehicle at launch. At the end of the highest swing arm, which provides a bridge for reaching the spacecraft, is the clean room

that precludes atmospheric contamination of the spacecraft while it is being prepared for launch. The design of this room had to be carefully worked out to assure contamination control while providing easy access to the spacecraft.

2. To accomplish the preflight checkout in the desired time, the input and output of the computers on board the space vehicle had to match exactly the output and input of the launch complex computers.

3. The design and shaping of optimum space vehicle trajectories involved a multitude of considerations, such as wind profiles and wind statistics during the month of the year, dynamic characteristics of the vehicle combination, minimum propellant consumption, optimum vehicle performance, guidance modes, minimum structural loads, and stage cutoff characteristics.

4. During the launch vehicle power phase of the flight, the spacecraft is subjected to an environment that has to be considered in establishing design criteria. The environmental conditions involved include: (a) acceleration loads; (b) vibration loads and patterns; (c) dynamic loads during the period of high aerodynamic pressure; (d) general dynamic behaviour of the space vehicle; (e) bending modes; and (f) acoustic noise. The level of each of these conditions is determined by vehicle dynamic tests, static firings of both the engines and the complete stages, computer theoretical analyses, and unmanned flight tests.

The total systems engineering task required continual monitoring of the interplay between the spacecraft and the launch vehicle systems throughout the design and development process. For this purpose, a number of interface panels were formed, organized generally according to scientific discipline. Each panel had representatives from both the Marshall Space Flight Centre, which had responsibility for the launch vehicle design, and the Manned Spacecraft Centre, which was responsible for the spacecraft design. These panels met on a regular basis to report progress being made by their respective centres and to exchange information on their systems' characteristics to preclude any incompatibilities. Most importantly, these meetings brought the best minds in each organization to bear jointly on the problems that arose in designing the launch vehicle, the spacecraft and the mission.

Development Testing

The development testing approach was strongly influenced by systems engineering as well as being based on a philosophy relatively new to the aerospace discipline. In the early years the testing procedure was to design, launch, fix and then launch again to verify the fix. As technical knowledge was gained, the number of launches needed was reduced. However, with the advent of manned spacecraft, which involved human lives and complex equipment that took months or years to perfect, design feasibility had to be thoroughly proven on the ground. The flight test was just the final verification.

The development tests were conducted on a step-by-step basis, beginning with the smallest units. For example, the recirculating blower for the Command Module was first thoroughly tested as a component. Next it was integrated into the environmental control system where it went through a series of further tests. It was then assembled into the Command Module for additional testing, and finally into the space vehicle for a completely integrated test.

The early development tests of the Command and Service Modules, prior to the qualification of flight-type hardware, used boilerplate spacecraft, a module simulating spacecraft in weight, shape, and centre of gravity. However, the later and major portion of the testing was conducted on flight-type spacecraft. Wherever necessary, both types of hardware were tested in two phases, designated Block I and Block II.

The Block I test hardware was configured and tested to verify the Command and Service Module design for the early Earth orbital flights, when it was not yet necessary to verify the Service/Lunar Module interface design. The Block II spacecraft incorporated the necessary features for Lunar Module docking and was used for the Earth orbital rendezvous flights, and lunar mission simulation. This block method of development permitted testing to begin almost immediately after the programme started. It also allowed a reduction in the number of intermediate spacecraft configurations by accumulating early design changes and incorporating them into the Block II configuration.

The Lunar Module development tests followed the same general pattern. Preceding the major tests was a broad-based programme that included early feasibility tests of materials and components, leading to qualification of components and complete subsystems. System and vehicle development testing followed this series. A large amount of hardware was required to support the development test programme for the Lunar Module—mockups, engineering test models, propulsion test rigs, and LM test articles. With these hardware items it was possible to perform a number of significant tests independently, yet overlapping timewise. This provided flexibility to absorb change, and shorten the failure, diagnostic, correction and verification cycle in subsystems development and vehicle ground testing.

With all the precautions taken and the intensive testing performed, American spacecraft development was not a story of uninterrupted progress and success. The most serious setback occurred in January, 1967, when a fire in a spacecraft undergoing ground tests took the lives of three astronauts. This caused a two-year delay in manned flights and seriously impaired confidence in the programme. In the thorough and painstaking investigation that followed, 1,500 technical experts examined every aspect of design, manufacture and operation. Besides correcting the causes of the fire, a myriad of improvements were made in components and systems that up to then had seemed perfected. Only then did the programme proceed, with even greater caution than before.

Conclusion

The final proof of the achievement of programme objectives was the manned lunar landings and safe returns. Some particular aspects of the first two lunar landing missions that illustrated the success of the Apollo spacecraft design and development are briefly recounted here:

1. In spite of the fact that a lightning discharge passed through the vehicle immediately after launch, the Apollo 12 successfully completed its mission. Spacecraft systems and astronauts continued on to a pinpoint landing, about 600 feet from the Surveyor spacecraft that had landed on the lunar surface several years previously.

2. Only two or three of the seven planned midcourse corrections were necessary in each of the lunar missions.

3. None of the "single point" flight critical items has failed during a mission. A single point item is one for which there is no redundancy or backup, and it must function if the mission is to succeed. An example is the Service Module engine.

4. Although the mission profile, including specific event time, is prepared months in advance of the mission, the reliability and capability of the Apollo hardware is such that in the lunar landing missions all actual times have been within seconds of planned times.

5. The extra safety and reliability built into the spacesuit and Portable Life Support System permitted the astronauts to stay on the lunar surface for longer periods than originally planned.

6. The capabilities of the communications system are such that during the Apollo 12 mission geologists in Mission Control discussed the results of the first lunar exploration period with the astronauts and outlined new goals for the second period of extravehicular activity.

Now that the Apollo spacecraft has met the first objectives of placing man on the moon and safely returning him to Earth, the future uses of the spacecraft will be to meet the remaining objective, to provide a vehicle that can be used to explore space, both for the advancement of scientific knowledge and the benefit of mankind. During the next few years it will be used to conduct extensive scientific investigations of the moon, and the spacecraft is already being modified for this purpose. The carrying space is being increased to accommodate experiments, exploration aids such as the Lunar Rover Vehicle, and larger supplies of consumables, and the Portable Life Support System is being modified to support longer stay time on the lunar surface.

Another future use for the Apollo spacecraft will be to provide a transportation system for the crews manning the Skylab. The basic modifications that will be needed for the Skylab programme involve two areas. One is the removal of systems not needed because the Apollo will be primarily carrying the crews to a relatively low orbit and returning them to Earth. Items that can be deleted are one of the fuel cells, and some of the propellant tanks for the Service Module engine. The other area is, of course, adding those systems necessary for the mission. Additional propellant tanks will be needed for the reaction control system to help maintain the Skylab in the correct attitude, and the spacecraft's thermal control system will require more heaters since the craft will be in the shadow of the Apollo Telescope Mount solar panels most of the time.

With the Apollo, the United States has developed a spacecraft that is capable of exploring space anywhere in a zone extending at least a quarter-million miles from Earth. What it represents is even more important: the Apollo spacecraft is but one element of a resource of incalculable and enduring value, the ability to explore and use space. With this capability, it will be possible to carry out a wide variety of missions of practical and scientific value. Future manned space missions can be conducted in Earth orbit, in lunar orbit, on the surface, and beyond.

The Mercury and Gemini were the initial building blocks in manned spacecraft development, and Apollo must be considered only an intermediate block. Other blocks will be added as spacecraft are evolved for exploration beyond the moon, for shuttling back and forth to supply manned space stations and rotating personnel, for missions not yet dreamed of and actualities unimagined.

CHAPTER FOUR

Astronaut Selection and Training

by G. Hage

Apollo 13, planned as the most challenging lunar landing mission to date, became instead the most crucial test in America's manned space flight programme... On the third day of the mission, some 200,000 miles from Earth, loss of the main liquid oxygen supply aboard the spacecraft caused the scheduled lunar landing in the hilly Frau Mauro formation to be cancelled. The return trip to Earth of the damaged spacecraft required the ability and ingenuity of the Apollo 13 astronauts, the ground controllers, and the trained minds of hundreds of engineers and technicians throughout the country to effectively carry out the established back-up plans of the Apollo programme for survival in space.

Apollo 13 was the United States' fifth flight to the Moon and the twenty-third space flight by American astronauts. The crew for the mission was commanded by veteran astronaut Jim Lovell, with his vast experience gained in three previous space missions including the first journey to orbit the Moon. The other members of the original prime crew, both assigned to their first space mission, were Ken Mattingly, the Command Module pilot, and Fred Haise, the Lunar Module pilot (*Figure 4-1*). However, during the final countdown of the Apollo-Saturn V space vehicle, Command Module pilot Mattingly was replaced when medical results indicated a possible infection of measles. His back-up pilot Jack Swigert had only two days to review and rehearse final mission plans. In the spacecraft simulator, Swigert demonstrated that he could function with unquestioned teamwork with the other two members of the prime crew as well as perform all of the manoeuvres required for the Command Module during the mission.

Figure 4-1. Apollo 13 crew: Left to right are James A. Lovell, Jr., commander; John L. Swigert, Jr., Command Module pilot; and Fred W. Haise, Jr., Lunar Module pilot.

Fifty-four hours after lift-off and well on their way to the Moon, the Apollo 13 crew entered the Lunar Module for the first time and checked all systems. At the completion of the checkout, the crew transmitted colour television of their activities to Earth. A few minutes after the completion of the television broadcast, the crew had the first indication of trouble aboard their spacecraft when Commander Lovell calmly relayed, "Houston, we've had a problem. We've had a main B bus interval . . . and we had a pretty large bang associated with the caution and warning there . . . it looks to me looking out the hatch that we are venting something. We are venting something out into space." Pressure in the number two super cold oxygen tank aboard the Service Module had dropped to zero, and fuel cells number one and three failed. The remaining oxygen tank began to lose its pressure at a slow rate and the decision was made to cancel the intended lunar landing mission.

Ground control carefully checked each of the three fuel cell systems in an attempt to locate the oxygen leak. In the existing configuration, all oxygen and electric power supplied by the Service Module to the Command Module would be expended within three hours and the Command Module would become virtually useless, except for its re-entry capability. The increased load on the remaining fuel cell in generating essential electrical power and manufacturing water, plus the diminishing pressure in the remaining oxygen tank led to the decision to activate the Lunar Module, turn off all equipment in the Command Module to conserve power, and use the systems of the Lunar Module for life support.

Two of the crew transferred to the Lunar Module and prepared the vehicle for an alternate procedure known as the "lifeboat" mode. The Lunar Module would now supply all of the oxygen, water, and power to operate Apollo 13 until the Command Module could enter the Earth's atmosphere.

A similar abort procedure calling for the Lunar Module to help return a crew to Earth had been worked out early in the programme, in 1964, and was practised on the Apollo 9 mission, a 10-day Earth orbital flight. The Apollo 9 mission commander, Jim McDivitt, was at Mission Control in Houston and helped direct activation of the "lifeboat" mode aboard the Lunar Module.

The possibility of returning directly to Earth was quickly eliminated because the required 6000-foot per second burn of the service propulsion engine could not be made. It was decided that the crew would move their spacecraft back into a free-return trajectory, fly around the back side of the Moon, and return to Earth. At sixty-one and a half hours into the mission, the Lunar Module engine was ignited. This was the first time that an engine built to land men on the Moon had been used to propel an entire spacecraft cluster and to place the spacecraft on a safer path to Earth.

Many of the systems aboard the Lunar Module were turned off to conserve electrical power and the water used to cool the electronic components. Until Earth entry, two of the astronauts worked and slept in the Lunar Module while the other crew member remained in the darkened and cold Command Module. A 10-foot long hose, taken from one of the lunar space suits, was used to supply oxygen to the Command Module.

As Apollo 13 emerged from behind the Moon, preparations were made to burn the Lunar Module descent engine again to speed the return to Earth. The successful four-minute burn of the Lunar Module engine reduced the travel time to Earth by 12 hours, and changed the alternate splashdown point from the Indian Ocean back to the South Pacific. The spacecraft cluster was then programmed into a slow rotating manoeuvre so that the Sun would uniformly heat the exterior surfaces. On the ground, fellow astronauts of the Apollo 13 crew worked many hours in simulators checking every proposed manoeuvre and the use of consumables aboard the Lunar Module. Mission Control also determined the best way to devise an air-purifying system in the Lunar Module to keep carbon dioxide at a safe level. Lithium hydroxide canisters taken from the Command Module were rigged with plastic, cardboard, a sock, and held together by tape; they helped keep the air clean in the Lunar Module.

Several hours after the engine burn, Mission Control radioed the astronauts that calculations showed that the crew would have sufficient oxygen and water to last through splashdown. However, life aboard the spacecraft was very uncomfortable for the crew.

The original Command Module pilot of Apollo 13 worked out mid-course manoeuvres and re-entry procedures with other astronauts and technicians on the ground and relayed instructions to the crew members in the spacecraft. A third burn of the Lunar Module engine was made for a slight correction in the flight path. A final correction on the last day of the mission assured landing on target in the Pacific Ocean. Seven hours later, Astronaut Swigert turned on the systems of the Command Module and prepared for separation from the damaged Service Module. After separation, the astronauts returned to the Command Module and the hatch connecting them with their lifeboat, the Lunar Module, was closed. Since the Lunar Module was not designed to enter the Earth's atmosphere, it was abandoned.

On April 17, 1970, the Apollo 13 spacecraft splashed down on schedule within four miles of the recovery ship. The six-day journey around the Moon—the shortest and most perilous Apollo lunar mission—was completed.

The Apollo 13 crew's performance was acclaimed throughout the world as a triumph of the human spirit, an exoneration of the human mind, a tribute to human perseverance, and a victory for all mankind. Most noteworthy was their calm, precise reaction to the emergency situation and their subsequent diligence in configuring and maintaining the Lunar Module for safe return to Earth. Despite lack of adequate sleep and low temperatures in the spacecraft, neither their performance nor their spirit ever faltered throughout the flight. Later, when asked about what contributed most to the crew's ability to sustain the rigours of the Apollo 13 mission, Commander Jim Lovell replied, "I think that the ability to keep working under the conditions that exist is the result, perhaps, of the many years of training in the business that Fred Haise, Jack Swigert, and myself are in . . . we expect at times to meet adverse conditions. In this business, you cannot expect complete success all the time."

With the saga of Apollo 13 and Astronaut Lovell's remarks in mind, let us now turn to a discussion of what makes an astronaut; the selection criteria for picking the right men; and the rigorous training they undergo to attain the skill and calm ability to isolate the cause of trouble in time of a crisis.

Astronaut Selection

The criteria used for selection of astronaut personnel are determined by the kind of training the astronauts are to receive and the requirements of the missions to which they will be assigned. Astronaut selection is a responsibility of the National Aeronautics and Space Administration (NASA) Manned Spacecraft Centre, Houston, Texas. Public announcements concerning the selection criteria for each astronaut training group and the periods during which applications will be received are issued and controlled from the Manned Spacecraft Centre.

Astronaut personnel must have an adequate background of experience and education to enable them to learn rapidly the many intricate details of space flight operations and to analyze quickly any unusual or unexpected circumstance and make a rapid decision on the correct procedure to follow. Astronauts must be in excellent health and have the physical and mental stamina to endure stresses and strains not commonly encountered by the average person. The demands of space flight require that they have sufficient self-control to remain steady and calm in a crisis and be able to remain alert and responsive after hours of cramped confinement and periods of extreme tension.

Early in 1959, prior to the beginning of manned space flight, the National Aeronautics and Space Administration asked the military services to search their records for men who met the following qualifications:

1. Less than 40 years of age.
2. Less than 5 feet 11 inches tall.
3. Excellent physical condition.
4. Bachelor's degree in engineering or its equivalent.
5. Qualified jet pilot.
6. Graduate of test pilot school.
7. At least 1500 hours' flying time.

The armed forces listed a total of 508 men who qualified under these criteria. The military and medical records of these men were examined, psychological and technical tests were administered, and personnel interviews were conducted by psychological and medical specialists. Following this screening, a large portion of the original 508 were eliminated and an additional number decided that they no longer wished to be considered.

Each man who remained on the eligible list was subjected to the most comprehensive physical examination the NASA medical staff could devise. The objectives of this portion of the selection programme were to provide crew members who: (1) would be free of intrinsic medical defects at the time of selection; (2) would have a reasonable assurance of freedom from such defects for the predicted duration of the flight programme; (3) would be capable of accepting the predictable psycho-physiologic stress of the missions; and (4) would be able to perform those tasks critical to the safety of the mission and the crew.

Figure 4-2. Left to right standing: Alan B. Shepard, Jr., Walter M. Schirra, Jr., John H. Glenn, Jr. Left to right seated: Virgil I. "Gus" Grissom, M. Scott Carpenter, Donald K. "Deke" Slayton, L. Gordon Cooper, Jr.

Following this screening, the first seven astronaut-pilots, known as the Mercury astronauts, were selected. The names of these men have since become household words in America. They were Alan B. Shepard, Jr.; Virgil I. "Gus" Grissom; John H. Glenn, Jr.; M. Scott Carpenter; Walter M. Schirra, Jr.; L. Gordon Cooper, Jr.; and Donald K. "Deke" Slayton (*Figure 4-2*).

In April, 1962, the Manned Spacecraft Centre issued a call for volunteers for a second group of astronauts to train for the Gemini and Apollo programmes. Minimum qualification standards were published and distributed to the Press, aircraft companies, government agencies, military services, and the Society of Experimental Test Pilots. These standards required that an applicant:

1. Be an experienced jet test pilot and preferably be engaged in flying high-performance aircraft.

2. Have attained experimental flight test status through the military services, the aircraft industry, or NASA, or have graduated from a military test pilot school.
3. Have earned a degree in physical or biological sciences or in engineering.
4. Be a United States citizen under 35 years of age at the time of selection, and be six feet or less in height.
5. Be recommended by his present organization.

These criteria were similar to those originally established for the Mercury space flight programme but allowed the candidates to be taller and opened the way for civilian volunteers. The maximum age limitation was also lowered, because of the long-range nature of the Gemini and Apollo programmes.

More than 200 applications were received from civilians and from volunteers in all four military services. Each candidate who met the five basic standards was asked to complete a variety of forms describing his academic background and his flight and work experience in detail. Each was also asked to take a medical examination and to forward the results of this examination to the Manned Spacecraft Centre.

A preliminary selection committee met and selected 32 of the most highly qualified applicants for further examinations, tests, and interviews. From this group, nine were selected as future astronauts. Included among the selectees were Frank Borman and James A. Lovell, Jr.—two of the three-man Apollo crew who first orbited the Moon—and Neil A. Armstrong—the first man to set foot on the Moon.

At the time of this second selection, Dr. Robert R. Gilruth, Director of the Manned Spacecraft Centre, pointed out that: (1) assignment of flight personnel to specific missions depends upon the continuing physical and technical status of the individuals concerned and upon flight schedule requirements; and (2) flight personnel have an important role in addition to any flight participation to which they may be assigned; this role includes contributions to engineering design, development of future spacecraft, monitoring of flights, and the development of advanced flight simulators.

In June, 1963, volunteer applications were requested for a third group of astronauts. In the requirements stipulated for this group, increased emphasis was placed on academic requirements, and emphasis on flight experience and test pilot work was reduced. The applicants were required to:

1. Be a citizen of the United States; no taller than six feet; not over 34 years of age.
2. Have a bachelor's degree in engineering or physical or biological science.
3. Have acquired 1000 hours' jet pilot time or have attained experimental flight test status through the armed forces, NASA, or the aircraft industry.
4. Be recommended by his present organization.

A total of 271 applications were received, 200 from civilians and 71 from military personnel. From these applicants, 14 were selected for astronaut training. Two of the 14 men selected were civilians; seven were Air Force pilots; four were Navy pilots; and one was a Marine pilot.

Extensive changes were made in the selection criteria for the fourth group of astronauts who were to be called scientist-astronauts. NASA called for applications for the selection of the first group of scientist-astronauts in October, 1964. The applicants were required to be aged 34 or under and have a bachelor's and doctor's degree or equivalent experience in natural sciences, medicine or engineering. No requirement was made for jet pilot experience. A total of 1492 letters of interest were received. Some were formal applications; some were informal inquiries. Of the applications received, 422 were selected as being qualified on the basis of the minimum criteria established and were forwarded to the National Academy of Sciences in Washington, D.C., for evaluation.

The National Academy of Sciences evaluated these applications on the basis of scientific criteria developed co-operately with the NASA Office of Space Science and Applications and selected 16 of the applicants as being highly recommended for consideration. Following thorough physical examinations and extensive testing at the Manned Spacecraft Centre, six from this group were selected for training as scientist-astronauts. The group selected was composed of one geologist, two physicians and three physicists. Selection was made primarily on the basis of scientific background, regardless of jet pilot experience; however, two of the group were jet pilots and needed no basic flight training prior to entering the regular astronaut training programme.

In September, 1965, NASA again issued a call from the Manned Spacecraft Centre for volunteer applicants for astronaut training. The eligibility requirements were basically the same as those set for the third astronaut group in June, 1963, except that the applicant must have been born on or after December 1, 1929. Applications were received from 351 persons, of whom 159 met the basic requirements. Of that number 100 were military personnel and 59 were civilians. Following the usual screening procedures, a total of 19 pilot-astronauts were selected on April 4, 1966. Four were civilians, seven were Air Force pilots, six were Navy pilots and two were Marine Corps pilots.

NASA requested that the National Academy of Sciences nominate a second group of scientists for selection and training as astronauts in September, 1966. The Academy was to seek experienced scientists of exceptional ability to conduct scientific experiments in manned orbiting stations and to observe and investigate the lunar surface and circumterrestrial space. Applications were invited from United States citizens and persons who would be citizens on or before March 15, 1967, no taller than six feet, born after August 1, 1930, and having a doctorate in the natural sciences, medicine or engineering. The applicants would also be required to meet physical qualifications for pilot crew members, but exceptions to any of the above requirements would be allowed in outstanding cases. Selection procedures were similar to those used in choosing the first group of scientist-astronauts in 1965. In its announcement for applications, the National Academy of Sciences stated:

> The quality most needed by a scientist serving as an astronaut might be summed up by the single word "perspicacity". The task requires an exceptionally astute and imaginative observer, but

> one whose observations are accurate and impartial. He must, from among the thousands of items he might observe, quickly pick out those that are significant, spot the anomalies, and investigate them. He must discriminate fine detail and subtle differences in unfamiliar situations, synthesize observations to gain insight into a general pattern, and select and devise key observations to test working hypotheses. He must have the good judgement to know when to stop a particular set of observations and turn to the next. The scientist as an astronaut must translate observations into verbal form and be able to generalize from observations to derive appropriate conclusions.

This recruiting effort, completed in 1967, yielded 11 civilian scientist-astronauts, two of whom were naturalized citizens of the United States—one having been born in Wales and the other in Australia. Following a brief period of general orientation activities at the Manned Spacecraft Centre, these astronauts began a programme of academic training. This included orbital mechanics, computers, spacecraft orientation and general math and physics refresher courses, as well as field trips for contractor facility orientation. In March, 1968, they began Air Force flight training to become qualified jet pilots.

When the Air Force discontinued its plan to place a Manned Orbiting Laboratory into Earth orbit in 1969, NASA assigned seven of the Air Force aerospace research pilots to its astronaut programme. The addition of these seven men brought the total number of active astronauts to 54 in August, 1969. No further astronaut selections have been made to date.

Although crew qualifications change and space hardware becomes more refined with each new programme, the importance of man as an integral part of the machine will not change. The ability of man to assess a situation and provide the necessary input is as important today in space as it was during earlier exploration. And, in retrospect, it is difficult to isolate a manned space flight during which the crew did not provide a contributing or decisive input. This capability was especially proven during the flight of Apollo 13.

Astronaut Training

The astronauts' part in the space programme is to provide the crews needed to man the space flight vehicle and to provide an operational input to design. This input is based on individual judgement, past experience and education, engineering simulations, and actual space flight experience.

In Project Apollo, man for the first time has left his orbit around the Earth and has landed on the Moon—the first stepping stone in the exploration of the solar system. The navigation required for this voyage demands that the astronaut be skilful, not only as a spacecraft pilot and in carrying out the routine computations on digital computers in the spacecraft, but also in using other more complex equipment such as propulsion control systems and fuel cells for electric power generation. In addition to operating scientific equipment the astronaut was required to make meaningful observation in order to select and interpret those phenomena which may be of significant scientific interest. Therefore, the basic objectives of the astronaut training programme are to train crew members (1) to operate the spacecraft in the best possible manner for the accomplishment of an assigned mission, and (2) to serve as competent observers who can conduct the scheduled in-flight experiments. The astronaut training programme is organized in two basic divisions: general training and specific mission training.

General Training

When one considers the complexity of the Apollo mission, it is clear that the training programme should be similar in some respects to graduate study activity in several disciplines concurrently. The flight crew must assimilate knowledge of space flight trajectories, lunar geology, and many spacecraft and launch vehicle systems operational details. Then, working with the design, procedures, flight planning, and simulation personnel, they must develop proficiency to fly not only the planned mission, but be able to fly the planned alternate missions under an almost infinite number of abnormal conditions.

The Apollo training programme is based on extrapolations from Mercury and Gemini training experience. Simulation hours can be used to compare the relative operational complexity of Mercury, Gemini and Apollo. Considering only the time spent in spacecraft simulators, which represent approximately 20 per cent of the training activity, the average totals are: Mercury—50 hours, Gemini—195 hours and Apollo—380 hours.

The general training programme for astronauts consists of an academic programme, an aircraft flight programme, environmental and contingency training, spacecraft design and development studies, and physical conditioning. Although the general programme is organized primarily for new incoming groups of astronauts, other astronauts also participate in certain phases of the programme for purposes of maintaining proficiency.

The academic programme is composed of courses conducted in a formal classroom situation, with accompanying support activities and field exercises. The programme generally consists of courses in basic science and technology, and familiarization with spacecraft and space flight operations.

The basic science and technology courses help to bring the flight crews to a common level of understanding in the prescribed subjects. Typical of such courses are geology, astronomy, digital computers, flight mechanics, meteorology, guidance and navigation, and physics of the upper atmosphere. With the exception of guidance and navigation, the courses are all fundamental in nature. The guidance and navigation course is primarily a functional description of the Apollo system but also covers the basic components of inertial guidance systems. Knowledge of these subjects enables the crews to understand the design and operation of the spacecraft and the launch vehicles used in manned space flight and to perform more effectively the assigned mission experiments.

From a training standpoint, the most significant difference in Apollo as compared to Gemini and Mercury is in the area of guidance and navigation. Learning the capability and gaining proficiency in the operation of the spacecraft guidance and navigation systems absorbs approximately 40 per cent of the training hours.

The crew must use onboard sextants and telescopes to align the spacecraft navigation systems to the stars, landmarks or horizon. They must also ensure that the navigation systems of their two spacecraft are tracking each other. For each mission phase such as launch, mid-course navigation, lunar descent, lunar ascent, rendezvous, and Earth re-entry there is a corresponding computer programme which must be activated by the crew. Within each individual programme are many routines and options which may be selected. The optimum integration of this guidance and navigation activity into the normal and emergency operation of the other spacecraft systems has been an iterative process where experience gained during simulations and previous flights are fed back to improve the procedures and arrive at the best concept for subsequent flights.

Training in spacecraft and launch design and development is accomplished through astronaut participation in spacecraft and launch vehicle engineering studies and mockup reviews of specific contractor-NASA design and development studies. The astronauts attend certain contractor-NASA meetings which are of concern to them, such as conferences on the development of pressure suits and personal equipment, on development of the pre-flight test programme of the spacecraft, and on launch vehicle ground test programmes.

However, in a training programme of this scope and complexity, it is impossible for all the astronauts to keep up with all the day-to-day progress and the ever-changing status of the launch vehicles and spacecraft and their many intricate systems. For this reason, the Flight Crew Operations Directorate has assigned one or more astronauts to each of a number of vital specialized fields. These astronauts closely investigate the activities in these fields and periodically report to the entire group of astronauts on the changes made and the progress effected. Specialized assignments are made in the following fields:

(a) Command and Service Modules.
(b) Lunar Module and cockpit layout.
(c) Launch vehicles.
(d) Control systems, communication systems, and instrumentation.

(e) Mission planning and guidance and navigation.

(f) Recovery systems.

(g) Trajectory analysis and flight plan.

(h) Training simulators.

(i) Spacecraft propulsion.

(j) Deep space network.

(k) Range operations and crew safety systems.

(l) Attitude and translation control systems and cockpit integration.

(m) Environmental control systems, and radiation and thermal protection.

(n) Pressure suits and extravehicular experiments.

(o) Future manned programmes and in-flight experiments.

(p) Electrical and sequential experiments, and monitoring of non-flight experiments (including experiments conducted on the lunar surface).

Environmental Training

The environmental conditions of space flight are either more extreme or occur for a longer period than the conditions met in aircraft flying. Since man's performance is affected by his familiarity with the environmental conditions under which he must operate, astronauts are given training to familiarize them with the conditions to be encountered in space flight, insofar as possible. Each man is exposed to the environmental conditions of weightlessness and launch and re-entry accelerations, and familiarized with the operation of the pressure suit.

Periods of weightlessness of approximately 30 seconds' duration are produced in a modified KC 135 aircraft by flying the aircraft in a parabolic trajectory. This enables the crew members to practise such activities as eating and drinking, free-float manoeuvring, self-rotation, and tumble and spin recovery.

A large centrifuge facility at the Manned Spacecraft Centre is used to familiarize astronauts with the expected acceleration profiles

Figure 4-3. Centrifuge facility at the Manned Spacecraft Centre.

of Earth orbit launch, launch aborts, and orbit re-entry (*Figure 4-3*). Personal familiarity with the forces encountered also gives an astronaut an opportunity to evaluate his own operational capability under these forces.

Pressure suit familiarization includes briefings on suit design and construction and demonstrations in donning and doffing the suit. The pilots also wear pressure suits during specific mission training exercises in order to become familiar with their operation.

Survival Training

During a manned space flight mission, an emergency requiring an immediate return to Earth could result in a landing in a remote and unplanned area. In the event of such an emergency the astronauts must be prepared to land and survive in all types of

Figure 4-4. Astronauts during desert survival training near Pasco, Washington.

terrain and climate. The contingency portion of the training programme includes water, desert and jungle survival exercises.

Water survival training is conducted by NASA and the U.S. Navy at the Pensacola Naval Air Station, Pensacola, Florida. Here the astronauts practise riding the Dilbert Dunker, where they experience the difficulty of getting out of a spacecraft or aircraft if it should become inverted in the water. They also practise other water survival techniques such as getting into a life raft.

Desert and tropic training makes use of existing U.S. Air Force Schools. For desert survival training the astronauts visit the deserts of Nevada and Washington State during the month of August, when temperatures are at their peak. The astronauts spend two days in the classroom hearing lectures and seeing demonstrations, and then three days on a field expedition into the desert. On this expedition they carry only survival kits, parachute and three and one-half quarts of water. They are taught to use a portion of their parachutes to make clothing to protect them from the heat, which can get as high as 110 degrees Fahrenheit. The rest

Figure 4-5. Receiving instruction at the Panama Jungle Survival School is Astronaut James B. Irwin in the foreground, and seated next to him is Scientist-Astronaut F. Curtis Michel.

of the parachute is used to make a tent with a life raft as the centre pole (*Figure 4-4*).

For tropic survival training the astronauts take a one-week trip to Panama. Again, the first two days are spent in the classroom learning survival techniques. In the morning of the third day they are taken into the jungle by helicopter. They are divided into three-man teams, like an Apollo crew. They hike off into the jungle until they find a suitable spot on which to construct a lean-to for protection from the tropical afternoon rains. They live "off the land" for three days, with nothing but a parachute and a survival kit (*Figure 4-5*).

So far, none of the desert and jungle emergency survival measures have been needed by a manned space flight crew. In the most serious emergency, that of Apollo 13, the crew was able to land in the Pacific Ocean and splashdown was performed in the manner practised for all missions.

Aircraft Flight Programme

Spacecraft flight readiness is maintained through a flight programme utilizing high-performance jet aircraft. In addition to local flights, these aircraft are used by the astronauts for cross-country flights in support of engineering and training activities. This permits greater flexibility in travelling as well as maintenance of the crew members' flying skills. A continuing programme of helicopter flying provides the astronauts with an opportunity to familiarize themselves with lunar landing trajectories.

Physical Training

The physical condition of an astronaut is very important, because his performance must not deteriorate under stress. Since individual schedules vary somewhat, each astronaut maintains his own physical fitness programme. Facilities are provided and physical training specialists are available to assist the astronauts in their individual programmes.

Egress Training

Egress training consists of preparing the crew for all pre-flight and post-flight nominal and emergency egress from the spacecraft while in the vacuum chambers, on test stands, on the launch complex and in the water.

Water Egress Training

Water egress training is divided into three phases. Phase I involves procedures familiarization utilizing a Command Module mockup. The crew receives a briefing and demonstration of the survival gear components and their operation. Then the crew reviews egress procedures (unsuited) using the trainer and survival equipment. During this review, they receive further instruction on hatch operation, survival equipment stowage, location and removal, and postlanding system activation. In Phase II—the freshwater tank exercise—the crew receives a briefing on the

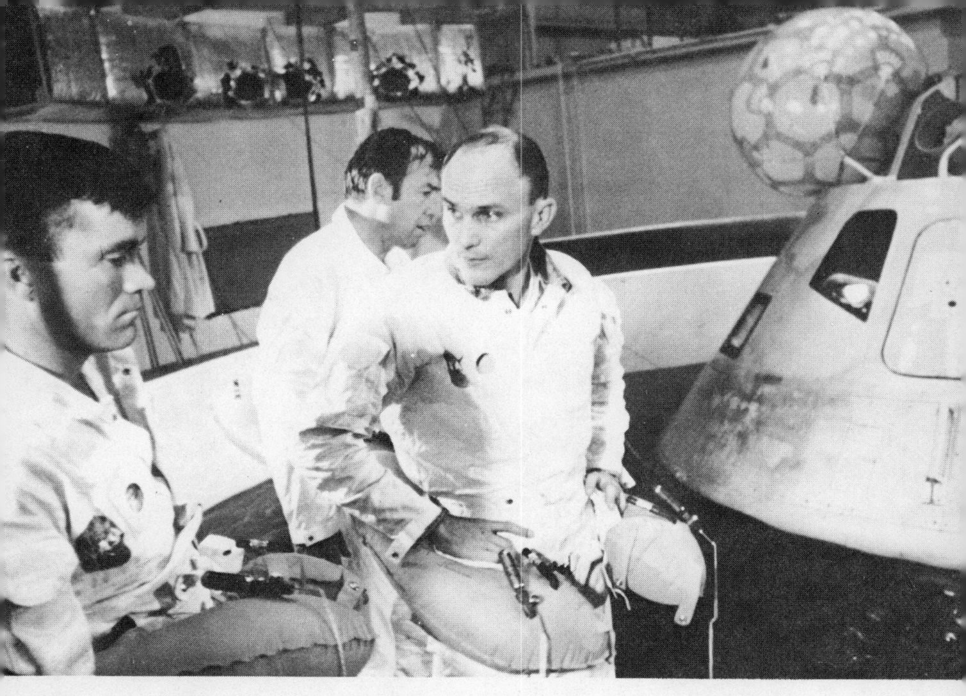

Figure 4-6. The crewmen of Apollo 13 prepare for water egress training at the Manned Spacecraft Centre. Left to right are Astronauts Fred W. Haise, Jr., Lunar Module pilot; James A. Lovell, Jr., commander, and Thomas K. Mattingly II, Command Module pilot.

differences between the spacecraft and the trainer. This is followed by crew egress (suited) practice from the egress trainer while floating upright (Stable I position) in the flotation tank, located at the Manned Spacecraft Centre. After completing the Stable I practice, the crew receives a briefing on apex down (Stable II) egress with subsequent practice in uprighting the training article and Stable II egress. Phase III is a full-scale recovery operation in the Gulf of Mexico. The crew receives a briefing on overall recovery operations and crew recovery procedures. They then practise uprighting the trainer to the Stable I position, egress into their rafts and are picked up by helicopter (*Figure 4-6*).

Spacecraft Prelaunch Egress Training

The crew uses egress mockups early in the training programme to develop familiarity with the operation of egress equipment and to practise egress procedures in conjunction with closed hatch

Figure 4-7. Astronaut Stuart Roosa, Apollo 9 support crew, prepares to descend a rope following the first manned run down a slide wire in a cab from the three-foot level.

spacecraft tests in vacuum chambers and on the launch pad. Emergency egress procedures cover fire, internal and external contaminants, electrical power failures and other emergencies.

At approximately 40 days before launch, the crew accomplishes a launch pad egress exercise at the Kennedy Space Centre designed to assure a rapid crew egress from the spacecraft and launch pad vicinity. This training consists of a briefing on the total launch pad operation, followed by a practice launch pad egress walk-through demonstration of the three evacuation modes: high speed elevator, slide tube and slide wire (*Figure 4-7*).

Specific Mission Training

Crew training following assignment to a flight is designed to prepare the crew to perform a specific mission in accordance with a detailed flight plan. Approximately 2300 hours of crew training is programmed to develop a highly skilled crew to fly the Lunar Landing Mission. In addition to the programmed training, each crew member spends many additional hours participating in other training activities, i.e., physical exercise, study, informal briefings and reviews and necessary mission support activities (air-craft flying, suit fits, pilot meetings, travel, physical examination, mission development).

Criteria used to develop the training programme and overall training objectives are as follows:

(a) The training programme encompasses approximately a 12-month period.

(b) Initially, crews are scheduled for a two- to four-day training week to permit incorporation during remaining time of essential mission development activities requiring crew participation early in the training programme.

(c) The prime and backup crews receive the same training. The support crew members (a third astronaut crew) primarily support crew training by substituting for the prime and backup crews in activities that are essential for mission development.

(d) Because of mission complexity and training time constraints, crew members only train for their specific inflight responsibilities with sufficient cross-training to assure a redundant capability for the more critical mission tasks.

Briefing

Flight crew training is initiated with a series of briefings. These briefings cover a wide variety of subjects, including:

(a) Command Service Module Systems—describing each major subsystem and emphasizing its operation to assure that each crew member has a thorough comprehension prior to initial crew simulator training or participation in spacecraft tests.

(b) Lunar Module Systems — describing each of the major systems and emphasizing its operation.

(c) Launch Vehicle—covering launch vehicle systems operation and performance and related aspects such as countdown

Figure 4-8.
Suited Scientist-Astronaut Harrison H. Schmitt participates in a simulation of using the Aseptic Sampler on the surface of the Moon. He is strapped to a partial gravity simulator.

techniques, range safety, failure modes, abort situations and vehicle flight dynamics and instrument unit operation.

(d) Guidance and Navigation Programme — covering Boost, Earth Orbit, Coast, Prethrust, Thrust, Midcourse Navigation, Entry, Backup and Service Programmes. In conjunction with these briefings, functional descriptions of the Inertial, Optical, Rendezvous Radar, Landing Radar and Computer Subsystems are presented.

(e) Lunar Surface — to provide an understanding of the purpose, results expected and the constraints under which lunar surface data is to be collected; and to describe the equipment and methods to be utilized on the lunar surface.

(f) Geology — verbal descriptions of geological features, observation techniques, geological sampling, information provided by studies by Ranger, Surveyor and Lunar Orbiter data and photography. Simulated geological missions are conducted using Apollo tools and instruments (*Figure 4-8*).

(g) Biology — to instruct the crew in techniques and methods for obtaining aseptic samples, and to give them an appreciation for the concern of back contamination. Lectures are conducted to develop a basic understanding of microbiology as it pertains to the lunar surface.

(h) Experiments — to familiarize the crew with the purpose of the various lunar surface experiments, the constraints involved, and the desired results. Whenever possible, these briefings are conducted by the principal investigator for the particular experiment. Examples of experiments covered are:
 (a) Solar Wind Composition.
 (b) Early Apollo Surface Experiment Package (EASEP).
 (c) Apollo Lunar Surface Experiment Package (ALSEP).

(i) Equipment — in conjunction with the experiment briefings, a series of training sessions is conducted to familiarize the crew with the operation of the experiment hardware. The exercises range from a table-top display of the equipment to a suited exercise of an entire EVA timeline (*Figures 4-9* and *4-10*). Some hardware items covered are the S-Band antenna, geological hand tools, sample return containers, cameras (*Figure 4-11*), modular equipment stowage assembly (MESA) (*Figure 4-12*), EASEP and ALSEP (*Figure 4-13*).

(j) Photography — early in the training programme, the crew receives several briefings by the Mission Operations Branch on the photographic requirements of the mission and operation of the involved photographic equipment. At this time cameras and film are given to the crew for their practice.

(k) Extravehicular Mobility Unit (EMU) — the crew receives a briefing and demonstration of the EMU, which consists of the Pressure Garment Assembly (PGA), the Portable Life Support System (PLSS), and the Oxygen Purge System (OPS). Additional knowledge is gained during subsequent training exercises requiring wearing of this equipment.

(l) Lunar Landing Site—lunar topography recognition for the three selected landing sites. Training encompasses landing site landmark sighting, powered descent track monitoring,

Figure 4-9. Apollo 13 prime crew participate in a walk-through of the extravehicular activity (EVA) timeline.

Figure 4-10. EVA timeline exercise.

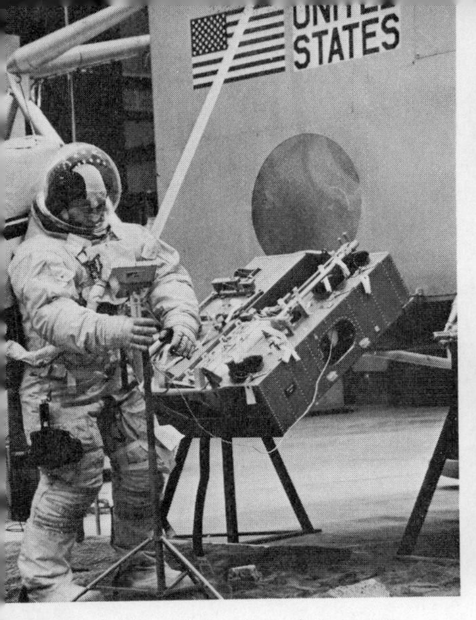

Figure 4-11. Scientist - Astronaut Harrison H. Schmitt wearing an Extravehicular Mobility Unit (EMU), goes through a simulation of deploying the lunar surface television camera.

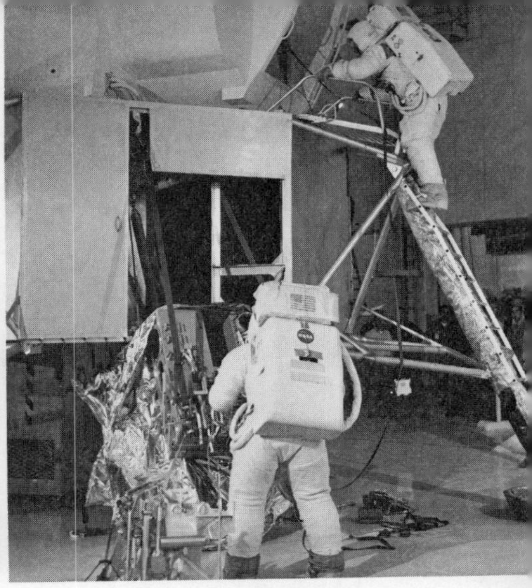

Figure 4-12. Apollo 12 Astronaut Charles Conrad, commander, left, releases equipment from the Modular Equipment Stowage Assembly (MESA) as Alan Bean, Lunar Module pilot, descends the ladder to the ground.

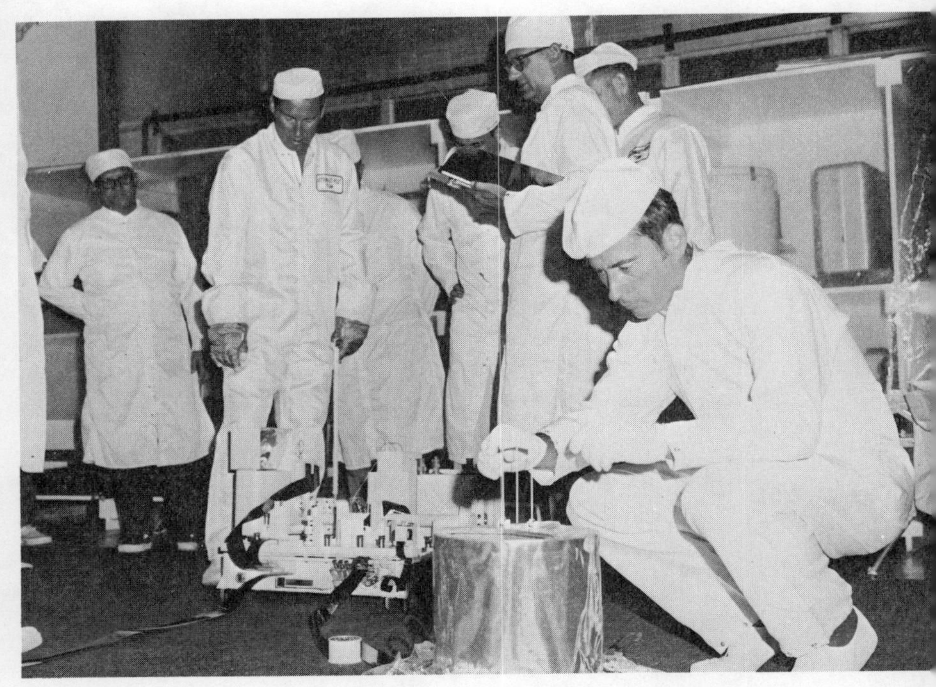

Figure 4-13. The Apollo 13 crew practise deploying the Apollo Lunar Surface Experiments Package (ALSEP) which would be left behind on the Moon. Working with ALSEP experiments are: Left background, James A. Lovell, Jr., prime crew commander, and in right foreground, John W. Young, backup crew commander.

LM pitch variation monitoring, post high gate landing point identification and selection and post landing location of the LM landing point.

(m) Mobile Quarantine Facility (MQF) — the MQF is designed to biologically isolate the flight crew and support personnel from recovery to delivery to the Lunar Receiving Laboratory at the Manned Spacecraft Centre in Houston. The MQF briefing emphasizes communications, oxygen and decompression, sanitation, emergency egress and crew safety. It also includes a familiarization tour of the Lunar Receiving Laboratory facilities and an explanation of the procedures to be followed during the quarantine period.

Mission Simulator Training

To simulate has been defined as "to assume the appearance of, without reality". This is exactly what NASA simulators are designed to do. Crews receive extensive practice in performing all of the nominal and contingency missions tasks on one or more of the simulators located at the Manned Spacecraft Centre in Houston, Kennedy Space Centre in Florida, and the contractor sites. Contractor simulators are primarily intended for engineering development. When the astronauts practise in the NASA simulators, they are made to feel that they are on actual space missions. As a result, when they head into space, they are already familiar with almost every detail of their mission and have had advance preparation for possible emergencies.

NASA has simulators for training Apollo flight crews at two of its facilities, mentioned above. The Apollo Mission Simulator (AMS) is the basic and primary device for the preparation of Apollo flight crews (*Figure 4-14*). This simulator can provide full simulation of a mission, in the appropriate spacecraft configuration, with the proper interface with the Mission Control Centre. Visual displays of the navigation devices in the cockpit and the out-of-the-window views will be complete for the total mission. The AMS training programme has been developed in four phases, to allow the crews to progress easily from single-member part tasks, through increasingly complex operations, to integrated crew and flight controller training.

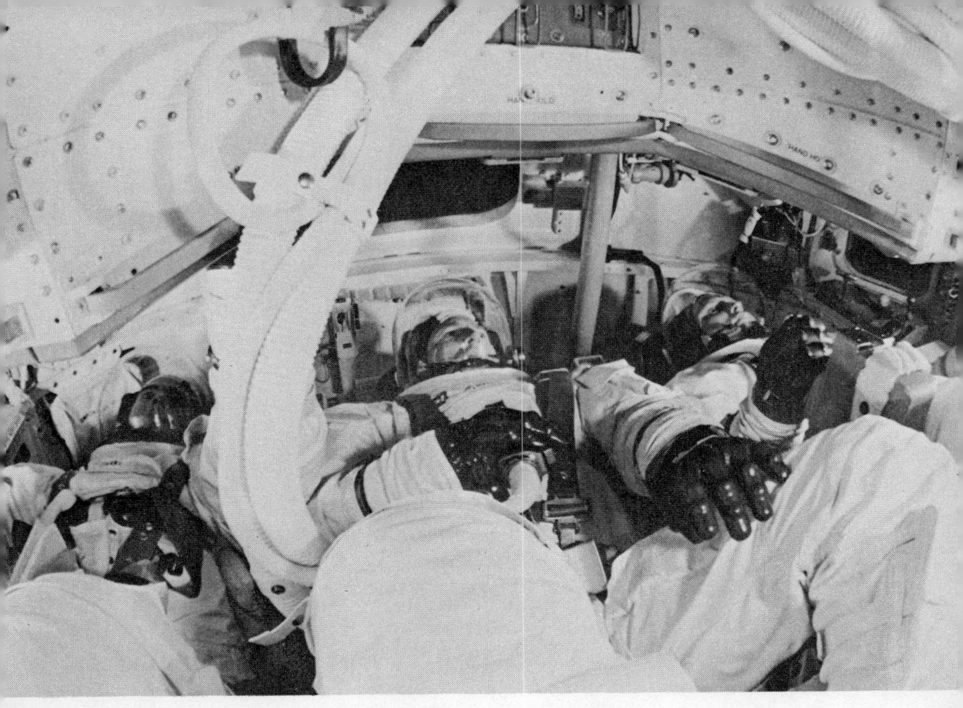

Figure 4-14. The Apollo 8 prime crew during training in the Apollo Mission Simulator at the Kennedy Space Centre. Left to right are Astronauts William A. Anders, Lunar Module pilot; James A. Lovell, Jr., Command Module pilot, and Frank Borman, commander.

Part task training is accomplished on an individual basis, and each man becomes familiar with the basic procedures and skills necessary for the mission. Since all crew members receive this training and all crew stations are considered, the crews are cross-trained during this phase.

Mission task training is a series of exercises in which tasks are combined to provide practice in a specific phase of the mission and the crews are introduced to a co-ordinated effort with a two- or three-man operation. It ties the crew tasks together in a complete mission concept in real time, without extending the simulator session beyond reason for crew comfort and effective training. Missions of several orbits are practised, along with sessions on entry and launch aborts.

Integrated mission training exercises tie the AMS with the Mission Control Centre and simulated remote site. This portion of the training is conducted in the last few weeks before a flight, to train the flight controllers and to give the flight controllers and astronauts the experience of working together. This training also exercises the full capability of the Manned Space Flight Network.

In addition to familiarizing crews with the spacecraft controls and all aspects of a mission, the mission simulator provides training in procedures to be followed during a systems failure. Malfunctions can also be simulated at random in all phases of the training.

Command Module Simulators

The most important as well as extensive Command Module crew training is accomplished on the Command Module Simulators, one located at the Manned Spacecraft Centre and two at Kennedy Space Centre. Crew training proceeds through three distinct phases, progressing from basic systems familiarization and individual crew tasks (Phase I), through mission simulations (Phase II) and concluding with integrated mission simulations (Phase III).

Typically, each Command Module Simulator (CMS) training exercise includes a briefing and debriefing. The briefing covers salient aspects of the training exercise to be accomplished and information relating to the system or operations exercised. Crew questions, performance and simulation discrepancies are reviewed during the debriefing.

In addition to the Command Module Simulator, the crew trains on three other types of simulators related to command module operations. The Command Module Procedures Simulator is used to familiarize the flight crew with the rendezvous and entry phase of the mission. The Dynamic Crew Procedures Simulator is utilized to obtain basic familiarity and subsequently maintain proficiency in crew procedures relative to the Launch Escape System and Service Propulsion System aborts with the attendant entry manoeuvre. This training also provides the crew with the opportunity to review with launch vehicle cognizant engineers nominal and abnormal booster systems operations and performance. Training exercises progress from normal runs with minor deviations to runs with one or more malfunctions inserted.

The Command Module Pilot also utilizes the Rendezvous Docking Simulator at Langley Research Centre, Virginia, for familiarization with Command Module active (Lunar Module rescue) docking procedures. Training is conducted under both day and night lighting situations.

Lunar Module Simulators

The most important and extensive Lunar Module crew training is accomplished on the two Lunar Module Mission Simulators, one each at the Manned Spacecraft Centre and the Kennedy Space Centre. These simulators are capable of simulating the entire LM mission in the appropriate configuration with external displays and interfaces with the CMS and the Mission Control Centre.

LMS training is phased similarly to that employed in the CMS. Simultaneous with Phase I and Phase II LMS training, the crew accomplishes a comprehensive rendezvous training programme on the Lunar Module Procedures Simulator. Phase I frequently requires only one crew member in the LMS, whereas Phase II and Phase III emphasize lunar descent, landing site identification, ascent and rendezvous with the CMS. Crew briefings and debriefings are similar to those for CMS training.

In the Lunar Module Procedures Simulator, the crew receives its basic familiarization and initial proficiency training in LM rendezvous procedures and techniques related to the Abort Guidance System and the Primary Guidance and Navigation Control System.

The Lunar Module crew uses the Translation and Docking Simulator to become familiar with the LM active docking technique (passive CM) (*Figure 4-15*). The crew commander develops proficiency in the LM docking task under various lighting and malfunction situations.

There are other simulators besides the Command Module and Lunar Module Simulators, which are important to manned space flight missions. Housed in a 135-foot diameter circular room at the Ames Research Centre is a Space Navigation Simulator, which duplicates every known factor of control and navigation during space flight. The facility contains two spacecraft models. One is a three-man model to conduct simulated lunar and interplanetary missions. The other is a one-man model for study of physiological and psychological factors of prolonged flight in space. The models are equipped with systems such as may be used on long-duration flights; for example, life-support (air, temperature control, food, etc.), navigation, guidance and power. A feature of the facility

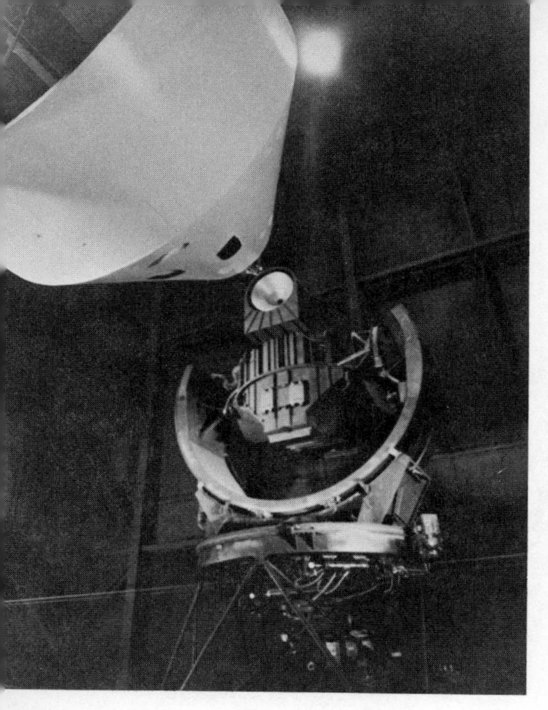

Figure 4-15. The Translation and Docking Simulator used to train astronauts in the rendezvous and docking of the Lunar Module ascent stage and the Command and Service Modules.

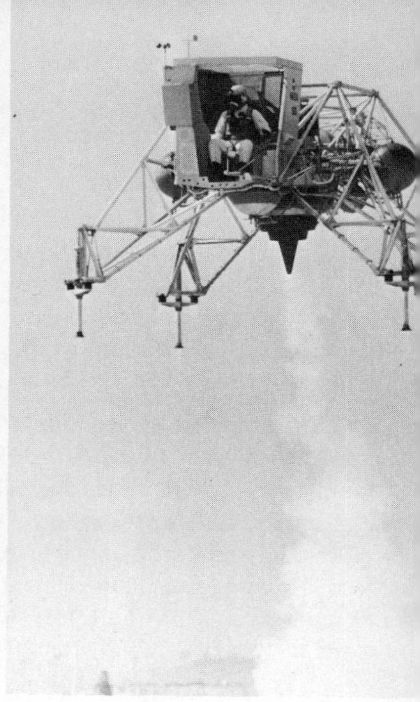

Figure 4-16. Astronaut Charle Conrad, Jr., commander of th Apollo 12, sits in the cockpit of Lunar Landing Training Vehic (LLTV) during a lunar simulatio flight.

is its motion generator — a carefully controlled centrifuge that provides the acceleration and deceleration forces associated with lift-off from Earth, firing of rockets during flight path adjustments (mid-course manoeuvres), and entry into the atmosphere of Earth on return from the Moon and planets.

Several simulators help perfect techniques for landing safely, using only rocket power, on the airless Moon. The Lunar Landing Research Vehicle at the Flight Research Centre in Edwards, California, has a jet engine that is automatically regulated and controlled to counterbalance five-sixths of the Earth's gravity (the vehicle's weight). The Lunar Landing Research Vehicle uses hydrogen peroxide gas jets to lower, raise and balance itself in the same manner as the Lunar Module is expected to operate as it descends to the Moon's surface. An improved version of this craft, looking like the LM, is called the Lunar Landing Training Vehicle (*Figure 4-16*).

Celestial Training

Celestial training is given to increase the crew's capability to orient and control their spacecraft by use of celestial information and to make certain astronomical observations. In missions involving long flights to the Moon, the star background remains constant and the planets do not change position appreciably due to man's movement from the Earth to the Moon. However, the Moon and Earth are objects of prime importance in the man-to-the-Moon missions.

Planetarium facilities are used for general reviews of the celestial sphere, including the recognition and location of prominent stars and constellations. The star patterns that will be visible during an actual mission can be simulated with planetarium equipment and with specially prepared out-of-the-window optical displays.

The planetarium's Apollo capsule simulator has four windows in addition to instruments designed for star observation and navigation. From his seat in the capsule, the astronaut operates a machine to bring a recognizable star pattern into the 60 degree diameter field of the Apollo telescope. Specific sightings may then be taken through the sextant which has a field of view of 1.8° (a full Moon is .5° across). A great deal of practice time is spent getting from one field of view to another by the shortest route in order to economize on imagined fuel.

Conclusion

When training for a specific mission is complete, the flight crews and all the support personnel have been thoroughly trained and rehearsed together in every aspect of the mission. For this reason, an actual mission often seems like a repeat of a fully integrated mission simulation. Of course, there are notable exceptions, such as the flight of Apollo 13.

A crew's training, however, is not terminated with the completion of a space flight. Each astronaut's flight experience is valuable in the evaluation of the current astronaut training programme. Flight experience helps in determining the need to minimize or strengthen certain aspects of the training programme. Also, future flights to which astronaut crews may be assigned will make new demands, creating the need for additional training. An astronaut must train and study continuously, since his occupation is one that places him constantly in the forefront of new knowledge.

It must be kept in mind that the space programme of the United States is very broad and includes a great deal more than the selection and training of astronauts. There are thousands of people whose jobs are just as vital to the success of manned space missions as those of the astronauts. These men behind the scenes represent almost every field of science and engineering, and they participate in a remarkable variety of activities.

Medical personnel study the reactions of the astronauts under physical and psychological stress. Biologists, physiologists, radiologists and doctors determine the physical conditions necessary to sustain human life during lunar exploration. Chemists and chemical engineers develop new materials necessary to absorb the tremendous heat of re-entry. Suit designers and engineers join with geologists to determine what the space suit should be like not only to sustain life but to perform the tasks required in space. Nutritionists design condensed foods which are carried on extended missions. Aerodynamicists, structural engineers, electrical engineers, physicists, thermodynamicists, metallurgists, meteorologists, data analysts, and people trained in various other disciplines are extremely important to the efforts of space exploration. So it becomes obvious that the space programme offers a challenge to many types of dedicated and trained individuals.

Many young people write to the astronauts asking how they can get into the space programme. When astronaut James Lovell, the Apollo 13 commander, was queried as to how he answers these letters, his reply was, "We answer them in a manner that suggests that of all things, they continue their education. We feel that one of the best ways that we can forward our programme is to have well-educated people in it. I think that their resourcefulness, their background, made it possible for this flight — Apollo 13 — to be completed. We say above all things, continue your education. We believe the space programme, if nothing else, is a stimulus to education and inspires young people to follow along."

CHAPTER FIVE

Apollo Missions 1 Through 10

by G. Hage

The spectacularly successful flight of Apollo 11 and the landing of Astronauts Armstrong and Aldrin on the lunar surface are unquestionably man's greatest technological achievement. Not only did this "Great Leap for Mankind" remove the fundamental barriers to our travel throughout our solar system and eventually the universe, but it demonstrated that men and resources, properly directed, can solve any problem, attain any goal, no matter how impossible it may seem.

When, in 1961, only nine years ago, President Kennedy announced the National intention of landing men on the Moon and returning them safely to Earth within the decade, only a handful of astute scientists really believed that it would be done, but it has been done, and the procedure has been well documented so that we will know precisely how it was done. The knowledge and experience gained in the Apollo programme can thus be applied to other large and complex problems.

I would like to review with you the major steps which preceded the lunar landing.

The rapid, but carefully planned progress which brought men to this high plateau of achievement began with the setting of the lunar goal and the establishment of milestones enroute to the Moon. Those which were set toward the end of 1963 have been attained and the lunar landing was accomplished within the time period specified and at the lowest cost estimated in 1961.

In 1960 serious doubt existed concerning man's ability to live in space. Although authoritative opinions differed somewhat, the consensus was far from optimistic. Opinions were based upon research and experimentation in man's tolerance to the environment known, or thought, to exist beyond Earth's atmosphere.

Categories of particular concern were: The effects of weightlessness; man's response to severe acceleration and deceleration loads; difficulties associated with the creation of a life-sustaining atmosphere; and physiological and psychological reactions to prolonged confinement and radiation hazards.

In the face of these unknowns, the six subjects and 54 hours of manned flight accumulated in Mercury could not be considered as an acceptable clinical sample. Therefore, the Gemini programme became the essential link in the chain which would carry men to the Moon.

In the Gemini programme some major steps toward Apollo were taken. The vital rendezvous and docking operations were performed, extravehicular activity was accomplished and it was proven that men could live and do useful work in space for up to 14 days with no ill effects. Training and conditioning of the astronauts which had begun with Mercury were carried to a high point in Gemini. It was learned that man was an essential part of this system designed to conquer a new environment, for he was able in numerous cases, by applying his knowledge and his ingenuity, to save a mission that would otherwise have had to be aborted.

Many of the procedures which were developed for Gemini were later applied to Apollo. Guidance Technique—rolling the spacecraft to get lateral control for entry—was one of these. Much of the manned spaceflight network was operational for the Gemini flights, although it had to be augmented for Apollo. Launch techniques and mission control were developed and reached a high state of readiness before the first Apollo flight.

Within 20 months, the Gemini flights, each carrying two men, had made significant accomplishments. They included: the longest manned flight, 330 hours; the highest altitude, 742 nautical miles; the accumulation of 1,993 hours of manned space flight by 18 different astronauts; 12 hours of extravehicular activity; the greatest number of miles travelled on one mission, 5,129,400, and the greatest number of revolutions about the Earth on one flight, 206. With this experience, the way was cleared for the Apollo programme.

ure 5-1. Apollo Spacecraft 009 ɔ the Saturn IB launch vehicle ing preparations for Apollo/ Saturn 201 test flight.

Three primary launch vehicles were developed in Apollo—Saturn I, IB and V.

The Saturn I was originally planned to be used for manned flight. However, when the reorganization of 1963 took place, and the "All-Up" test concept was instituted, it was decided to discontinue the manufacture of the Saturn I.

The orderly process of the Apollo programme is best revealed by a description of the 10 flights which preceded the flight of Apollo 11. Therefore, I will describe these to you in their chronological order.

An orderly build-up of objectives was necessary to maximize the returns from the flight programme. First, the unmanned tests qualified the booster and the spacecraft. Then the manned tests proceeded in gradually increasing complexity, finally culminating in the landing on the Moon.

APOLLO/SATURN 201

Apollo/Saturn 201 was the first flight of the uprated Saturn launch vehicle, the Saturn IB. It carried, for the first time, both the S-IB first stage and the S-IVB second stage. Flight separation of the launch vehicle and the spacecraft took place for the first time in a non-orbital mode. The Command Module was recovered, the Service Propulsion System burned and restarted and the Block I Apollo spacecraft was tested in flight, all for the first time (*Figure 5-1*).

Many objectives were assigned to this flight and all of them were successfully accomplished. In the interests of economy of both funds and time, as many tests were conducted as possible on each flight. In the event that all could not be accomplished, others could be substituted, and those not performed, put forward to the next flight.

As the first flight test of the launch vehicle and the Block I spacecraft, this flight demonstrated their structural integrity and their compatibility. Separation of the S-IVB stage Instrument Unit and spacecraft from the S-IB stage was performed according to plan as was the separation of the launch escape system and the boost protective cover of the Command Service Module. The Command Module made a smooth removal from the Service Module.

Flight operation information was obtained on the launch vehicle, the propulsion, guidance and control, and electrical systems. The heat shield was tested for entry from low Earth orbit at approximately 28,000 feet per second. The emergency detection system was checked out in an open-loop configuration.

This flight was also the first flight to operate under the newly installed Mission Director concept, and the first to be controlled from the new Mission Control Centre. It proved not only the competence of the hardware but also the viability of the support facilities for launch, mission conduct and Command Module recovery (*Figure 5-2*).

Figure 5-2. A Navy helicopter hovers over the Apollo Spacecraft 009 Command Module as a frogman team prepares it for recovery. The flotation collar increases the spacecraft's buoyancy.

All of the foregoing tests as well as many others which would be performed on later flights, were designed to ascertain the readiness of the equipment for manned missions. Particularly important on this flight were the performance of the spacecraft Control System; the Service Propulsion System in operation for a minimum of 20 seconds after at least two minutes in the space environment, its ability to restart and its ability to operate for 200 seconds and shut down; the Reaction Control Systems of the Command Module and the Service Module; the total environmental control system; the communications system.

The recovery operation including the parachute recovery subsystem and all other recovery aids were exercised for the first time with complete success.

Unmanned, the flight took only 37 minutes, yet with a single flight proved the integrity and compatibility of the systems and hardware. The launch vehicle placed the Command Service Module in an Earth intersecting orbit with an apogee of 266 nautical miles. The spacecraft engine was then fired twice to increase velocity and the Command Module landed in the Atlantic Ocean and was recovered by U.S.S. Boxer on February 27, 1966.

APOLLO/SATURN 202-203

The two following flights, with Apollo/Saturn 202 and 203, had similar objectives, but it was the function of each to do more detailed testing. 203, for instance, was used to check out the systems needed to restart the S-IVB stage and to determine the action of liquid hydrogen in orbit. The heat shield was subjected to additional testing on 202 as the flight apogee of 617 nautical miles brought it back into the atmosphere at a velocity of 28,500 feet per second.

The continuous venting system of the S-IVB LH_2 was evaluated as was the chilldown and recirculation system of this stage, and the fluid dynamics of the tank. Heat transfer into liquid oxygen through the tank wall was examined and data was obtained for a propellant thermodynamic model. Both the S-IVB and the Instrument Unit were checked out in orbit.

The orbital operation of the launch vehicle attitude control and thermal control systems was tested. Tests were also made of the ability of the launch vehicle guidance system to insert a payload into orbit (*Figure 5-3*).

Unusual features of the 203 flight included the simulation of the S-IVB engine restart in orbit and the use of hydrogen continuous vents to accelerate payload in orbit for settling the LH_2 of the S-IVB stage. This was the first orbital flight of the S-IVB and of the redesigned, lighter weight S-IB stage. The most weight to that date was inserted into orbit by the U.S.—28 tons. However, no spacecraft was carried aboard AS-203 which was launched July 5, 1966, and performed four revolutions of the Earth at

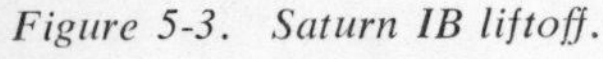

Figure 5-3. Saturn IB liftoff.

Figure 5-4. Apollo/Saturn Mission 203 liftoff. Its primary objective was the orbital behaviour of liquid hydrogen.

approximately 101 nautical miles distance. An aerodynamic fairing weighing 3,700 lbs. was attached to the instrument unit and contained a cryogenic experiment.

Since recovery of the vehicle was not planned, a pressure test above the design value was conducted and the vehicle broke up.

AS-202 flew after 203, and was launched August 25, 1966. It carried a spacecraft which went through rigorous testing of its separation qualities and of its subsystems.

These three flights man-rated the Saturn IB as well as the heat shield and the Command and Service Modules. Additional tests conducted on 203 looked far ahead at the examination of the venting of spent stages; an important element in potential future programmes (*Figure 5-4*).

Fuel cells were first used on Apollo/Saturn 202 as was the Apollo Guidance and Navigation System. This was the first test of the unified "S" band high gain antenna, and the first recovery of an Apollo space craft in the Pacific Ocean, after a one-hour 33-minute flight.

Figure 5-5. Apollo/Saturn V liftoff.

APOLLO 4

The flight of Apollo 4 was undoubtedly the most important flight of the Apollo test programme; this was the first flight of the Saturn V launch vehicle, and it accomplished every one of its objectives flawlessly. From liftoff at Cape Kennedy to recovery eight and one-half hours later in the Pacific, all systems behaved perfectly.

Before the "all-up" flight test programme was instituted, four flights were scheduled to perform the tests which were accomplished on the single flight of Apollo 4. All three stages of the vehicle were tested at once. The spacecraft thermal qualities were tested at an entry velocity of 36,537 feet per second, equal to lunar return velocity. The new hatch, designed after the spacecraft fire at the beginning of that year, 1967, was also tested; restart of the S-IVB as well as burns of the Service Module propulsion system were successfully conducted (*Figure 5-5*).

Figure 5-6.

The mission of Apollo 4 was unusual in many ways. It saw the first launch from Pad 39A at Cape Kennedy, the site especially designed to launch the huge Saturn V to the Moon. As well as being the first flight of the Saturn V it was also the first of the S-IC stage, the S-II stage and of a Lunar Module test article. The S-IVB stage was restarted in orbit and the Service Propulsion System performed the first no-ullage start. The return velocity from the Moon was successfully achieved for a full test of the simulated Block II heat shield. The Command and Communications system had its first flight test as did the Apollo Range Instrumentation Aircraft. And for the first time Apollo-configured ships were used in the communications net.

The Lunar Module test article was a "boiler-plate" test article instrumented to measure vibration, acoustics, and structural integrity at 36 points in the spacecraft-LM adaptor. Data was telemetered to the ground stations during the first 12 minutes of flight. The test article used a flight-type descent stage without landing gear. Its propellant tanks were filled with water/glycol and freon to simulate fuel and oxidizer, respectively. The ascent stage was a ballasted aluminium structure containing no flight systems.

This flight was launched November 9, 1967, and lasted for 8 hours 37 minutes and 8 seconds; it was recovered in the Pacific by the U.S.S. Bennington (*Figure 5-6*).

This was a "textbook" flight, perfect in every way, and it confirmed not only the integrity of the hardware and systems, but also the value of the "all-up" test philosophy. This single flight moved the lunar landing almost a year closer to accomplishment.

APOLLO 5

The flight of Apollo 5 was primarily for the verification of the Lunar Module and its ascent and descent engines.

The Saturn IB carried the Lunar Module into its first flight unmanned. This fantastic mechanism, the Lunar Module, designed

Figure 5-7. Apollo 5 liftoff.

to operate only outside Earth's atmosphere, performed well. An unscheduled hold of 3 hours 48 minute occurred during the countdown at T-2 hours 30 minutes. The hold was caused by two problems: a failure in the freon supply in the environmental control system ground support equipment, and a power supply failure in the digital data acquisition system.

The flight of the SA-204 Launch Vehicle was according to plan. The LM-1 spacecraft also performed according to plan until the time of the first descent propulsion engine burn. The engine started as planned but was shut down after slightly more than four seconds by the LM guidance subsystem when the velocity did not build up at the predicted rate. The problem was analyzed and was determined to involve guidance software only, and the decision was made to go to an alternate mission plan that provided for accomplishing the minimum requirements necessary to meet the primary objectives of the mission (*Figure 5-7*).

The major difference between the planned and alternate missions was the deletion of a long (12-minute) descent engine burn and the substitution of Programme Reader Assembly control for primary

Figure 5-8. Complex 37 blockhouse during the Apollo 5 countdown.

guidance control during the propulsion burns. During all burns thus conducted there was no attitude control; only rate damping was provided. The alternate plan was successfully executed by the flight operations team (*Figure 5-8*).

Sufficient data were obtained to proceed with the mission schedule and to man-rate the Lunar Module. The flight was launched on January 23, 1968, and lasted 7 hours and 50 minutes. No recovery was planned.

APOLLO 6

Apollo 6, the second unmanned flight of the Saturn V, was a complex and interesting one from which a great deal was learned. It was planned to be the same as Apollo 4, to be sure that the perfection of that flight was not a "random" success. The launch vehicle developed a massive longitudinal oscillation and the S-IVB did not restart. A high apogee had been designed into this flight in order to simulate re-entry from the Moon. Upon the failure of the third stage to restart, the service propulsion system of the spacecraft, operating on a contingency plan, pushed the spacecraft to an apogee of 12,020 nautical miles. Although only 32,000 instead of 36,000 feet per second entry velocity was attained, this speed was considered adequate for a heat shield test.

The spacecraft, however, worked very well throughout the mission and as telemetry was returned concerning the activity of the launch vehicle, one of the world's greatest technological detective stories unfolded. Although no part of the launch vehicle was recovered, film and telemetry information were available. Within three months it had been established that four anomalies had occurred in flight and a number of fixes were instituted (*Figure 5-9*).

First, a longitudinal oscillation descriptively called "POGO" had occurred. Second, a film revealed that a piece of the shroud covering the Lunar Module had fallen away. The third anomaly, a premature shutdown of two of the five second stage engines, was found to have been the result of first a fire in one engine and then of faulty wiring in an interconnection. The fourth, the failure of

igure 5-9. Apollo 6 prior to liftoff.

Figure 5-10. Apollo 6 liftoff.

the third stage to restart, was traced to a fire in this engine during its first burn.

The Marshall Space Flight Centre undertook a study of "POGO" and established that while oscillation had been noted in some of the five engines operating on Apollo 4, and while the statistical probability of all five engines operating in a destructive resonance pattern at the same time was highly unlikely, this appears to be what actually did happen on Apollo 6. In order to prevent a recurrence, bubbles of helium were inserted in the fuel line's accumulators in order to change the resonant frequency of the lines and thus prevent any oscillation from building up. The shroud was found to have blown apart from internal pressure — vent holes cured that problem. The engine fires resulted from the breakage of a flexible line from vibrations that could only occur in space. Flex lines were replaced with solid lines. These were the first and, so far, the last problems with the Saturn V launch vehicles. The Command Module landed within 50 miles of the targeted landing point and was recovered in good condition by U.S.S. Okinawa (*Figure 5-10*).

Figure 5-11. Apollo 7 mating. Apollo Spacecraft 101 Command Service Modules being moved into position for mating with Spacecraft Lunar Module Adaptor (SLA)-5.

APOLLO 7

Apollo 7 was the first manned flight in the Apollo Programme. Its overwhelming success generated enough enthusiasm to carry men to the Moon within nine months. This Saturn S-IB vehicle was launched successfully from the Cape on October 11, 1968, carrying Wally Schirra, Don Eisele and Walter Cunningham on a 10.8-day journey.

Accomplishments included eight successful burns of the service propulsion system. A rendezvous of the Command Service Module with the S-IVB stage brought the two spacecraft within 70 feet of each other. On this mission we enjoyed our first live TV shows from space — seven from Earth orbit (*Figure 5-11*).

All primary Apollo 7 mission objectives were successfully accomplished. In addition, all planned detailed test objectives plus three that were not originally scheduled were satisfactorily accomplished.

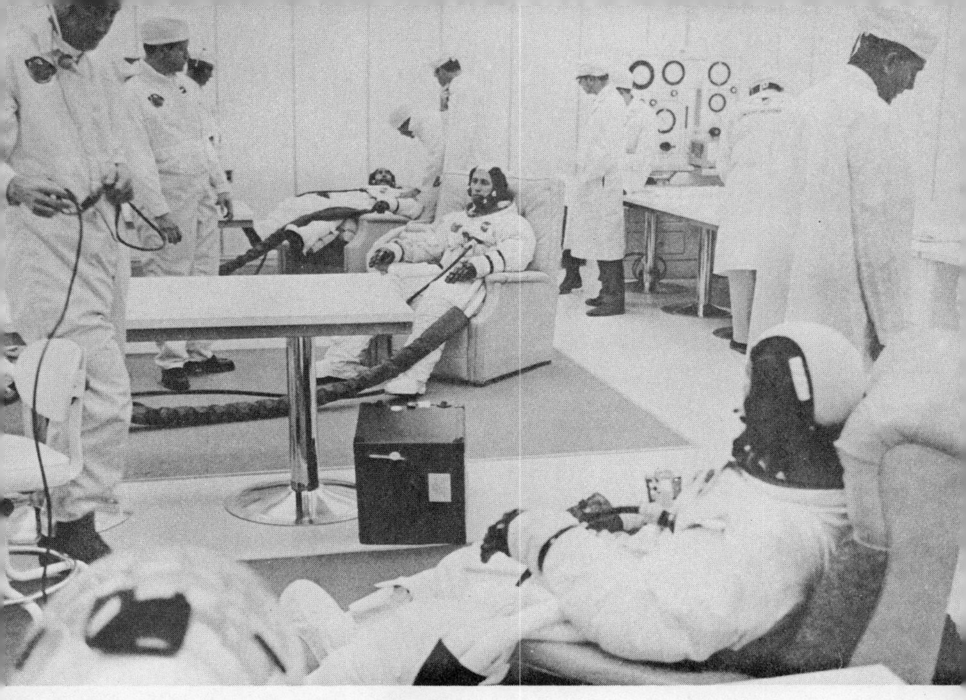

Figure 5-12. Apollo 7 astronauts relax during suiting up. Front to rear: Walter M. Schirra, Jr., Donn F. Eisele and Walter Cunningham.

As part of the effort to alleviate fire hazard prior to liftoff and during initial flight, the Command Module cabin atmosphere was composed of 60% oxygen and 40% nitrogen. During this period the crew was isolated from the cabin by the suit circuit, which contained 100% oxygen. Shortly after liftoff, the cabin atmosphere was gradually enriched to pure oxygen at a pressure of 5 psi.

Hot meals and relatively complete freedom of motion in the spacecraft enhanced crew comfort over previous Mercury and Gemini flights. The Service Module main engine proved itself by accomplishing the longest and shortest manned burns and the largest number of inflight restarts. This engine is the largest thrust engine to be manually thrust vector-controlled (*Figure 5-12*).

Manual tracking, navigation and control achievements included full optical rendezvous, daylight platform realignment, optical platform alignments, pilot attitude control of launch vehicle and orbital determination by sextant tracking of another vehicle by the spacecraft. The Apollo 7 mission also accomplished the first

Figure 5-13. Apollo 7 astronauts, left to right, Walter M. Schirra, Jr., Donn F. Eisele and Walter Cunningham, leaving the recovery helicopter on board U.S.S. Essex.

digital auto pilot-controlled engine burn and the first manned S-Band communications.

All launch vehicle systems performed satisfactorily throughout their expected lifetimes. All spacecraft systems continued to function throughout the mission with some minor anomalies. Each anomaly was countered by a backup subsystem, a change in procedures, isolation or careful monitoring so that no loss of system support resulted. Temperatures and consumables usages remained within specified limits throughout the mission (*Figure 5-13*).

During this mission we gained an appreciation, not only of the skills but also of the humour of the astronauts. The crew developed colds while they were in orbit and there was considerable discussion during the mission of the possibility of ear damage on landing, but this did not occur. The crew, the launch vehicle and the spacecraft performed faultlessly throughout the mission.

APOLLO 8

While Apollo 7 returned the excitement of manned flight to the programme, it was Apollo 8, with man's first departure from his own planet to fly around the Moon, that caught the imagination of the people of the world. The first manned flight of the Saturn V, Vehicle 503, proved conclusively that the detective work that had been done after the flight of Apollo 6 had been successful. Apollo 8's primary objectives were to check out the vehicle with modifications to the J-2 engines, and the changes which had been occasioned by "POGO" and its consequences (*Figure 5-14*).

The mission was launched from Launch Complex 39A on December 21, 1968. It was recovered in the Pacific Ocean on December 27, 1968, by U.S.S. Yorktown.

All primary Apollo 8 mission objectives were completely accomplished. Every detailed test objective was accomplished as well as four which were not originally planned.

igure 5-14. Erection and mating Apollo/Saturn 503, S-1C stage.

Figure 5-15. Apollo 8 on launching site.

The AS-503 Space Vehicle featured several configuration details for the first time, including: a Block II Apollo Spacecraft on a Saturn V Launch Vehicle, an O_2H_2 gas burner on the S-IVB for propellant tank repressurization prior to engine restart, open-loop propellant utilization systems on the S-II and S-IVB stages, and jettisonable Spacecraft Lunar Module Adaptor panels.

For this first Apollo flight to the lunar vicinity, Mission Operations successfully coped with lunar launch opportunity and launch window constraints and injected the S-IVB into a lunar "slingshot" trajectory to prevent recontact with the spacecraft or impact on the Moon or Earth. Apollo 8 provided man his first opportunity to personally view the back side of the Moon, view the Moon from as little as 60 NM away, view the Earth from over 200,000 NM away, and re-enter the Earth's atmosphere through a lunar return corridor at lunar return velocity.

All launch vehicle systems performed satisfactorily. All spacecraft systems continued to function satisfactorily throughout the mission (*Figure 5-15*).

Figure 5-16. The crater Langrenus as photographed from Apollo 8 at an altitude of nearly 150 nautical miles. The small, sharp circular crater nearby is Langrenus C.

Figure 5-17. Thirteen-frame sequence of the Moon from Apollo 10. The darker maria contrast with the lighter highlands. The small circular, more completely surrounded by highlands, is the Sea of Crises. The largest terraced crater shown is Langrenus.

Lunar photography was an important part of the assignment of Astronauts Frank Borman, James A. Lovell Jnr. and William A. Anders. Although not a primary objective, astronaut photographs of the Earth, taken from above the Moon, were one of the treasures collected on this flight. For the first time we saw ourselves as a whole world (*Figures 5-16* and *5-17*).

Only five non-critical parts of the complex hardware which performed Apollo 8 failed — a percentage of reliability which is about 99.999996. The "Awareness Programme" carried out by all of the major contractors who manufactured the millions of parts and thousands of systems for Apollo has proven its effectiveness (*Figure 5-18*).

APOLLO 9

The Earth orbital flight of Apollo 9, with Astronauts James McDivitt, David Scott and Russell Schweickart on board, was actually as innovative and critical a flight as Apollo 8 had been, for the Lunar Module was flown manned for the first time. Beyond the imaginings of any of ou. science fictioneers, this strange craft went through its paces with a perfection rarely achieved with

ure 5-18. Apollo 9 crew, from left, nes McDivitt, David Scott and Russell Schweickart.

Figure 5-19. View of the docked Apollo 9 Command / Service Module and Lunar Module "Spider" during Astronaut David R. Scott's extravehicular activity on the fourth day of the Apollo 9 Earth-orbital mission.

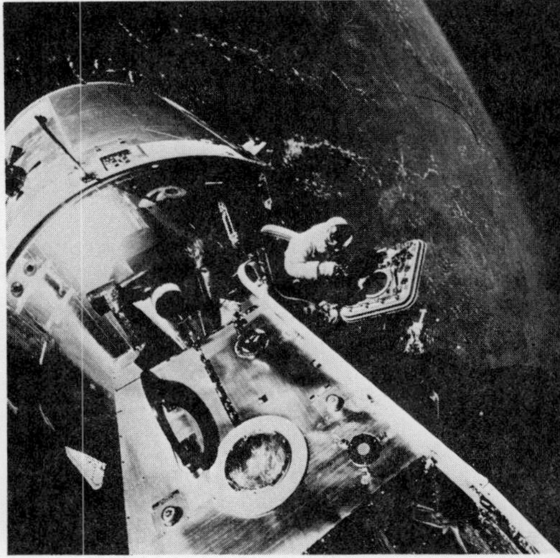

anything so revolutionary—so completely new. The Lunar Module had gone through five changes as it evolved on the drawing board, and changes were incorporated even after its first unmanned test.

A mild virus respiratory illness which infected all of the Apollo 9 crew members was the primary factor in the decision to reschedule the launch from February 28 to 11:00 EST, March 3, 1969. This decision to reschedule was made February 27, 1969, in order to assure the good health of the astronauts. The countdown was accomplished without any unscheduled holds and was the fourth Saturn V on-time launch.

The Apollo 9 launch was the first Saturn V/Apollo Spacecraft in full lunar mission configuration and carried the largest payload ever placed in orbit. Since Apollo 9 was the first manned demonstration of Lunar Module systems performance, many firsts were achieved. These were highlighted by Command and Service Module and Lunar Module-active rendezvous and docking, the first Apollo extravehicular activity, and intervehicular transfer in shirt sleeve environment. This flight also contained the first demonstration of S-IVB second orbital restart capability (*Figure 5-19*).

In the third day of the mission, Lunar Module Pilot Schweickart was struck by nausea and this illness caused a small delay from the normal timeline in the donning of pressure suits and in the transfer to the Lunar Module. It also caused shortening of the proposed extravehicular activity plan. Later the next morning, Commander McDivitt assessed Lunar Module Pilot Schweickart's condition as excellent and with ground control concurrence decided to extend his extravehicular activities.

The Apollo 9 crew had remarkable success in sighting objects using the Crewman Optical Alignment Sight. Their success seems to confirm the thesis that the visual acuity of the human eye is increased in space. One example is their sighting of the Pegasus II satellite at a range of approximately 1,000 miles (*Figure 5-20*).

All primary objectives were successfully accomplished on the Apollo 9 flight. All mandatory and principal detailed test objectives were accomplished, except two, and these two were partially accomplished. One secondary detailed test objective, the S-IVB propellant dump and safing, was not accomplished.

Figure 5-20. The Pegasus II satellite sighted from Apollo 9 at a range of approximately 1,000 miles.

All launch vehicle systems performed satisfactorily with the exception of inability to dump propellants following the third S-IVB burn. All spacecraft systems continued to function satisfactorily throughout the mission. No major anomalies occurred. Those minor discrepancies which did occur were primarily procedural and were corrected in flight with no mission impact, or involved instrumentation errors on quantities which could be checked by other means.

An experiment of considerable importance in future programmes was carried aboard this flight, the multi-spectral camera equipment for experiment S065. Using filters of different wave lengths, photographs of the Earth were taken by the astronauts while the same Earth areas were being photographed from high and low flying aircraft. Apollo 9 was landed in the Atlantic after 10 days, 1 hour and 53 seconds of flight and was recovered by the U.S.S. Guadalcanal on March 13, 1969 (*Figure 5-21*).

Figure 5-21. Apollo 9 recovery. Astronaut Russell L. Schweickart on open spacecraft prior to joining David R. Scott in life raft.

Figure 5-22. Apollo 10 crewmen, from left, Eugene Cernan, John Young and Thomas Stafford, with vehicle in background.

APOLLO 10

The dress rehearsal for the lunar landing, Apollo 10 was launched on May 18, 1969, from Cape Kennedy with Saturn V 506 carrying the Command and Service Modules and the Lunar Module on a flight around the Moon. For the first time, all of the equipment destined for the lunar landing was tested in lunar orbit, and everything worked. It is not possible to estimate the worth of the confidence which has been generated by this series of Apollo successes in one of man's most difficult and complex endeavours.

Astronauts Eugene Cernan, John Young and Thomas Stafford orbited the Moon at 60 nautical miles. Stafford and Cernan entered the Lunar Module and flew it down to within 47,000 feet of the lunar surface (*Figure 5-22*).

The most complex mission yet flown in the Apollo Programme was performed in the full lunar landing configuration, paralleling as closely as possible the lunar landing mission profile and timeline. Extensive photographic coverage of candidate lunar landing sites provided excellent data and crew training material for subsequent missions. This was the fifth on-time Saturn V launch.

Nineteen colour television transmissions (totalling 5 hours 52 minutes) of remarkable quality provided a world audience the best exposure yet to spacecraft activities and spectacular views of the Earth and the Moon. The Lunar Module pericynthion of 47,000 feet was the closest man had come to the Moon, and the crew reported excellent visual perception of the proposed landing areas (*Figure 5-23*).

The mission was nominal in all major respects. Translunar and transearth navigational accuracy was so precise that only two of seven allocated midcourse corrections were required, one each during translunar and transearth coast periods. Significant perturbations in lunar orbit, resulting from differences in gravitational potential, were noted. All launch vehicle systems performed satisfactorily (*Figure 5-24*).

Figure 5-23. Photograph taken minutes before docking of Apollo 10 shows ascent stage of the Lunar Module. Rendezvous radar dish is located in the corner of photograph.

Figure 5-24. View from Ap 10. Bruce, the prominent cr near the bottom of this scene about six kilometres in diame

Spacecraft systems generally performed satisfactorily throughout the mission. One exception was the No. 1 fuel cell which had to be isolated from the main bus, but work-around procedures made it available for load sharing, if required. Another problem was the occasional difficulty with direct Lunar Module to Earth communications. Two incidents of unexpected motion occurred prior to and during Lunar Module staging. Data indicated an unscheduled transfer of the abort guidance system mode from "Attitude Hold" to "Automatic" (*Figure 5-25*).

A number of minor discrepancies occurred which were either primarily procedural and were corrected in flight with no mission impact, or which involved instrumentation errors on quantities that could be checked by other means. Two cameras that malfunctioned were returned to Earth for failure analysis. All detailed test objectives were met, except for two secondary spacecraft objectives that were partially accomplished. Five other major activities not defined as detailed test objectives were fully accomplished.

re 5-25. International Astronomical on Crater No. 302 on the lunar side. Photographed from Apollo 10.

Figure 5-26. View of Moon from Apollo 10. The large dark area near the centre of photograph is the Sea of Tranquillity.

Flight crew performance was outstanding. Their health and spirits remained excellent throughout the mission. Unexpected bonuses from the mission were several sightings of individual spacecraft Lunar Module adaptor panels, three sightings of the jettisoned descent stage as it orbited the Moon at low altitude, and a few sightings of the receding S-IVB stage with the naked eye, once from nearly 4000 miles as it tumbled and flashed in the sunlight (*Figure 5-26*).

The years of hard work, innovation and zealous performance by almost half a million people in government, science and industry were soon to be rewarded, when Apollo 11 landed on the Moon and returned safely to Earth.

These 10 flights of the Saturn launch vehicle, and the four flights of the manned spacecraft had thoroughly tested the men, machines and systems which would bring mankind to the demonstration of his highest efficiency — and fulfil the ancient dream of walking on another body in our solar system.

CHAPTER SIX

The Lunar Landing

by G. Hage

"HOUSTON. TRANQUILLITY BASE HERE. THE EAGLE HAS LANDED."

These words proclaimed that the impossible dream of man for countless generations had come true. For the first time in history, man had left his planet Earth and landed on another celestial body.

In 1961, the President of the United States, John F. Kennedy, stated that America should commit itself to achieving the goal of landing a man on the Moon and returning him safely to Earth before the end of the decade. Thus, the manned exploration of the Moon became the national goal of the emerging government-industry space team of the United States. And in less than a decade man's footprint on the surface of the Moon was an accomplished reality.

The flight of Apollo 11, the first manned landing on the Moon, began on the morning of July 16, 1969. Exactly on schedule, Apollo 11 lifted off from Launch Pad 39A at Cape Kennedy, Florida, to start the trip to the Moon (*Figure 6-1*). Atop the 363-foot, 7.6 million-pound Saturn V launch vehicle, the astronauts were strapped to their couches in the Command Module. These three dedicated men had been selected by the National Aeronautics and Space Administration as the crew of Apollo 11. Neil Armstrong, the mission commander, was to be the first man in history to set foot on the surface of the Moon. Edwin E. Aldrin, the Lunar Module pilot, was the second member of the lunar landing team. Michael Collins, the Command Module pilot, would remain

Figure 6-1. Apollo 11 liftoff.

alone in his spacecraft orbiting above the Moon during the 28 hours his fellow astronauts descended to the lunar surface and returned.

Watching the launch of Apollo 11 was a world-wide television audience and an estimated one million eyewitnesses. Three and one-half miles away on the sandflats of the Kennedy Space Centre or seated in grandstands were half the members of the United States Congress and more than 3,000 newsmen from 56 countries.

After several static seconds for thrust buildup, Apollo 11 lifted from the launch pad and moved up past the retracted gantry, gaining speed as it climbed. The events of the pre-orbital phase—the roll sequence, jettison of the launch escape tower, first stage rocket cut-off, second stage burn, second stage cut-off and third stage burn—took place with clockwork precision. With the shut-down of the third stage rocket engine, both spacecraft and the third stage entered a 103 nautical mile circular orbit. All systems operated satisfactorily.

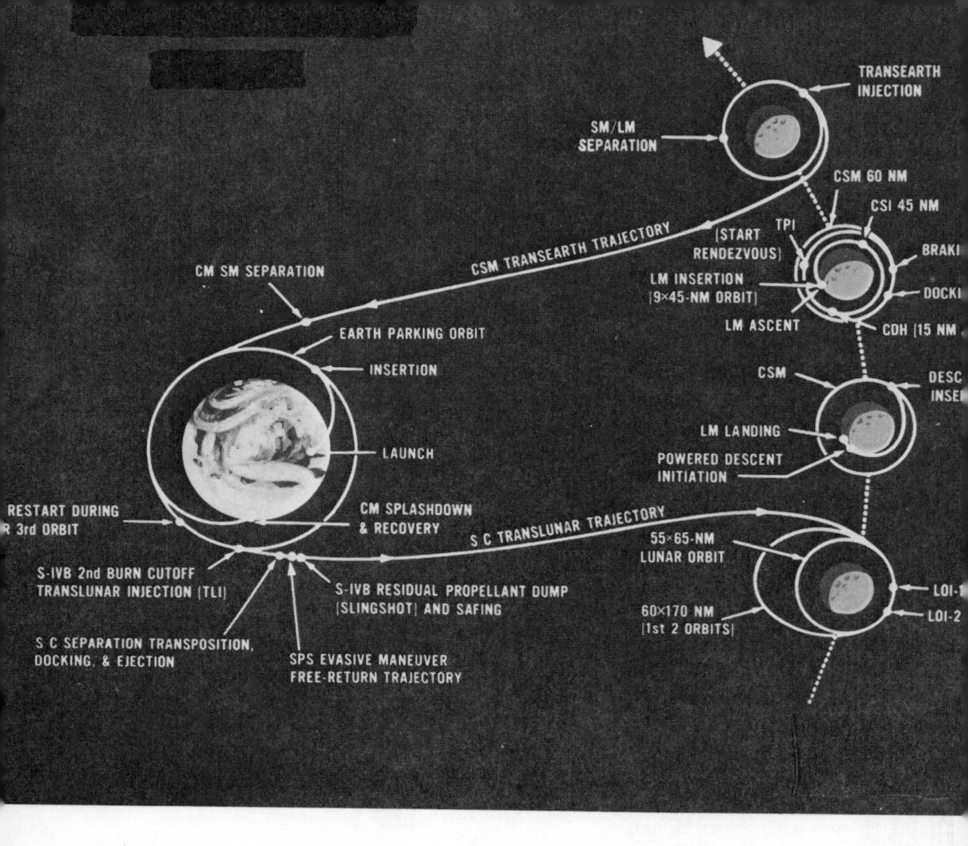

Figure 6-2. Apollo 11 Flight Profile.

The crew spent the next full orbit and part of the second in an engineering, communication and equipment checkout. Over a point north-east of Australia, Mission Control at Houston, Texas, gave them "go" for insertion into their translunar course (*Figure 6-2*). Re-firing the third stage engine increased velocity to roughly 24,200 miles per hour, sufficient to break out of low-Earth orbit into a free-return trajectory. This trajectory was an elliptical course that, if undisturbed, would loop the spacecraft around the Moon and bring it back to Earth.

Once on course and moving farther and farther from Earth, the crew set about separating the Command Service Module (CSM) from the third stage which still housed the Lunar Module (LM) in the protective shelter of the panelled adaptor section. First, the astronauts fired explosive bolts which caused the main spacecraft, called Columbia, to separate from the adaptor and blow apart the four panels that make up its sides, exposing the LM, code-named Eagle. Then, the CSM was pitched 180 degrees and flown slowly back for docking with the LM. The LM separated from the third stage, docked with the CSM, and the newly joined components moved a safe distance from the stage. Mission Control then ordered the third stage to dump its remaining fuel. This manoeuvre, which reduced its weight, caused the stage to propel itself into a course around the Moon and on into solar orbit. The mated Columbia and Eagle continued on toward the Moon.

With the flight on schedule and proceeding satisfactorily, the astronauts were informed by Mission Control later in the day that the first scheduled mid-course correction was considered unnecessary.

On the morning of July 17, Mission Control gave the Apollo crew a brief review of the morning news and informed them about the progress of the Russian spacecraft Luna 15. Shortly after noon, a mid-course correction was made with a three-second burn, sharpening the course of the spacecraft and testing the engine that would be used to get it in and out of lunar orbit. That night, the astronauts began their first scheduled colour telecast, showing a view of the Earth from a distance of about 128,000 nautical miles.

The third day into the mission, July 18, was spent doing housekeeping chores, such as charging batteries, dumping waste water, and checking fuel and oxygen reserves. Once again, Mission Control informed the astronauts that course corrections scheduled for that afternoon would not be necessary. At 4.40 Eastern Daylight Time one of the clearest television transmissions ever sent from space was begun, with the spacecraft 175,000 nautical miles from Earth and 48,000 from the Moon. With the telecast in progress, the hatch to the Lunar Module was opened and Armstrong and Aldrin squeezed through the 30-inch-wide tunnel to inspect it.

Though it left Earth orbit speeding at more than 24,000 miles per hour, relative to Earth, the gravitational pull of Earth steadily slowed the spacecraft until its velocity had been cut to slightly over 2000 miles per hour on the night of July 18. At this low point, the Apollo spacecraft was approximately 34,000 nautical miles from the Moon, a zone where the pull of the Moon's gravitational field is stronger than that of Earth and the spacecraft, accordingly, began to pick up speed.

Early afternoon of the next day, July 19, permission was given by Mission Control for lunar orbit insertion. As the spacecraft passed completely behind the Moon and out of radio contact with the Earth for the first time, the 20,500-pound-thrust engine was fired for about six minutes to slow the vehicle so that it could be captured by lunar gravity. The resulting orbit ranged from a low of 61.3 nautical miles to a high of 168.8 nautical miles. Later, a second burn of the main engines was employed for 17 seconds to stabilize the orbit at 54 by 66 nautical miles and the spacecraft began circling the Moon every two hours.

On the morning of July 20, Armstrong and Aldrin crawled into the Lunar Module and powered-up the spacecraft. At 1.46 p.m. EST, the LM was separated from the Command Module in which Michael Collins continued to orbit the Moon.

An hour and 22 minutes later, the descent manoeuvre began with a retrograde burn of the LM's descent engine that placed the LM in an elliptical orbit with a low point 8.5 nautical miles above the lunar surface. When the orbital low point was reached, the powered-descent stage started. This involved dropping the LM out of orbit into an arching glide with a terminus on the Moon's surface. The glide path had two check points: one called "hi-gate" at an altitude of 7,600 feet and 26,000 feet laterally from "lo-gate", 500 feet in altitude and adjacent to the landing zone. During the glide the spacecraft's velocity would be cut from 342 miles an hour to about 50 miles an hour and eventually to almost zero. The descent went as planned and as the LM reached "lo-gate", its attitude approached the vertical to the Moon's surface. As the LM dropped below 500 feet in altitude, the crew transmitted a staccato numerical report to Mission Control on its rate of drop and lateral movement.

Just seconds from touchdown, there was a break in communications providing what can only be described as very tense moments. The next word heard from the crew was that the Eagle had landed.

The full story of what actually happened became known after the astronauts returned to Earth. When Neil Armstrong first saw the landing site through the window of the LM, he was not absolutely sure where he was. Most of the landmarks he had studied and memorized were actually behind him and of no help. Because of the previous training Armstrong had received in the simulators, however, he knew exactly what must be done. Taking over partial control from the LM's autopilot, he ordered the computer to keep the spacecraft at a steady altitude and gave the LM a heading, reducing the braking effect of the descent engine and letting the craft surge forward at high speed. After clearing a large crater which appeared extremely rugged with boulders of five to 10 feet in diameter and larger and then a second, smaller crater 100 feet in diameter, Armstrong brought the descent engine's braking power into full play again in order to land at a level, relatively clear site.

During the last 40 feet of descent, the descent engine exhaust sent up a cloud of Moon dust engulfing the LM. The particles flew at low angles and high velocity with no atmosphere to buoy or impede them. As soon as the engine was cut off, however, the view from the window of the LM was clear again.

This manoeuvre took the astronauts more than 1,000 feet beyond where the computerized autopilot would have set them down. When the descent engine was cut off, there was only 30 seconds worth of fuel remaining.

After touchdown, Armstrong began running through his post-landing check list and told the flight controllers at Mission Control that he had shut down the engine of the descent stage and that he had removed his controller from a neutral position and had programmed an attitude instruction into his onboard computer which would cause a reaction control thruster to fire as an audible cue that the Lunar Module might be tipping over. Both his primary guidance and navigation systems and his abort guidance system were ready to be operated by computer if any equipment malfunctioned.

The controllers on the ground and the crewmen in the module were very busy for the next few minutes as critical "stay-no-stay" times were relayed. Finally, a "stay" for one revolution of the Command Module, orbiting 60 miles above the lunar surface, was given. Mission Control radioed instructions to reset the mission timer as the LM crew continued to prepare for their exit from the Lunar Module to the lunar surface. Concurrently, preparations were made to prepare the spacecraft for ascent from the lunar surface. By this time, the crew had been on the surface for more than an hour.

Aldrin began describing the view from his window of the Lunar Module: ". . . it looks like a collection of just about every variety of shapes, angularities and granularities, every variety of rock you could find. The colours vary pretty much depending on how you're looking. . . . There doesn't appear to be much of a general colour at all; however, it looks as though some of the rocks and boulders, of which there are quite a few in the near area, are going to have some interesting colours to them."

A few moments later, he told of seeing numbers of craters, some of them 100 feet across, but the largest number only one or two feet in diameter. He saw ridges 20 or 30 feet high, two-foot blocks with angular edges, and a hill half a mile to a mile away.

Finally, in describing the surface, Aldrin said: "It's pretty much without colour. It's grey and it's a very white, chalky grey, as you look into the zero phase line, and it's considerably darker grey, more like ashen grey as you look up 90 degrees to the Sun. Some of the surface rocks close in here have been fractured or disturbed by the rocket engine and are coated with light grey on the outside, but when they've been broken they display a dark, very dark grey interior, and it looks like it could be country basalt."

As the Command Module came around the backside of the Moon, Collins was asked to try to accurately pinpoint the location of the Lunar Module. He was informed by Mission Control the LM had landed about four miles beyond its targeted area. Despite repeated efforts, Collins was never able to spot the LM and its

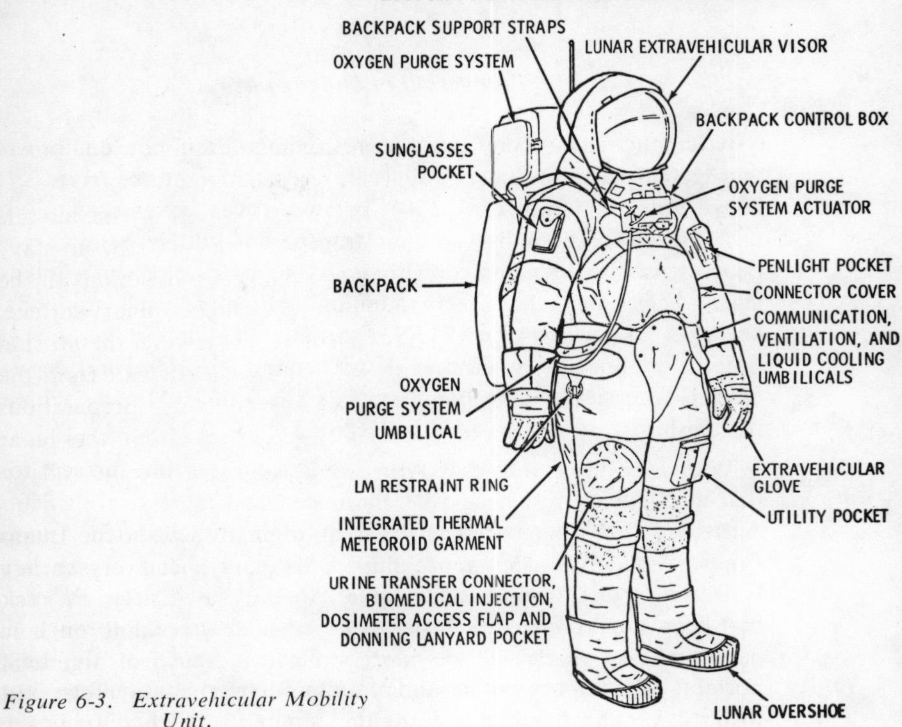

Figure 6-3. Extravehicular Mobility Unit.

actual landing site was defined only after lift-off from the Moon's surface when the rendezvous radar was utilized.

Another "stay" time was given to the LM and the crew prepared to power down their spacecraft and end the simulated countdown for an earlier than scheduled lift-off. Then the first lunar explorers put on space suits for their excursion onto the lunar surface.

The space suits that they donned were designed for ingenious protection against the hazards of the Moon's environment (*Figure 6-3*). From the skin out, the basic pressure garment consisted of a nomex comfort layer, a neoprene-coated nylon pressure bladder and a nylon restraint layer. The outer layers, known as the extravehicular integral thermal/metheoroid cover, consisted of a liner of two layers of neoprene-coated nylon, seven layers of Beta/Kapton spacer laminate, and an outer layer of Teflon-coated Beta fabric. These outer layers are designed for protection against micrometeoroids travelling at 64,000 miles an hour, 30 times the speed of a rifle bullet.

After the suits were on, the astronauts then put on heavily corrugated plastic overboots that can resist temperatures from 250° above zero Fahrenheit to 250° below, gloves covered with fine metal mesh, and hoods for their transparent bubble helmets with double visors (both coated with gold) to block the sun's intense glare, heat, and ultraviolet radiation. Finally, each donned a backpack, known as the PLSS (portable life-support system) to provide cooling water, electric power, communications and enough oxygen to last for four hours outside the Lunar Module without replenishing.

With everything in order, Armstrong radioed a recommendation that a scheduled rest period be eliminated and the Extra Vehicular Activity (EVA) be started earlier than originally scheduled. Mission Control concurred and, more than five hours ahead of schedule, Neil Armstrong opened the Lunar Module hatch after the craft had been depressurized. Armstrong began to squeeze through the hatch, a task made all the more difficult because of the bulky portable life support system and an unfamiliar gravitational environment. About 12 minutes from the report of the hatch opening, Armstrong was out of the spacecraft and on the porch of the Lunar Module's ladder.

The next task for Armstrong was to unstow the module equipment stowage assembly (MESA), a pallet on the descent stage of the Lunar Module containing fresh batteries for the portable life-support system, a television camera, sample bags and tools for obtaining lunar samples. This task was accomplished without difficulty and shortly thereafter Mission Control reported picking up a signal from the television camera. The television transmissions were fuzzy and scored with lines but nonetheless held the world spellbound as Armstrong prepared to descend to the surface of the Moon.

As he descended the ladder, Armstrong inspected the footpads of the Lunar Module and took his first close look at the surface of the Moon. He reported that the LM footpads were only depressed in the surface about one or two inches and that the surface appeared to be very fine grained and almost like powder. Armstrong continued down the ladder, put his left foot on the Moon

Figure 6-4. Lunar surface footprints.

and said into his radio microphone, "That's one small step for man, one giant leap for mankind". In less than two seconds, this message was received at the huge telescope dish at Honeysuckle Creek, near Canberra, Australia, bounced to the COMSAT satellite over the Pacific Ocean, then to the switching centre at the Goddard Space Flight Centre outside of Washington, D.C., and finally to the Manned Spacecraft Centre at Houston, Texas, and the rest of the world (*Figure 6-4*).

Armstrong's attention was first directed toward the nature of the surface material and observed that the exhaust of the descent engine had not cratered the area directly below the LM engine nozzle. He then proceeded with his scheduled task of collecting a contingency sample consisting of several pounds of lunar surface material which he stowed in a space suit pocket. This collection was assigned as a first task to make sure that there would be samples aboard in case an early abort of the mission was necessary.

Figure 6-5. Aldrin makes the descent to the lunar surface.

Once the LM inspection and the contingency sample collection were completed, Aldrin came out of the LM and climbed down the ladder, with Armstrong providing guidance and taking photographs (*Figure 6-5*). The two astronauts then unveiled the plaque mounted on the strut behind the ladder of the LM. They read the inscription for the benefit of their world audience:

HERE MEN FROM PLANET EARTH
FIRST SET FOOT UPON THE MOON
JULY 1969 A.D.
WE CAME IN PEACE FOR ALL MANKIND

The plaque was signed by the astronauts and United States President Richard Nixon.

Figure 6-6. Astronaut on lunar surface.

Armstrong then removed the TV camera that had covered his first steps on the Moon and placed it in a position so that the LM and surface activities could be observed.

After surveying the surroundings, Armstrong and Aldrin began moving about, testing themselves in the gravity environment one-sixth of that on Earth (*Figure 6-6*).

There was no problem in collecting samples of lunar rocks and soil. The astronauts bagged upwards of 50 pounds of the dark, loose surface material and representative samples of the lunar rocks. The men used a specially made aluminium scoop on an extension handle and a pair of long aluminium tongs to perform this task since the space suits were too cumbersome to facilitate bending over. The samples were put into two boxes, each formed from a single piece of aluminium. A ring of indium, a soft metal, lined the lip of each box. When the box was closed and the

Figure 6-7. Lunar rock sample collected by Apollo 11 crew.

straps drawn tight around it, a knifelike strip around the edge of the lid bit deeply into the indium which sealed the samples in a vacuum and protected them against contamination.

Armstrong and Aldrin noted that the Moon was strewn with rock fragments of a wide range of size, angularity, and texture (*Figures 6-7* and *6-8*). Although some rock fragments were obviously lying on top of the surface, it was not always possible to judge their depth of burial. In the course of using the scoop, rocks buried under several inches of soil were encountered.

The major textural types of rock fragments observed were plain, even-grained basaltlike rocks, vesicular basaltlike rocks; basaltic-appearing rocks with one to five per cent small white minerals; and rocks consisting of aggregated smaller fragments. In some instances, loosely aggregated clods of soil were difficult to distinguish from the rock fragments until they were disturbed and broken up.

Figure 6-8. Lunar rock samples.

Smaller pieces of material that had a metallic lustre were noted. These pieces of material were concentrated in scattered aggregates at the bottoms of three or four-feet diameter craters. Several examples of lunar material that seemed to be transparent crystals were also observed. These crystal-like materials resembled quartz crystals and appeared to be opaque from some views and translucent from other views. There were also fragments that resembled biotite but these fragments were not examined closely.

The astronauts described the fine surface material as a powdery, graphitelike substance that seems to be dominantly sand to silt size and when the material is in contact with the rocks, it makes them slippery. This phenomenon was checked on a fairly smooth, sloped rock. When the powdery material was placed on the rocks, the boot sole slipped easily on the rock and the slipping was sufficient to cause some instability of movement. Otherwise, the astronauts reported that traction was generally good in the loose powder.

Figure 6-9. Aldrin takes lunar core sample. Solar wind experiment in the background.

Armstrong and Aldrin found unexpected differences in the consistency and softness of the top layer of the surface material at locations having minor changes in surface topography. These differences were manifested in significantly different footprint depths. The depth differences indicated that there may be different depths of surface material covering the more resistive subsurface, particularly on the rims of small craters.

Surface penetrability decreased quickly within the first few inches of the surface. When specifically probed more than four or five inches, the surface was found to be quite firm. Surface penetration by using core tubes was no greater than eight or nine inches, even when the sampler extension was hammered hard enough to be significantly dented (*Figure 6-9*). However, there were no rocks

Figure 6-10. Aldrin sets up solar wind experiment.

under the core tubes during the driving operations. The material at the bottom of the core sample appeared to be darker than the surface material, and this material packed in and adhered to the sides of the tube in the same manner as wet sand or silt.

As scheduled, the astronauts then set up three planned experiments, on one of which was mounted a dust detector experiment.* From an outside storage compartment in the Lunar Module, Aldrin removed the solar wind experiment. The solar wind is an ionized, or electrified, gas which constantly streams away from the Sun at speeds of 200 to 400 miles a second. The wind is not normally detected on Earth because the magnetosphere

---

* The Dust Detector Experiment, developed by Dr. Brian J. O'Brien of the School of Physics, University of Sydney, is designed to monitor dust accumulation on three solar cells.—Ed.

Figure 6-11. American flag planted on lunar surface.

deflects the gas. Its effects can only be seen when a small amount of the solar wind enters into the magnetosphere in the polar regions, accelerated by some unexplainable process, and causes the aurora high in the atmosphere.

Since the Moon does not have a strong magnetic field, a steady barrage of atomic particles, carried by the solar wind, is battered against it. Scientists believe that these particles may slowly erode the lunar rocks. A simple device was deployed by Aldrin to trap these particles which consisted of a strip of aluminium foil about a foot wide and four and a half feet long that was unfurled and hung on a mast stuck into the Moon (*Figure 6-10*). This sheet was left exposed to direct sunlight for an hour and 17 minutes. Then it was rolled up and stored inside one of the lunar sample boxes. During this exposure, scientists hope the sheet received the full blast of the solar particles. This experiment was sponsored

Figure 6-12. Deployment of Early Apollo Scientific Experiment Package.

by a team of Swiss researchers at the University of Bern and the Federal Institute of Technology in Switzerland where they hope to find isotopes, or varieties, of these elements in the returned foil. Knowledge about the proportions of such isotopes is expected to enhance the understanding of the origin of the solar system and how the Earth and its atmosphere were formed.

With the solar wind experiment implanted, the astronauts next deployed a three-by-five foot American flag. They joined together its two-piece aluminium staff and fitted a support along its upper edge so that it would remain unfurled in the windless Moon environment (*Figure 6-11*). While implanting the flag, the men discovered a strange phenomenon. When they pushed the flag in the lunar soil, they had to press hard to force the staff down, yet it would fall over easily. The soil showed great resistance downward, but little sideways.

After the completion of this task, Mission Control put through the longest long-distance telephone call in history to the astronauts.

The call originated in the White House in Washington and was relayed by the facilities at Mission Control to one of the giant dish antennae handling ground-Moon communications, and on to Armstrong and Aldrin. When the conversation was completed, the astronauts faced the TV camera and saluted.

The remaining two experiments were taken out of the LM and set up approximately 70 and 80 feet away from the LM. These distances were a safeguard against damage to the instruments by the ascent engine exhaust at take-off. The seismometer, designed to record and report events affecting the physical structure of the Moon, such as moonquakes, meteorite impact or volcanic eruption, began returning data to Earth immediately. The laser reflector, which was to provide very precise information on the Moon's distance from Earth and its orbital path, did not function immediately. However, a few days later it began operating correctly. Together, these two experiments form the EASEP or Early Apollo Scientific Experiment Package (*Figure 6-12*).

The seismometer consisted of a mechanical combination of booms, hinges, and springs that respond to vibrations, and electronic devices to record the intensity of these vibrations and transmit them to Earth. Two solar panels provide the necessary electric power during the two-week long lunar day. During the Moon's night, the instrument was designed to be silent with nuclear heating to keep the transmitter warm. The seismometer was 10 to 100 times more sensitive than those used on Earth and could detect the impact of a meteorite the size of a small pebble half a mile away on the Moon. It was designed to record tremors about one million times smaller than the vibration level that a human being can feel.

The laser reflector, unlike the seismic package, had no moving parts and required no power supply. It consists of a hundred fused-silica prisms, set in an aluminium frame 18 inches square. The prisms form the most accurate reflectors ever made in any quantity. Knowing the speed of light and timing the round trip to an accuracy of one billionth of a second, the distance to the laser reflector can be calculated with an exactness never before possible. It is expected that the distance between the Earth and Moon will be measured to an error of only six inches. Within a decade, scientists from all over the world hope to use the laser reflector to check on how fast the Moon is receding from the Earth, examine the wobble of the Earth on its axis, and test new theories of gravity.

Leading up to the flight of Apollo 11, scientists had reservations about man's ability to move around freely in the lunar environment. The space suit that the astronauts wore with its backpack of life support and communication equipment had an Earth weight of over 180 pounds. Adding to this problem, the suit's internal pressure inflates it and substantially reduces its flexibility. Some scientists had also expressed doubts about the human ability to adapt to the one-sixth gravity of the Moon, and thought that disorientation would make movement awkward.

As Armstrong and Aldrin demonstrated, these fears were groundless. The astronauts reported that movement on the Moon was easier than it had been in the one-sixth gravity simulator in which they had practised. The earthbound audience watching Aldrin perform a series of leaps and bounds will attest to the apparent agility he demonstrated. However, as he pointed out, it was important to know where the body's centre of mass was and to keep a foot under it. The "kangaroo lope" worked quite well as a method of lunar locomotion but not as well as the time-tested earthly method of putting one foot in front of the other. The fact that the astronauts ignored the rest periods that had been scheduled for them during the Moon walk confirms the ease of movement on the Moon's surface. At no time during the extravehicular activities was any heavy breathing detected. For the most part, the astronauts' heartbeat was lower than expected. Pulse rates for both men were within the acceptable range throughout.

Even though movement on the Moon was relatively simple, certain other unusual physiological effects were noted by the astronauts. They found that distances on the lunar surface were deceiving. A large boulder field located north of the LM did not appear to be very far away when viewed from the LM cockpit. However, on the surface they did not come close to this field, although they traversed about 100 feet toward it. The flag, the television camera, and the experiments, although deployed a reasonable distance away from the LM and deployed according to plan, appeared to be immediately outside the window when viewed from the LM. Because distance judgement is related to the accuracy of size estimation, it was concluded that these skills may require refinement in the lunar environment.

The astronauts also discovered that the lunar gravity field had differing effects on Earth-learned skills. Although the gravitational pull on the Moon is known to be one-sixth of the gravitational pull on the Earth, they found that objects seemed to weigh approximately one-tenth of their Earth weight. The mass of an object made the object easy to handle in the reduced lunar atmosphere and gravitational field. Once moving, objects continued moving, although their movements appeared to be significantly slower in the lunar environment.

The absence of any natural vertical features, coupled with the poor definition of the horizon and the weak gravity indication at the feet of the astronauts, caused difficulty in identification of level areas when looking down at the surface. The ability to discern level areas was further complicated by the fact that, when wearing a space suit, the centre of mass of the astronaut is higher and farther back than the normal centre of mass of a man on Earth.

Armstrong and Aldrin found that walking in the up-Sun direction posed no problem, although the light was very bright with the Sun shining directly into the visor. While walking in the down-Sun direction, most objects were visible, but the contrast was not vivid. Varying shapes, sizes, and glints were more easily identified in the cross-Sun directions.

During the exercises on the Moon, the astronauts used a series of cameras to document the lunar landing. They used a Hasselblad lunar surface camera extensively during the Moon walk to photograph each of their major tasks. Additionally, they made a 360-degree overlapping panorama sequence of still photos of the lunar horizon, photographed surface features in the immediate area, made close-ups of geological samples and the area from which they were collected and recorded on film the appearance and condition of the Lunar Module after landing. A stereo close-up camera permitted the Apollo 11 landing crew to photograph significant surface structure phenomena which would remain intact only in the lunar environment, such as fine powdery deposits, cracks or holes and adhesion of particles.

Radioing Mission Control to ensure that all assigned tasks had been completed, experiments set up, and photographs taken, the astronauts prepared to re-enter the LM. Aldrin climbed back the ladder first and Armstrong handed him the lunar samples and film packs. Then they both entered the LM. Two minutes later the hatch was secured. Armstrong had walked on the surface of the Moon for two hours, 31 minutes and 37 seconds; Aldrin 40 minutes less.

While entering the Lunar Module, one incident aroused some apprehension. One of the backpacks, which barely cleared the hatch entrance, struck a circuit breaker just inside and snapped its end off. This was a circuit breaker that was needed to arm the ascent engine, a necessary step before the engine could be fired to lift the astronauts off the Moon. However, the circuit breaker was not damaged so badly that it could not be pushed back in. More important, there were other ways in which the engine could be armed. As in most all Apollo systems, redundant or back-up features are provided.

After entering the LM, the astronauts removed their portable life support systems that had sustained them on the lunar surface and began answering a number of questions concerning the geology of the Moon. Then Mission Control informed the astronauts that they were to sleep. Armstrong rigged himself a makeshift hammock and Aldrin curled up on the LM floor. However, due to the excitement of the Moon walk and the cramped quarters, neither slept well.

Before leaving the Moon, Armstrong and Aldrin opened the hatch of the LM once more and jettisoned their portable life support systems and other items which could not be returned to Earth because of weight restrictions during lift-off. Left on the Moon was an Apollo shoulder patch commemorating the three American astronauts, Gus Grissom, Ed White, and Roger Chaffee, who died on January 27, 1967, in a spacecraft fire. Medals honouring two Soviet cosmonauts who lost their lives, Yuri Gagarin and Vladimir Komarov, were also left behind. A final item carried messages of goodwill from leaders of 73 nations. Etched on a 1½-inch disc of silicon by the same process used for manufacturing miniaturized electronic circuits, the messages were reduced in size 200 times.

The Lunar Module lift-off from the lunar surface occurred at 1.34 p.m. EDT using the descent stage as a launch pad. The total lunar stay time was 21 hours and 36 minutes. All lunar ascent and rendezvous manoeuvres were nominal and the LM rendezvoused with the Command Service Module and docked about four hours later while circling the back side of the Moon. After transfer of the crew, samples and film to the Command Service Module, the Lunar Module ascent stage was jettisoned and will remain in lunar orbit for an indefinite period of time. Subsequently, a small burn of the Service Module propulsion system placed the Command Service Module in a 62.6 by 54.7 nautical mile orbit about the Moon.

Shortly after midnight on July 22, while on the back side of the Moon with the LM trailing 20 miles behind, the CSM was injected into a transearth trajectory after a total time in lunar orbit of 59 hours and 28 minutes or 30 revolutions.

Compared to the events of the preceding days, the trip back to Earth was relatively routine. Shortly after noon on July 22, the spacecraft passed the point in space, 33,800 nautical miles from the Moon and 174,000 from the Earth, where the Earth's gravity took over and began drawing the astronauts homeward. Sometime later, a mid-course correction was made to readjust the flight path of the spacecraft and 18 minutes of live television was transmitted to Earth was relatively routine. Shortly after noon on July 22, the effect of weightlessness on food and water and brief scenes of the Moon and Earth. The final colour television broadcast was made on the following night after the spacecraft had passed the midway point of the journey to Earth, 101,000 nautical miles from splashdown.

Figure 6-13. Recovery of Apollo 11 astronauts from the Command Module.

Figure 6-14. The astronauts on board the recovery ship U.S.S. Hornet. They are wearing Biological Isolation Garments.

Early on the morning of July 24 the crew of Apollo 11 awoke and began preparations for the splashdown. Because of deteriorating weather in the nominal landing area, the aim point had been moved downrange 215 nautical miles where the weather was excellent. Visibility was 12 miles, wave height was three feet, and the wind was blowing at 16 knots. At 12.21 p.m. EST the Command Module of the spacecraft was separated from the Service Module with no complications and about 15 minutes later the Command Module entered the Earth's atmosphere. At 12.51 p.m. EDT, the spacecraft splashed down 825 nautical miles southwest of Honolulu and 13 nautical miles from the prime recovery ship, the U.S.S. Hornet. Flotation bags were deployed to right the spacecraft and the crew reported that they were feeling fine and were in good condition.

Following splashdown, the recovery helicopter dropped swimmers who installed a flotation collar to the Command Module. Then a large, seven-man raft was deployed and attached to the flotation collar. Biological Isolation Garments (BIGs) were lowered into the raft. One swimmer put on a BIG while the astronauts donned BIGs inside the Command Module.

These biological isolation garments were worn by the astronauts until they entered the Mobile Quarantine Facility aboard the recovery ship. The garment is fabricated of a lightweight fabric which completely covered the wearer and served as a biological barrier. Built into the hood area was a face mask with a plastic visor, air inlet flapper valve, and an air outlet biological filter. The suits were one of many precautions to ensure that there were no adverse effects of lunar material upon terrestrial life.

Within the Command Module, after the suits were on, the postlanding ventilation fan was turned off, the spacecraft was powered down, and the astronauts egressed and assisted the swimmer in closing the hatch (*Figure 6-13*). The swimmer then decontaminated all garments, the hatch area, the collar and the area around the postlanding vent valves. The helicopter which had been hovering overhead lowered a specially designed seat for the Apollo 11 crew and departed for the U.S.S. Hornet (*Figure 6-14*).

Figure 6-15. Mobile Quarantine Facility.

After landing aboard the recovery ship, the astronauts and a physician entered the Mobile Quarantine Facility. This facility was equipped to house six people for a period up to 10 days. The interior was divided into three sections — a lounge area, kitchen and sleep/bath area. It was powered through several systems to interface with various ships, aircraft and transportation vehicles. The shell was air and water tight and air was filtered as it came through the vent. A negative pressure differential for biological containment in the event of leaks was provided. Specially packaged and controlled meals were passed into the facility where they were prepared in a microwave oven. Medical equipment to complete immediate post-landing crew examination and tests was provided (*Figure 6-15*).

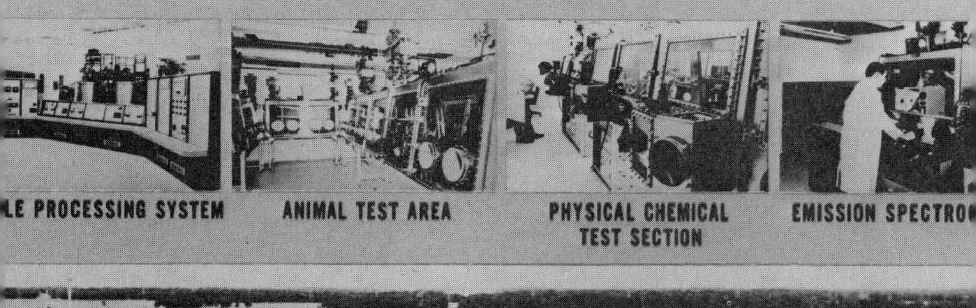

Figure 6-16. Lunar Receiving Laboratory.

President Nixon was aboard the recovery ship to greet the astronauts. He spoke to the Apollo 11 crew members by intercommunications in the quarantine facility and congratulated them for their extraordinary feat.

The lunar samples that the astronauts collected were taken from the Command Module aboard the recovery ship, flown by helicopter to Johnston Island, put aboard an airplane and flown directly to the Lunar Receiving Laboratory at Houston, Texas. The flight crew, remaining in the Mobile Quarantine Facility, was transferred from the U.S.S. Hornet to an airplane in Hawaii which flew directly to Houston, arriving July 27. The astronauts were taken from the airplane to the Lunar Receiving Laboratory to begin a two-week quarantine period.

The Lunar Receiving Laboratory was the final phase of the back-contamination programme. The laboratory had as its main functions the quarantine and testing of lunar samples, spacecraft and flight crews for possible harmful organisms brought back from the lunar surface, and the protection of the lunar samples from contamination on Earth (*Figure 6-16*).

Detailed analyses of returned lunar samples were done in two phases — time-critical investigations within the quarantine period and post-quarantine scientific studies of lunar samples repackaged and distributed to participating scientists. Thirty-six scientists and scientific groups were selected in world-wide competition on the scientific merits of their proposed experiments. These scientists represented some 20 institutions in Australia, Belgium, Canada, Finland, West Germany, Japan, Switzerland, United Kingdom and United States. Major fields of investigation were mineralogy and petrology, chemical and isotope analysis, physical properties and biochemical and organic analysis.

The crew recreation area of the Lunar Receiving Laboratory served as quarters for the Apollo 11 flight crew and attendant technicians for the quarantine period during which the astronauts were debriefed and examined. Other occupants of this area were physicians, medical technicians, housekeepers and cooks. The crew reception area also served as a contingency quarantine area for people accidentally exposed to spills or vacuum system breaks. Both the crew reception area and the sample operations area were contained within biological barrier systems that protected the lunar materials from Earth contamination and protected the outside world from any possible contamination by lunar materials.

Analysis of lunar samples was conducted in the sample operations area which was made up of vacuum, magnetics, gas analysis, biological test, radiation counting and physical-chemical test laboratories. Lunar sample return containers, or "rock boxes", were first brought to the vacuum laboratory and opened in the ultra-clean vacuum system. After preliminary examination, the samples were repackaged for transfer, still under vacuum, to the gas analysis, biological preparation, physical-chemical test and radiation counting laboratories. The gas analysis laboratory measured amounts and types of gases produced by lunar samples, and geochemists in the physical-chemical test laboratory tested the samples for their reactions to atmospheric gases and water vapour. Additionally, the physical-chemical test laboratory made detailed studies of the mineralogic, petrologic, geochemical and physical properties of the samples.

Other portions of the lunar samples travelled through the Lunar Receiving Laboratory vacuum system to the biological test laboratory where they underwent tests to determine if there is life in the material that may replicate. These tests involved introduction of lunar samples into small germ-free animals and plants. Some 50 feet below the ground floor of the laboratory, technicians in the radiation counting laboratory conducted low-background radioactive assay of the lunar samples using gamma ray spectrometry techniques.

On the morning of August 10, Neil Armstrong, Edwin Aldrin and Michael Collins stepped out of the Lunar Receiving Laboratory with no evidence of any contamination that could harm people on Earth, thus ending the adventure that had begun nearly four weeks before.

The remarkable and unerring voyage of Apollo 11 to the surface of the Moon and return was acclaimed by people throughout the world and will undoubtedly change the course of history of man just as other great discoveries of the past. Following in the wake of Apollo 11, eight more lunar landings were planned — each to a different and more challenging site. Perhaps these follow-on flights will not share the success of the first lunar mission. There may be setbacks and disappointments. However, now that man has learned to live and work on the Moon, there will be no turning back from the exploration and discovery of this new domain.

CHAPTER SEVEN

Scientific Results of Apollo 11 and 12 Missions

By G. E. Mueller

Scientists from many parts of the planet Earth who, several months previously, had received the first specimens ever brought back from another celestial body met in Houston, Texas, in January, 1970, to compare notes on their findings.

In the dust-covered rocks taken from the Moon they had discovered extraordinary beauty. In the drab dust itself they had found a galaxy of colours; seen under the microscope, roughly half the lunar dust particles resembled tiny spheres, dumb-bells, teardrops and other globules of glass, ranging in colour from purple, green and yellow to wine red. By using the most sophisticated measuring techniques known to modern physics they had demonstrated that the Moon is an archive of information dating back long before the Earth took its present form.

Perhaps the most challenging discovery was that the catastrophic event that coated the Sea of Tranquillity with molten rock coincided with formation of the oldest rock on the surface of the Earth. Yet, the conferees were told, even the very ancient rocks on Earth, some 3.5 billion years old, contain chemical hints that life already existed on this planet.

What then was the relationship — if any — between the outpouring of lava on the Moon and events on the Earth? Is it possible that life originated on the Earth during the first billion years of its existence, was wiped out (or almost wiped out) by a fearsome meteoric bombardment that also caused extensive melting on the Moon, and then was born again?

It was the prospect of being able to answer questions such as these which sent man to the Moon on July 20, 1969. The activities

Figure 7-1. Deployment of Early Apollo Scientific Experiment Package.

of Neil Armstrong and Edwin Aldrin (the Apollo 11 astronauts) on the lunar surface consisted of collecting samples of lunar material, including two core samples from depths of from six to eight inches below the lunar surface, and discretely selected surface samples. They deployed the Early Apollo Scientific Experiment Package (*Figure 7-1*). This package included a solar cell powered seismometer designed to measure the seismic activity of the Moon, and to detect meteoroid impacts, lunar oscillations and tidal effects. The package also included a Laser Retro-Reflector, a passive device consisting of an array of 100 precision reflectors which is presently being ranged upon by Earth-based laser systems. This experiment will enable precision measurements to be made of the Earth-Moon distance, the motion of the Moon about its centre of gravity, lunar size and orbit, changes in the gravitational

Figure 7-2. Aldrin takes lunar core sample. Solar wind experiment in background.

constant and the distance between continents on Earth. The astronauts also deployed an experiment to determine the composition of the solar wind (*Figure 7-2*).

The second manned lunar landing mission, Apollo 12, was launched on November 14, 1969. The Lunar Module *Intrepid* landed on the Moon four days later, and its crew of Commander Charles "Pete" Conrad and Lunar Module Pilot Allen Bean began to accomplish the objectives of the mission: perform geological inspection, survey and sample in a mare (sea) area; deploy and activate an Apollo Lunar Surface Experiment Package (ALSEP); develop man's capability to work in the lunar environment, and obtain photographs of candidate exploration sites.

After taking a contingency and deploying the S-Band erectable antenna, the Solar Wind Composition experiment and an American

Figure 7-3. Apollo 12 crewman deploying Apollo Lunar Surface Experiments Package.

flag, the two astronauts removed the ALSEP from the Scientific Equipment Bay and deployed it about 400 feet from the Lunar Module (*Figure 7-3*). The ALSEP is a much more sophisticated array of experiments than those deployed on the Apollo mission. The ALSEP contains its own energy source — a Radioisotope Thermoelectric Generator which supplies nuclear electrical power for the six experiments in the array. The experiments included in the ALSEP were: (1) a Passive Seismometer to measure seismic activity; (2) a Magnetometer (*Figure 7-4*) to measure the magnetic field; (3) a Solar Wind Spectrometer to measure the strength, velocity and direction of the electrons and protons which emanate from the Sun and reach the lunar surface; (4) a Suprathermal Ion Detector to measure the characteristics of positive ions near the lunar surface; (5) a Cold Cathode Ion Gauge to determine the density of any lunar ambient atmosphere; and (6) a detector

Figure 7-4. Al Bean emplacing the Lunar Magnetometer.

to measure the amount of dust accretion on the ALSEP to provide a measure of the degradation of thermal surfaces. The ALSEP array is expected to transmit scientific and engineering data on Earth for at least a year. After gathering additional samples and taking photographs the astronauts re-entered the Lunar Module, concluding the first extravehicular activity (EVA) of four hours and one minute. The second EVA, which lasted for three hours and 49 minutes, took the crew to the ALSEP deployment site, several craters, Surveyor III (an unmanned lunar lander which had been on the Moon for 31 months) and back to the Lunar Module.

Let me now present some of the scientific results of the Apollo 11 and 12 missions.

It is the responsibility of the lunar science programme to assure that the best possible science work is carried out on the Apollo missions. These efforts have to be reconciled with the necessity to assure safety and operational success. Consequently, the landing sites of the first two missions, Apollo 11 last July and Apollo 12 in November, were selected after consideration of safety, operation simplicity and scientific interest. Both sites are near the Moon's equator in the "seas" or *maria*, which are flat areas. The highlands offer lunar explorers the best chance of capturing the richest prize in planetary science — the record of the missing billion years in the solar system's history. But they are not close to the top of the list of targets for Apollo landings, because they present a treacherous terrain of jumbled rocks and steep slopes in which the radar of the Landing Module might be confused by multiple echoes, or the craft might come to rest in a dangerously canted position. If the selection had been guided by scientific considerations alone, one of the sites probably would have been in the more rugged highland areas. But for the first missions, any site promised a quantum jump in our knowledge of the Moon.

Nevertheless, we are both surprised and extremely pleased that the scientific results of Apollo 11 and 12 are so very significant. In early January, 1970, a symposium was held at Houston to discuss the results of the first analyses of the data returned and the laboratory study of the samples. Taking part were more than 1000 scientists, including 97 from 24 other countries. The results reported at that symposium generated considerable scientific interest.

Before we ever began manned exploration of the Moon, we had accumulated a body of knowledge about its major characteristics. It is about one quarter the diameter of the Earth. No other planet in the solar system has a satellite so near it itself in size. Some scientists infer from this that the Earth and Moon bear a very close relationship, that perhaps they were once part of the same body, or that they were formed at the same time and in the same manner.

The Moon has no atmosphere, as we know it. As a matter of fact, even with our best vacuum chambers we cannot approach the vacuum that exists naturally on the Moon. The surface of the Moon reaches some 270 degrees Fahrenheit at high noon, and sinks to about minus 270 degrees in the long lunar night.

As it rotates about the Earth, the Moon always keeps the same face towards the Earth. It is only in the last decade, from

Figure 7-5. View of a near full Moon, photographed from the Apollo 13 spacecraft.

lunar missions, that we have had any knowledge at all of the far side of the Moon. As can be seen, there is a very distinct difference in appearance between the near and far sides (*Figures 7-5* and *7-6*).

The far side surface is marked by craters of all sizes. The light areas are highlands or mountainous regions. The highlands are known to be older than the lunar seas, for photographs clearly show that the material of the seas fill the natural basins in the rocks out of which the highlands are formed and lap up against the "shores" of the highlands. For this to have happened, the highlands must have been present before the seas existed. The conclusion is strengthened by the fact that the highlands have a greater density of craters than the Sea of Tranquillity, indicating that they have been bombarded by meteorites for a long time.

The dark regions on the lunar surface are called maria or

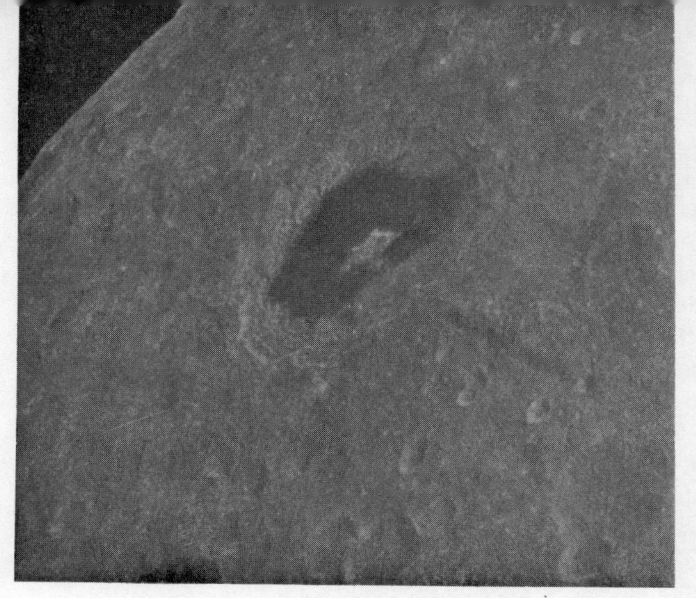

Figure 7-6. View of the lunar farside from Apollo 13.

seas. These maria are flat, plain-like regions in which our two landings to date have occurred. They may be material that has flowed from the interior of the Moon when the highlands were punctured by giant meteorites.

The major areas of interest in lunar exploration were established over the past decade, while we were proceeding through the automated lunar programmes — Ranger, Surveyor and Lunar Orbiter. From the Surveyors we learned that the surface would support manned landings and obtained initial data on the material and composition of the surface. With Lunar Orbiters we photographed the entire surface and discovered peculiar mass concentrations associated with the circular seas from studying their effects on the Orbiter trajectories. During this time, National Aeronautics and Space Administration was consulting regularly with the wider scientific community.

The following table lists the principal scientific objectives of the exploration of the Moon:

- Determine Age of Moon and Dates of Principal Events.
- Determine Chemical and Mineral Composition.
- Investigate Major Body Properties of the Moon.
- Study Dynamic Processes — Past and Present.

The first objective is to determine the age of the Moon and to date the principal events that have affected it. From the beginning of our space programme this question stood high on our list because we hoped to find on the Moon materials older than the 3½-billion-year age of the oldest rocks on Earth. That is, we hoped to uncover evidence of the lost history of the Earth. The early record of the Earth has been completely erased by our weather and atmosphere. We knew that changes occurred far more slowly on the Moon, due to its lack of atmosphere, so we hoped clues to our origin and early history might be preserved. The ages of materials taken from various parts of the Moon can supply evidence on its origin and perhaps even on the origin of the solar system.

The second objective is to determine the composition of the highlands, the seas and, if possible, the interior. This composition can tell us about the environment in which the rocks were formed. We hope to correlate the results of the chemical and mineralogical analysis of materials with the age dating of samples from various places to understand the sequence of events in the ancient history of the Moon and how they fitted together. Thus we plan to obtain rock and soil samples from each site visited.

The third objective is to understand the Moon's major structural body properties, such as whether it is layered like the Earth. The principal instrument here is the seismometer, which records moonquakes and other vibrations of the lunar surface. We need to obtain data from several sites simultaneously to trace the way seismic waves travel through the Moon's interior and thus establish its structure. Another experiment in this area is the magnetometer.

The fourth objective is to understand the dynamic processes that have acted upon and continue to act upon the Moon. The principal source of understanding here is the examination of the Moon's geology, at first hand or by study of photographs. In addition, the seismometer, the heat-flow measurement, the magnetometer and instruments measuring the solar wind can help considerably in this area.

With these four objectives in mind, let us now consider the Apollo 11 and Apollo 12 results. The Apollo 11 landing site was in the Sea of Tranquillity in the eastern hemisphere, which is the left eye of the man in the Moon. The Apollo 12 landing was in

Figure 7-7. Apollo 12 Lunar Module "Intrepid" descending to landing site in the Ocean of Storms.

Figure 7-8. Apollo 12 landing site.

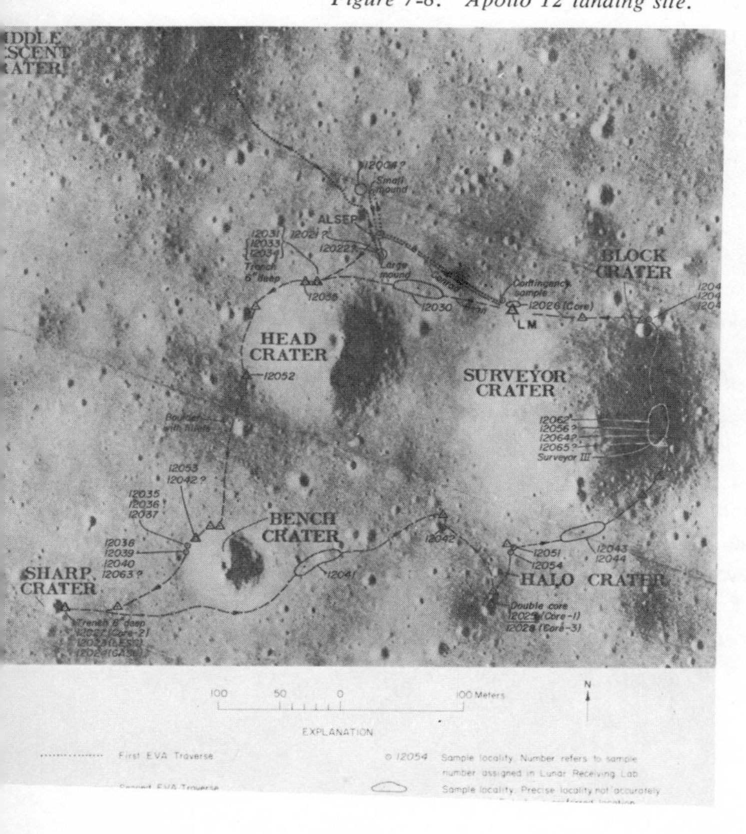

Figure 7-9. Apollo 12 crewman Pete Conrad inspecting Surveyor III.

the Ocean of Storms in the western hemisphere. The Ocean of Storms forms a part of the misshapen nose of the man in the Moon (*Figures* 7-7 and 7-8).

On the Apollo 11 mission, Armstrong and Aldrin collected about 44 pounds of rock and soil samples, took several hundred high-quality photographs and emplaced three devices—a seismometer, a laser retroreflector and a solar wind collector. On the Apollo 12 mission, Conrad and Bean brought back about 75 pounds of lunar material and emplaced a much more sophisticated scientific station that included five instruments. They also brought back some parts of the Surveyor III spacecraft, which landed on the Moon in April, 1967, 31 months earlier (*Figure* 7-9) — the complete TV camera and the shroud; the painted and unpainted aluminium tubing which was a good indicator of micrometeorite impact; electrical cable and the soil scoop. The shroud on the camera changed colour from white to tan. Scientists say that the colour is a dust on the surface of the shroud, and that they have not found any micrometeorite impact activity on the shroud itself. It is expected that meteorite impacts on the shroud would be found, but it now appears there was a low ebb of meteorite activity during the time that the Surveyor was on the Moon.

Aluminium tubing does show some indication of pitting, but it is all on one side of the tubing, and indicates that it might be dust impact from the landing of the Lunar Module itself. Engineers and scientists will inspect the optical parts of the camera for any micrometeorite impact and look for radiation effects on the internal parts of the camera.

The glass parts, on initial inspection, do not show any breaks at all, rather just some slight warping.

The engineering analyses now under way will contribute substantially to the design of future long-life spacecraft, such as the space station and automated planetary craft.

The Ocean of Storms samples may be contrasted with those from Tranquillity Base in several ways: (1) While still old by terrestrial standards, the Apollo 12 rocks are about one billion years younger than those of Apollo 11. (2) About half of the Apollo 11 material was microbreccia, as opposed to only two of the 45 rocks of Apollo 12. (3) The regolith mantlerock at the Apollo 12 site is about one-half as thick as at the Apollo 11 site. (4) The amount of solar wind material in the Apollo 12 fines is considerably lower than in the Apollo 11 fines. (5) The crystalline rocks in the Apollo 12 collection display a wide range of both modal mineralogy and in primary textures, in contrast to the uniformity of the Apollo 11 rocks. (6) The "nonearthly" chemical character of the Apollo 11 samples (high in refractory and low in volatile element concentrations) is shared by the Apollo 12 samples, but to a lesser degree. (7) The chemical composition of the fine material is the same as that of the crystalline rocks; this is not as pronounced in the Apollo 11 collection.

Age Dating: Perhaps the most surprising, interesting and important results concern the very old age of the samples. The Tranquillity Base area is truly ancient. First, the dating of Apollo

11 soil samples by measurement of products of radioactive decay has established that the Moon may have been formed about 4.6 billion years ago. Second, similar measurements for the returned Apollo 11 rocks indicate that many of them solidified about 3.7 billion years ago. The highland areas are expected to be more than 3.7 billion years old. By contrast, as noted previously, the oldest material we have found on Earth solidified 3.5 billion years ago.

Evidence collected prior to the Apollo landings had indicated that the solid matter of the Earth and the meteorites was formed about 4.7 billion years ago from a whirling cloud of gas. Thus the Moon appears to have been formed at about the same time as the Earth and the meteorites. We now expect to find on the Moon a record of events in the first billion years of the Earth and the other planets — a record that has been obliterated on Earth.

The three major contending theories of the origin of the Moon are: (1) that it was formed at the same time, the same place and in the same fashion as the Earth; (2) that it was captured by the Earth; and (3) that it originated by breaking away from the Earth.

The early evidence seems to decrease the probability of this last theory. If the Moon did break away from the Earth it did so at a very early stage of the Earth's development and by a very complex process. All three theories are still in the running, however.

The Apollo 12 rock and soil samples were released from the Lunar Receiving Laboratory for the detailed study that was previously given to the Apollo 11 samples. The Lunar Sample Preliminary Examination Team has reported that some of the Apollo 12 rocks are about 2½ billion years old — about a billion years younger than those from Apollo 11. Thus, such an age difference would indicate a total of at least three periods of major activity, each separated by about a billion years, in which the seas were solidified from lava-like liquid. We can anticipate that other areas of the Moon will show still other ages and thus permit the development of a chronology of events in the early history of the Moon.

Chemical and Mineral Composition: In the second area of interest, the composition of the materials found at the Apollo 12 landing site is also revealing much about the Moon and its history. Minerals identified in the Apollo 12 samples are similar to those observed in the Apollo 11 materials. Glass, plagioclase, pyroxene, olivine, low cristobalite, ilmenite, sanidine, troilite and iron metal have been positively identified. Spinel, tridymite, metallic copper and the iron analogue of pyroxmagite were tentatively identified by optical methods. Four new minerals have been discovered thus far. There is much more titanium and other heavy elements in the Apollo 11 samples than we find in Earth rocks. But for the most part, the mineralogists have been dealing with types of rock familiar to them. The significance of the compositions is what they tell about the environment in which the rocks were formed. Very little oxygen and little or no water were present when the rocks at the Apollo 11 site were solidified.

In looking for clues to the origin of life, extremely sensitive techniques were employed. To date no positive, unambiguous identification of life forms has been made. In view of the fact that the rocks and finely particulate remnants ("dust") of the lunar regolith from Tranquillity Base exhibited an extremely low carbon content, there seemed little or no possibility that the Moon had ever evolved a biosphere during the course of its history. This *a priori* conclusion has been confirmed by careful examination of rock chips (macrobreccia), thin sections (microbreccia) and dust (*Figure 7-10*). Observations were made by bright field and dark field reflected light, by transmitted light and by scanning electron optics. Morphology and optical properties of discrete objects in the lunar material, at all levels of observation employed in this study, show total absence of structure that can be interpreted as biological in origin. The lunar fines examined in this work were virtually devoid of terrestrial contaminants.

The dryness of the Moon is another factor which makes it unlikely that any form of life, primitive or advanced, exists on the Moon. Because of the weak gravitational pull of the Moon, any water which was trapped in the interior of the Moon in liquid or vapour form would have diffused to the surface and escaped, so that in the course of billions of years the outer layers of the Moon

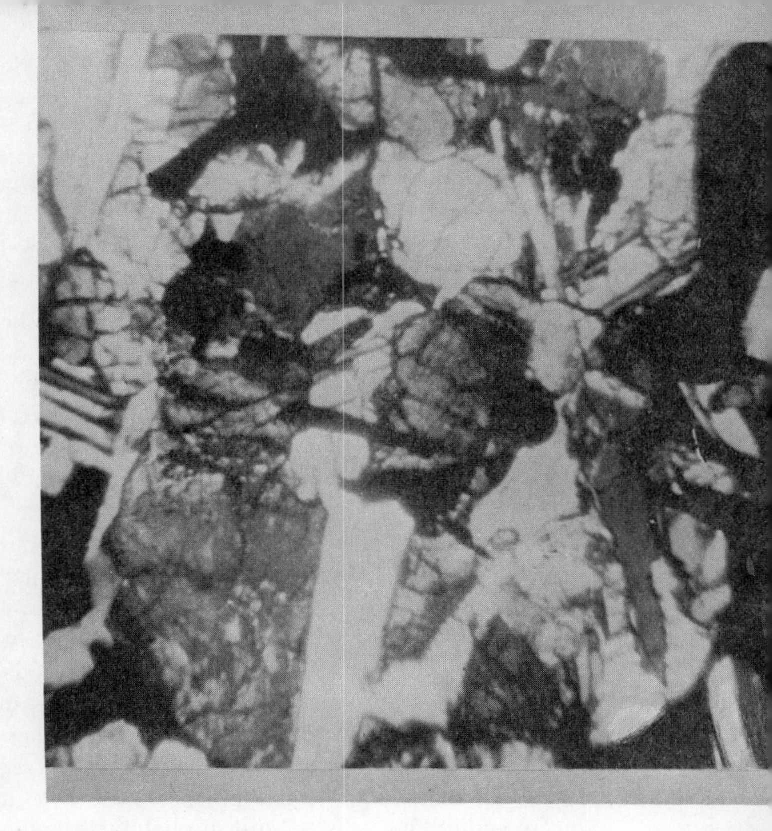

Figure 7-10. Typical lunar basaltic thin section.

probably became thoroughly dehydrated. Water is essential for the development of life as we know it. When the Earth was a young planet its surface contained an abundance of water, also an abundance of the basic molecular building blocks out of which all forms of life are constructed. The molecules — amino acids and nucleotides — immersed in the waters of the Earth, collided ceaselessly; now and then, the collision linked them into the large molecules — proteins, DNA and RNA — which are the essence of a living organism. All the basic chemicals of living matter might be spread out in a thick paste on the Moon, and yet they could never unite to form the simplest living organism because they would be unable to move about and collide on its dry surface.

The loose sediment seems the least likely place to discover any pre-existing forms of life due to the hostile conditions of soil turnover, vacuum, temperature and radiation present on the Moon.

The breccia might provide some protection; however, no organized forms were observed here either.

Future Apollo missions call for drilling down to 10 feet beneath the surface. Material from that depth will be subjected to careful analysis to search for life forms.

A wide variety of biological systems are now undergoing tests with lunar material to determine if there is any toxicity, microbial replication, or pathogenicity. Histolical studies are being made to determine whether or not there is any evidence of pathogenicity. Other activities involve extensive in vitro study of the early biosample and of the regular lunar samples.

Since we are dealing with unknown materials, quite elaborate quarantine procedures were established on the first two missions. We exposed a variety of plants and animals to the lunar soil to check for deleterious effects. As you know, there were none. An unexpected result was the manner in which certain of the plants grew far better when lunar soil was added. This unexpected turn of events is being carefully analyzed. The lunar soil may be acting to neutralize some growth-limiting factor secreted by the plant. The stimulation could be the result of a beneficial agent induced within the cells. Most probably, the stimulation is due to the lunar soil providing certain nutrients in a form optimal for plants to utilize.

The "melting pot" in which the rocks were formed was very hot — from 1800 to 2200 degrees Fahrenheit. These temperatures are comparable to those of the volcanic chambers of Earth. The rocks are similar in many respects to the basalts — solidified lava — found on the floors of ocean basins on Earth. The rocks returned from the Apollo 12 site in the Ocean of Storms seem to be cousins to the Apollo 11 rocks, but they displayed a wider range of mineral constituents and size of crystals. We expect that detailed study of these differences will indicate more about the internal composition of the Moon.

The mineralogists have found several other differences between the samples returned from the Apollo 11 and Apollo 12 sites. About half of the Apollo 11 rocks were of the type known as breccias — which have been deformed by shock and partial melting.

Only two of the 45 Apollo 12 rocks were of this type. The Apollo 12 samples also display a high content of titanium and other heavy elements, but to a lesser degree than those of Apollo 11. Finally, the Apollo 12 soil and microbreccias were of similar chemical composition to one another, but both were different from Apollo 12 rocks. All of the Apollo 11 samples were similar.

Chemical analyses of the samples were carried out mainly by optical spectrographic techniques conducted inside the biological barrier. The major constituents of the samples are, in general order of decreasing abundance, Si, Fe, Mg, Ca, Al and Ti. Gold, silver and the platinum group elements were not detected in any samples.

The chemistry of the Apollo 12 samples is not identical with that of any known meteorite, nickel in particular being strikingly depleted. The Apollo 12 material is enriched in many elements by one to two orders of magnitude in comparison with our estimates of cosmic abundances, and the maria material is strongly fractionated relative to our ideas of the composition of the primitive solar nebula.

The Apollo 12 site appears to be less geomorphologically mature than Tranquillity Base, with a thinner regolith. The lower amount of solar wind material in the fines, compared to those from Tranquillity Base, also suggests that Oceanus Procellarum (Ocean of Storms) mare material is younger than in Mare Tranquillitatis.

The chemistry at the two mare landing sites is clearly related. Both sets of lunar samples show the distinctive features of high concentrations of refractory elements and low contents of volatile elements which most clearly distinguish lunar material from other materials.

Thus, these materials were once exposed to extremely high temperatures. Now the question arises, is this typical of the whole Moon? Was the material from which the Moon formed baked by intense heat from the newly formed Sun? Is that why the Moon rocks seem virtually devoid of water and very low in hydrogen?

Or is this a result of processes that took place on the Moon itself? Some scientists argue that the Moon has undergone sufficient internal heating and churning to generate a dense core. The surface material of the Moon is depleted in heavy substances, like gold. On Earth, where this is also true, the explanation is that the gold, along with iron and nickel, has largely sunk into the core.

Other scientists point to evidence that the Moon cannot have an iron core like that of the Earth. In fact some proposed that the Moon, essentially, is a giant clump of debris, most of which has not undergone extensive alteration since gravity pulled it all together.

Another proposal was that the Moon was formed from an early atmosphere of the Earth, rich in silicates vaporized when the Sun was far hotter than it is today. As these silicates cooled, they formed a ring, like that around Saturn, whose particles were finally drawn together to form the Moon.

The chief objection to this theory was that the resulting Moon should have been in orbit around the Earth's equator, like all other large moons of the solar system. Instead, the orbit of the Moon is considerably tilted to the equator. Differing opinions such as these will undoubtedly undergo revision as further missions to the Moon's surface are undertaken.

In detail, there are numerous and interesting differences between the Apollo 12 and the Apollo 11 rocks. These include:

1) Lower concentration of Ti, both in the rocks and the fine material of Apollo 12.
2) Lower concentrations of K, Rb, Zr, Y, Li and Ba in Apollo 12 rocks.
3) Higher concentrations of Fe, Mg, Ni, Co, V and Sc in the crystalline rocks from Apollo 12.
4) A significant variation in the Apollo 12 rocks among the elements which favour ferromagnesian minerals. The range of abundance was not nearly as great in the Apollo 11 rocks.
5) The fine material at the Apollo 12 site differs from that at the Apollo 11 site in containing about half the titanium content, more Mg, and possibly higher amounts of Ba, K, Rb, Zr and Li.

The soil found at the Apollo 11 site presents a real puzzle. Some of it consists of small particles obviously broken from the larger rocks found there. But some of the particles are entirely different. A few are pieces of meteorite debris. Others seem to be older than the large crystalline rocks and their composition is quite different from either meteorites or the rocks. A solution to this puzzle must await more detailed analysis of material already collected or to be obtained from future sites.

The soil layer found at the Apollo 12 site, in the Ocean of Storms, appears to be about half as thick as that at the Apollo 11 site. The Apollo 12 landing site lies within a ray of material ejected when the crater Copernicus was formed by impact. This crater is several hundred miles to the north.

Lunar Soil Mechanics: The Apollo 12 regolith is generally similar to the regolith at Tranquillity Base. Similar penetrations of the Lunar Module footpads were observed under similar conditions, and bootprint depth was about the same. Also similar are colour, grain size, adhesion and cohesion of most of the soil samples.

The mechanical behaviour of the lunar soil can be summarized as follows:

1) Confinement of the loose surface material leads to a significant increase in resistance to deformation, which is characteristic of soils deriving a large portion of their strength from interparticle friction. The relatively small Lunar Module footpad penetrations of 2.5 to 7.5 centimetres and footprint depth of up to 5 centimetres correspond to average static-bearing pressures of 0.6 to 1.5 pounds per square inch.

2) The soil possesses a small amount of cohesion. This was evidenced by the following observations: (a) it possesses the ability to stand on vertical slopes and to retain the detail of a deformed shape; the sidewalls of trenches dug with the scoop were smooth with sharp edges; (b) the fine grains stick together, and, in some cases, it was hard for the astronauts to distinguish soil clumps from rock fragments;

(c) the holes made by the core tubes were left intact upon the removal of the tubes; and (d) the core tubes did not tend to pour out when the core bit was unscrewed.

3) Natural clods of fine-grained material crumbled under the astronauts' boots. This behaviour may be indicative of some cementation between the grains although in Lunar Receiving Laboratory (LRL) tests the soil grains were found to cohere again to some extent after being separated.

4) Most of the footprints at the low leads imposed by the astronauts caused compression of the lunar surface soil, although in a few instances bulging and cracking of the soil adjacent to the footprint occurred. The latter observation indicates shearing rather than compressional deformation of the soil.

5) At the LRL, the specific gravity of lunar soil was measured as 3.1, considerably higher than the average value (about 2.7) for terrestrial soils. Based on the value obtained for the lunar soil and the measured bulk densities, the void ratio of the material in core 1 is 0.87 and in core 2 is 1.01. The respective porosities are 46.5 and 50.1 per cent. Because of the disturbance involved in sampling, these values may not be representative of the material's properties in place.

6) In the LRL, material finer than one millimetre size obtained from the lunar bulk samples was placed loosely in a container and the bulk density of the material was found to be 1.36 grams per cubic centimetre. In a second test, the soil was compacted to a dense state with a bulk density of 1.80 grams per cubic centimetre. In the compact state, the bearing capacity of the material was determined by a small penetrometer. From these tests the cohesion of the material was estimated to be in the range between 0.05 and 0.20 pound per square inch. The above experiments were performed in a nitrogen atmosphere.

Two significant differences were observed by the Apollo 11 and 12 crews. The Apollo 12 astronauts experienced greater loss of visibility due to soil erosion during the Lunar Module landing than did the Apollo 11 astronauts. This was due to different soil conditions or to a different descent profile or to both at the two landing sites. The Apollo 12 astronauts were able to drive the core tubes to the full depth, approximately 70 cm for the double core tube, whereas the Apollo 11 astronauts were able to drive the tubes only about 15 cm. The Apollo 12 trenches were dug to a depth of 20 cm, whereas the Apollo 11 astronauts could only dig down about 10 cm.

Investigations indicate that although the lunar soil differs considerably in composition and range of particle shapes from a terrestrial soil of the same particle size distribution, it does not appear to differ significantly in its mechanical behaviour.

Body Properties: The third principal area of interest is the major structural properties of the Moon. The primary means of obtaining such information is by use of the instruments emplaced by the astronauts in the Apollo Lunar Surface Experiment Package (ALSEP).

The Apollo 12 ALSEP as well as one experiment from the limited Apollo 11 package are functioning at this time. Additional packages will be emplaced on later flights.

The Apollo 11 seismometer, which may be compared to an extremely sensitive microphone, provided the first definite indication that the Moon is very quiet in comparison with the Earth. During its 21 Earth days or two lunar days of operation, this instrument recorded over 80 natural events, which may have been moonquakes or meteoroid hits. But this was much less than what would have resulted from seismic activity comparable to that of Earth, or from the anticipated rate of meteorite impact. In both missions, seismometers were able to record the footsteps of the astronauts as well as small thermal shifts within the Lunar Module and the outgassing of residual fuels. The seismometer emplaced on the Apollo 12 mission is hearing vibrations that indicate events similar to those heard by the Apollo 11 instrument.

A major surprise of the Apollo 12 seismic experiment was the complex vibrations that followed the planned impact of the Lunar Module ascent stage, 45 miles from the landing site. The impact caused the Moon to reverberate for almost an hour, rising to its peak seven minutes after the impact and then gradually subsiding. In our experience on Earth there is no precedent for such prolonged vibrations or for vibrations increasing in intensity for several minutes following a single sharp impact of that magnitude on a passive object. This is another bit of evidence pointing toward a very complex lunar structure, at least in the region near the Ocean of Storms.

On the Apollo 13 mission, the S-IVB booster rocket impacted the Moon about 80 miles from the seismometer left by Apollo 12, with an equivalent force of 11.5 tons of TNT. The overall character of the seismic signal was similar to that of the LM impact signal from Apollo 12, but the combination of higher energy and greater distance between point of impact and seismometer gave a seismic signal 20-30 times larger than the LM impact and four times longer in duration. The impact signal was automatically radioed to Houston starting 30 seconds after the rocket casing hit. It rapidly built up from a modest level to its maximum, a pattern that surprised scientists. This part of the signal, at least, cannot be satisfactorily explained by scattering of seismic waves in a rubble material as was thought possible from the earlier LM impact data. It may indicate that the unknown events that melted the Moon's surface around 3½ billion years ago were so great that they melted material at least 35 to 40 miles in depth. The sound waves from the crash apparently penetrated at least that deep before returning to the surface. Scattering of signals may explain the later part of the signal. Several alternate hypotheses are under study, but no firm conclusions have been reached. One possibility is that the expanding cloud of material from the impact produces seismic signals continuously as it sweeps across the lunar surface. Whatever the explanation turns out to be, it will be of fundamental importance in explaining the origin and evolution of the Moon.

The signals also seemed to show that if the Moon has a molten or once-molten inner core, it must be deeply buried. The seismometer recorded no variation in signals to indicate any boundary or area of differing material.

To understand such phenomena fully, however, the geophysicists will need to examine the record of signals received at more than

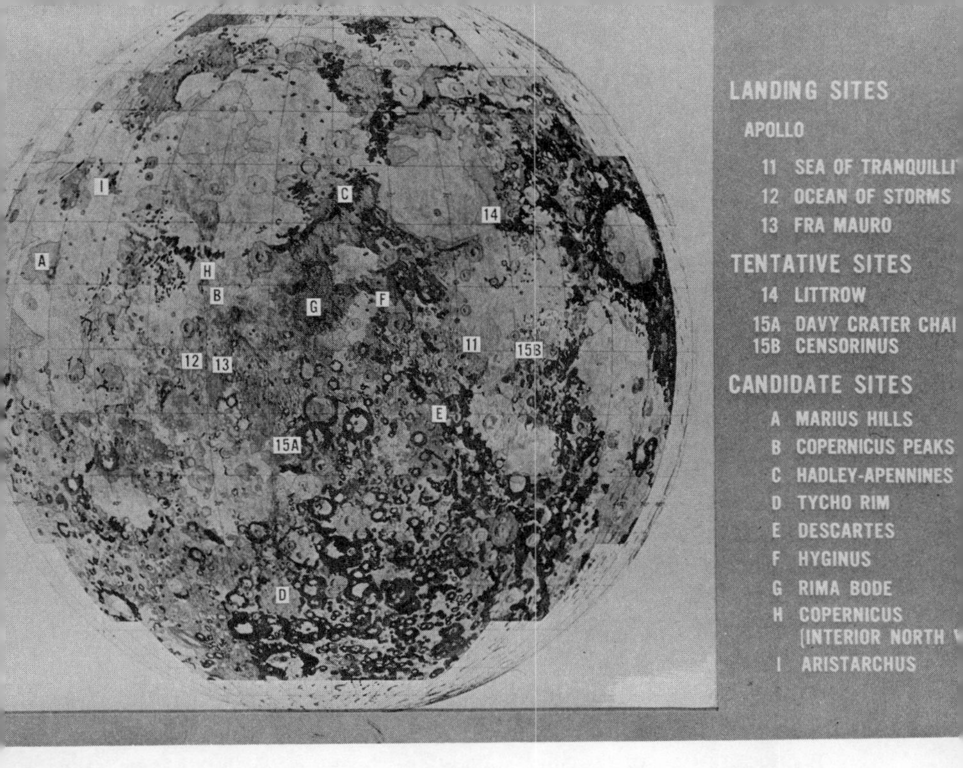

Figure 7-11. Geologic map of Moon with Candidate Future Landing Sites.

one site from such an impact. The study of data from a network of seismic stations may provide the understanding of how these vibrations travel through the upper layers of the Moon, making it possible to determine the internal structure (*Figure 7-11*).

With this understanding, it may also be possible to explain the mystery of the so-called mass concentrations or mascons that have been found associated with some of the lunar seas. These were discovered by analysis of Lunar Orbiter trajectories.

Lunar Orbiter 5, in particular, circuiting the Moon once every three hours and 11 minutes over a 10-day period, displayed unexpected orbital characteristics that have led to this most important discovery. Data received at the Deep Space Network tracking system on the Earth revealed the presence of a strong but initially unidentified influence that modified the observed orbit to an appreciable degree. An analysis of the orbital data by scientists has revealed strong gravitational anomalies in much the same manner that a sensitive mine detector locates its concealed prey. It has become clear that very large concentrations of mass (or mascons) are situated at no fewer than seven sites on the lunar nearside. When the spacecraft passed close to these buried masses, the resultant lunar gravitational attraction it experienced was abruptly and drastically modified, thereby changing the subsequent path in much the same manner as if engine thrust had been applied.

The gravitational variations caused by the mascons amount to about one per cent of total lunar gravity. This suggests that individual mascons may represent as much as 1/50,000 of the Moon's total mass. The full extent of the mascons has yet to be determined.

Another unexpected result was the density of the Apollo 11 rock samples, ranging from 3.2 to 3.4 times the density of water. This is very close to the average density of the Moon as a whole, which is 3.34 times the density of water. If these samples are representative of surface densities throughout the Moon, they indicate that heavier material is not greatly concentrated at the centre as is the core of the Earth. The density of Earth rocks varies from about 2.5 times that of water at the surface to an estimated 10 at the core.

A related surprise was the discovery of an unexpected magnetic field by the Apollo 12 magnetometer. The field is faint — about 30 gammas, a few thousandths of that of the Earth, but five to 10 times that predicted by theory. No such field has been detected by our Explorer 35 spacecraft launched in 1967 and placed in an orbit that approaches within 475 miles of the Moon. Nor was the field discovered by the Soviet lunar landers, possibly because it was below the sensitivity of their magnetometer. If the field encompassed the entire Moon, it should have been detected by Explorer 35. Thus it seems likely to be local to the Ocean of Storms.

Remanent magnetization has been found in some of the rocks returned from the Moon. This suggests that the ancient Moon may have had a magnetic field at least 10 times that detected by the Apollo magnetometer, or that the Earth and the Moon were close enough at one time for the Earth's magnetic field to be induced into the still-hot lunar rocks.

Dynamic Processes: The fourth area of interest is that of dynamic processes that have been and are at work on the Moon. It is now evident that impact has played a major role in forming the craters and soil of the Moon. In the returned rocks and soil particles we have found considerable evidence of strong shock and deformation by external forces. Analysis of photographs shows that impact is the most ready explanation of the formation of craters that range in size from as large as the Imbrium Basin, 600 miles in diameter, all the way down to tiny pits a fraction of an inch in diameter on individual grains of lunar soil. Constant bombardment has been responsible for creating much of the lunar soil and for ploughing and turning over the top few feet of the surface at a very slow rate. The initial results from Apollo 11 indicate that the turnover and erosion at Tranquillity Base were even slower that previously believed. Individual rocks have lain near the surface there as long as 500 million years, while erosion progresses at about one inch in 25 million years.

Of course our information on when most of these impacts occurred is still inadequate. When we do learn it will be of considerable assistance as we attempt to put together the pieces of the jigsaw puzzle that will disclose the Moon's early history and this history will have significance on Earth. If the Moon underwent a severe bombardment by large objects early in its history, it is likely that the Earth was also affected. Recognizable remains of such a bombardment cannot be found on Earth, but our oceans, continents and atmosphere may have resulted from such events.

A major question still remains in determining how to recognize those craters on the Moon that have been formed by volcanic processes. We now know these processes have had an important influence at some time in the Moon's history. The indications thus far are that the dynamic history of the Moon is quite complex, with both "hot" and "cold" periods and perhaps times when it was hot in some regions and cold in others.

Still another dynamic process is the solar wind (a steady rain of solar particles) streaming from the Sun and striking the Moon's surface at the rate of 36 million helium atoms per second on every square inch. This solar weathering is thought to contribute to the darkening of the lunar soil. The results of analysis of samples returned on the two missions differ in that the solar wind composition of the Apollo 12 soil is considerably lower than that of Apollo 11. This difference is consistent with the age differences determined by radioactive dating. It is hoped that the analysis of the solar wind content of lunar material will provide a record of the increases and decreases in the Sun's activity over the last three or four billion years. If so, it may be possible to understand the influence of solar activity on the Earth's climate through correlation with past ice ages and tropical periods. This will require considerable effort, however. The experimenters are still in the initial phase of their work to interpret the results of the solar wind composition experiment on Apollo 11.

The Moon as a Space Platform: Finally, there is one area of experimentation of particular relevance to our environment here on Earth. The laser retroreflector left on the Moon at Tranquillity Base has functioned perfectly. By continued ranging on the reflector array, three observatories in this country have been able to determine the relative Earth - Moon distance to far greater accuracy than ever before. One of these observatories has made relative Earth - Moon distance measurements to an accuracy of one foot. Further ranging is expected to reduce the uncertainty to about six inches. Distance measurements of this accuracy will supply precise information on lunar orbital motion and librations — periodic oscillations of the Moon. They can also lead to understanding of fluctuations in the Earth's rotation rate, measurement of the drift of the continents of the Earth and study of the variations in the wobbling of the Earth's rotational axis. There are indications that the wobbling of the Earth's axis is associated with major earthquakes. Thus the understanding of this relationship may make it possible to understand their causes and to develop a mechanism for predicting them.

One other point is worth mentioning with regard to the laser retroreflector. It is there for anyone to use. No special permission is needed from the United States government to operate it. It is simply a mirror device that will reflect certain types of laser beams back to any place on Earth from which they may be transmitted. Thus it is conceivable that observatories elsewhere in the world may be conducting experiments with this device without informing us. Of course, NASA is happy to provide information on how to utilize the reflector to scientists of other countries who wish to do so. We would hope that they would publish their results in the open literature, just as scientists in this country are doing.

Other International Aspects: A number of distinguished foreign scientists have been active in the sample analysis. Since the Apollo 11 mission we have received many other requests from overseas to join in the work. A number of additional countries are included in the allocation plan for Apollo 12 samples. We have carried one Swiss experiment on Apollo 11 and Apollo 12 and we anticipate carrying other experiments from other countries on other missions.* The lunar exploration programme lends itself especially well to international co-operation, both in terms of data analysis and in participation as an experimenter. We hope to make this programme an outstanding example of how the National Aeronautics and Space Act can foster international scientific co-operation.

---

* An agreement was reached in 1969 between the School of Physics of the University of Sydney and NASA on their joint participation in the Apollo Dust Detector Experiment (DDE) and the Charged Particle Lunar Environment Experiment (CPLEE). The DDE was put on the Moon on Apollos 11 and 12 and will also be placed there by Apollos 14, 15 and 16. CPLEE, which measures the energy distribution, time variations and direction of proton and electron fluxes on the Moon, will be placed there by Apollo 14. The data will be sent to the School of Physics for the one- to two-year active life of the experiments.—Ed.

CHAPTER EIGHT

The Impact of Space on Planet Earth

By G. E. Mueller

Throughout recorded history, the great civilizing influences have been advances in transportation and communications and the discovery of new territory. Our expedition into space is essentially the combination of all three—the operation of man-carrying rocket powered vehicles, the perfection of communications between Earth and deep space, the opening of the solar system to manned exploration and the new-found capability to walk upon the Moon.

Some idea of the magnitude of the potential results might emerge if we could contemplate the mind-boggling thought of the readiness of the airplane, and television at the time of the discovery of the Western Hemisphere! Using that assumption as a point of reference, we can reasonably expect that the eventual resulting changes in our way of life as a consequence of space activity may be more fundamental than any which have occurred in the history of man. For the multiplier effect of these factors operating together can be expected to produce results which are difficult to imagine.

If one would extrapolate from these evidences which are presently apparent in each of the sciences and technologies which have been substantially altered by our venture into space, the conclusions would be truly incredible.

Experience teaches that while near-term prognostication is usually extravagant, our glimpses of the more distant future are frequently far too conservative.

It is also hazardous to try to set dates for specific developments for a whole regime of activity may be delayed by the lack of one element, one breakthrough. However, we have learned, on our

Figure 8-1. Thirteen-frame sequence taken of the Moon after Apollo 10 trans-Earth insertion.

way to the Moon, how to discover on schedule, so that the lags in development which we have experienced in the past may not occur in the future. On this premise it may not be unreasonable to anticipate the most spectacular results of the intensive scientific and technological labours of the past decade.

Two new territories are immediately available for investigation—near-Earth space and the lunar surface (*Figure 8-1*). These can and are being explored with equipment prepared for the Apollo Programme.

Continuing landings on the lunar surface are impacting the sciences of astronomy, geology, geodesy, radiology, meteorology, volcanism, biology and physics. Over a thousand scientists gathered at the Manned Spacecraft Centre in Houston in January of this year to discuss their preliminary findings after cursory examination of the samples returned by Apollo 11. Samples from Apollo 12 differed so that additional expeditions are of vital importance to the scientists.

In near-Earth space over 50 experiments are scheduled in the Skylab Programme which will give us fundamental information about this planet as well as the essential knowledge concerning the actions of certain earth processes, and the behaviour of some materials in the weightless vacuum of space. This Programme, making use of equipment manufactured for Apollo, will be a pilot programme for the Space Station of the future. The inter-relationship of men and machines in space will be studied over significant periods and fundamental information will be applied to the Space Station.

However, today I will talk about those impacts of the space adventure which are already evident.

The Apollo Programme marshalled the largest work force ever gathered to execute a peaceful project. Over the eight years of its preparation, from the announcement of the national goal of a lunar landing by President Kennedy in May, 1961, until the launch of Apollo 11 on July 16, 1969, more than half a million people were engaged in this Programme at one time or another.

This strong and effective team was composed of men and women from government, industry and the academic community. Practically every science and technology was involved, from botany to zoology—from electrical engineering to glass making. And all of them were forced to think, to grow, to change because of the new and stringent requirements which the building of the capability to operate in space demanded.

There emerged from the necessary new mixture of disciplines a kind of cross-fertilization between engineering and science, a new kind of rapport between people whose experience and backgrounds were completely dissimilar—a melting pot, if you will, which has produced a new breed of scientist, technician and manager. Some of the remarkable results are well illustrated by the co-operative work being done today as engineers and doctors work together to design revolutionary medical equipment, of which I will speak later.

Because of the high cost of all space activity, let us first look at economic results. It is important to begin by realizing that all of the money was spent on Earth — there is not, as yet, any monetary system in space. Most of the funds were used for wages and salaries which were, in turn, used for food, clothing, shelter, education, travel, recreation and investment. Quite a lot was spent for fundamental research, and went eventually toward the maintenance of our schools and universities. Another portion bought the

Figure 8-2. Australian tracking station.

facilities which house the various elements of this many faceted programme. And these facilities, in many cases, can and are being used for purposes other than space support.

While all funds came out of tax dollars, many of them did, in fact, return to the national treasury in the form of tax payments made by companies and individuals. Although most of the highly qualified technologists and scientists who worked on the space programme would in any case be gainfully employed, there was a fairly large number of people who were called upon to man the Centres across the South who would, except for the creation of these facilities, have remained as subsistence farmers — not significantly contributing to their own or the national good. This fact is particularly pertinent in the states of Florida, Louisiana and Mississippi.

Major contractors for the Apollo Programme included 22 large aerospace and electronics manufacturers. They subcontracted work to over 20,000 companies in every State in the nation. Some funds were used internationally, especially, as here in Australia, for tracking stations (*Figure 8-2*). The manufacturing companies found it necessary to create new and very sophisticated facilities in order to produce the exotic hardware which would carry men to the Moon and back. As a direct result of these needs, the whole aerospace industry was updated and upgraded. New factories and new equipment replaced less sophisticated installations, so that the United States now has the most modern aerospace industrial plant in the world.

The expertise which was demanded by the space programme very quickly began to be applied to other products as factories adapted their experience with new systems and equipment to both civilian and military products. Profits from this kind of activity showed up in our international balance of payments, when, during a period when the balance was weighted against the U.S., technological exports were 10 times higher than imports. The items included in such a list are patent licence fees, highly specialized equipment, management services and computer rentals.

Another direct consequence was the creation of new industry. One of the first and most closely related to the space programme is the business of manufacturing and handling the cryogenic fuels which power the launch vehicles. The demands of space for these fuels has caused that industry to grow very rapidly. Because these chemicals were needed in such large quantities, they were produced economically and are now being used in many industries. New companies have also been formed to produce and market hundreds of individual items which stem from the technologies developed in response to space needs. They include paints, medical instrumentation, specially treated foods, fireproof fabrics and other non-metallic materials, sports equipment, hand tools, lubricants, many new metal alloys and other composites.

The pervasiveness of this new space-generated technology is demonstrated by the range and variety of the thousands of products which have been described. I shall discuss only a few.

The Saturn V launch vehicle, the world's largest flight article, has been called the most sophisticated engineering project in the history of man (*Figure 8-3*). Containing over 5.6 million parts in its systems, it produces 7½ million pounds of thrust at lift-off and sends a payload of 98,000 pounds into Earth orbit.

This vehicle is the most significant step in the direction of the

Figure 8-3.
The Saturn V launch vehicle.

continuing exploration of our solar system, and may well provide us with the first stage of the eventual transportation system which will permit our investigation of the planets.

At the other extreme, space has produced one of the world's smallest manufactured items, the integrated circuit. A silicon chip less than a tenth of a square centimetre in area can now hold a complete logic circuit or register. Its intricacy can only be seen through a powerful microscope, but its utility is almost unlimited. In addition to radio and computer components, the reliability of these circuits is enabling designers and manufacturers to apply electronic circuits to new and old products. There are those who think that these tiny chips may have as much influence upon our economy and upon our daily lives as the giant booster which allows us to leave the Earth.

Presently most important to the total society are the effects of the space expedition upon the educational system. Although no one would have said so at the time, Sputnik probably did more to revolutionize educational philosophy than any other single event. It focused attention upon the fact that most elementary and secondary schools in the United States were still geared to the nineteenth century while the world was racing into the twenty-first.

There was an almost immediate reaction which brought science and mathematics to the fore, along with astronomy and physics. Second-graders were being introduced to the "new Maths" while their siblings in secondary schools learned calculus and physics which had formerly been confined to college curricula. Engineering assumed a more important role in secondary schools and universities.

While it can generally be stated that all major research and development programmes raise the educational level, there is no doubt that in the U.S.A. the space programme surpasses anything previously done in this regard. There are specific reasons for this. First, it was understood at the outset that in order to develop the capability to operate in space, a great deal of new knowledge would be required. Therefore an orderly procedure for enabling universities to search for this new knowledge was set up. Second, the kinds of information essential to the programme had to come from a greater number of scientific and technological regimes. Therefore a broad spectrum of studies was affected which included the fundamental sciences concerned with man and his behaviour. In addition these other subjects would be involved: nuclear power, metals, lubricants, propellants, sealants, gases, aerodynamics, communications, navigation, geology, geodesy and astronomy.

From 1959 through 1968, Project Research, which is one aspect of the NASA university programme, received $479.9 million while the Sustaining University Programme received $203.9 million, for a total of $683.8 million. Work with the universities is broken down into these two major sections. About 70 per cent of NASA funds allocated to universities has been by the Project Research method. This system which supports the research of principal investigators within universities is serving both NASA and the schools well. These programmes also involve large numbers of faculty and graduate students and generate about three out of four of the space science publications for all NASA programmes.

In the decade of the 1960s Research Laboratories were built under NASA grants at 34 institutions. Facilities include Space Sciences Laboratories, Materials Research Centres, Biomedical Laboratories and Propulsion Research Laboratories.

Important research has already been carried out in these facilities. Equally important, research opportunities are being provided in these institutions for advanced work toward doctoral degrees in many disciplines. A substantial portion of young men and women who have already received their doctoral degrees under the NASA pre-doctoral training programme did their graduate research in these facilities.

More than 10 per cent of all funds supporting Project Research has been invested in these installations which are available for continuing education and research.

Scientific experiments performed in space represent an area of significant accomplishment. More than 98 per cent of balloon-borne experiments, more than 40 per cent of sounding rocket experiments, and more than 50 per cent of satellite experiments flown on NASA vehicles, had principal investigators or co-investigators in our universities. A large share of the significant discoveries in space science were made in university originated experiments.

Presently NASA supports about 13,000 project oriented research grants and contracts in universities. For example, 32 universities in 21 States are now working with NASA on various aspects of the Earth resources satellite programme.

The Sustaining University Programme accounts for about 30 per cent of NASA funds allocated to universities and provides support to institutions rather than to principal investigators. Many of its objectives, increasing the supply of trained manpower, increasing university involvement in aeronautics and space, broadening the base of confidence and consolidating closely related activities, have been achieved.

Over 4000 doctorates have been earned by trainees in the Sustaining University Programme for pre-doctoral traineeship grants. These have been made to 152 universities. More than half of these highly trained scientists and engineers are remaining in the universities and will continue to make a meaningful contribution to the nation through education and research for years to come. This very specific work with universities has had a significant influence on the level of education throughout the United States.

For primary and secondary schools, teachers are supplied with relevant information in useful formats, which the teacher may use at his discretion. In addition, publications are provided which relate aerospace results to various subjects. These are useful to curriculum planners as well as textbook writers. NASA's teacher educational services are used by approximately 25,000 teachers annually. Congresses of student scientists as well as Science Fairs encourage young people in their scientific interests. One of the most exciting of all educational projects is the Spacemobile, a travelling exhibit, manned by qualified speakers, which has been seen by millions of people either live or on TV throughout the U.S. Improvement of educational facilities and curricula is among the more visible benefits from the space programme.

People everywhere have already come to accept, without surprise, the communications satellites which bring us live TV from everywhere, and have improved and reduced the cost of overseas communications.

But we have not yet seen the use of the many-channelled large broadcast satellite in synchronous orbit. This development will permit tailored educational programmes for urban centres and for remote rural areas. It gives promise of surmounting the barriers now associated with communications and education in many parts of the world. India has completed negotiations for the experimental use of a satellite for educational purposes. Beginning in 1972, 5000 Indian villages will receive direct broadcast from a satellite hovering above the sub-continent. Each village in the experiment will have its own inexpensive receiving equipment. If this test is successful, India may expand the system to reach the rest of her 100,000 villages.

Arthur Clarke, author and scientist, estimates the cost of educating people by this method at slightly more than $2 per pupil per year.

Since 1958 when America initiated its space activity, more than a billion children have been born into the world—the first generation of the space age. Because of the space programme you young people are learning a new science, a new cosmology, and you will have a new view of man and his destiny in the universe. Although the dramatic flights to the Moon appear to us a revolutionary victory of man over the gravity of the Earth and the vacuum of space which have confined him to his native planet, space flight will become a commonplace to you and your colleagues.

You young people can look ahead confidently to new opportunities and great advances which will be made in the 21st century when you will be in your 'thirties and 'forties. You will be the first generation to know the Earth as a whole and you will be able to relate technology, science and philosophy as a unified experience, common to all men of the Blue Planet, Earth.

But perhaps the most significant aspect of the space adventure to you is its promise that you will, yourselves, participate in the pioneering stages of this great exploration. For men will be going into space in increasing numbers and for a vast number of reasons as this new frontier opens.

One of the areas of interest to everyone which has been most forcefully impacted by the space programme, particularly Apollo, is the biomedical. We at NASA had to know on a real-time basis, while it was happening, how fast the hearts of the astronauts were beating, how much oxygen they were using, how their muscles were reacting to the stresses imposed by their tasks in a weightless environment. And these evaluations had to be made without restricting the activity of these space test pilots. So medical doctors, biologists and engineers worked together to perfect a new system to relay information from a biosensor attached to the body of the astronaut to a computer, to data screening equipment, and through the Apollo communications network to the medical team at the Manned Spacecraft Centre at Houston, Texas — from 200 miles — or 800 miles or from a quarter of a million miles out in space.

A recent adaptation of this system is being used in many cities to increase the efficiency of hospitals. Radio-equipped ambulances transporting a heart attack victim use spray-on electrodes to attach biosensors which transmit an electrocardiogram to the hospital emergency staff (*Figure 8-4*). When the patient arrives both staff and equipment are ready to administer the indicated treatment.

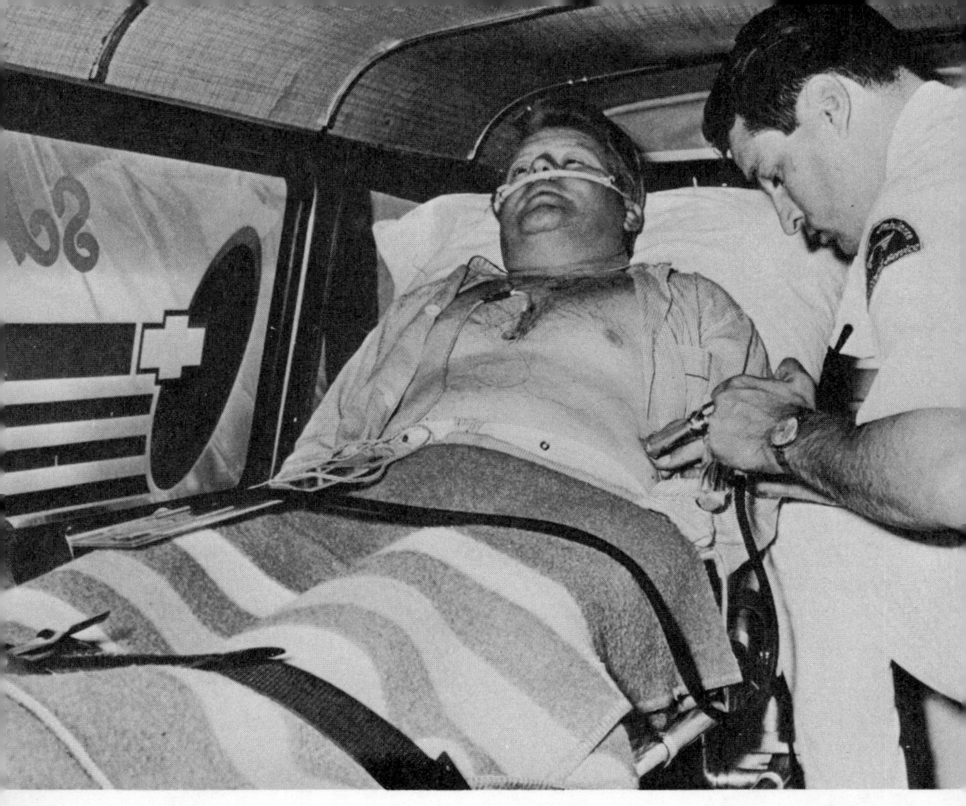

Figure 8-4. Biosensor attached to heart attack victim in ambulance transmits an electrocardiogram to hospital emergency staff.

Since more than 60 per cent of deaths resulting from heart attacks occur within one hour, the value of speed is evident. The spray-on electrodes can be used to monitor a patient after he has returned home, so that he can communicate an electrocardiogram to his doctor by telephone.

An exciting new development is the result of the work of scientists at Stanford University and NASA researchers. A system for studying the actions of the heart from the exterior by means of sonar, or ultrasound, replaces another NASA developed procedure in which a thin catheter is inserted into one of the heart chambers. The new system is still experimental but is expected to accomplish important diagnostic work without either danger or discomfort to the patient.

For the first time well men are being examined over long and continuous periods while they perform complex tasks, while they eat, sleep, rest and relax. I venture to say that we already know more about the well human being today than we have ever known before. And we are only at the beginning of our exploration of men in space.

Very significant information will be coming from the Skylab Programme when men stay in space for as long as 56 days at a time. But even our relatively few manned flights have produced enough knowledge to have created new concepts of medical procedures and equipment. You have heard of some of them.

The precision which we associate with the flights of Apollo is the result of precise testing and measuring. In our search for exact analysis of rockets and thrusters, we developed some instruments so sensitive that they have naturally found application in medicine. One of the remarkable adaptations is the breathing sensor. This is a delicate instrument which examines the breath coming through a tracheotomy tube which is inserted into the wind-pipe in certain operations. Formerly a nurse would need to be in constant attendance to monitor the tube. Now, integrated circuitry notes infinitesimal differences in temperature of the air, and actuates a visual or audible alarm at the nurse's station if any change occurs.

Electronic sensors that were used to monitor astronauts during Mercury and Gemini flights have been adapted to continuously and simultaneously measure the pulse and respiration rates, temperature and blood pressure of up to 64 patients in a hospital and provide continuous display of this information at a central control station. This single development promises to revolutionize hospitals throughout the world. It is being incorporated in new hospitals everywhere.

Anyone who has stayed in bed for two weeks or more knows how difficult it is to walk again with unused muscles. It is much worse for those who have been bedridden for months or years. If these people could get to the Moon they would find it easy to move about in the lunar gravity as their muscle strength returned. That not being possible yet, they can use the Lunar Gravity Simulator at a NASA Centre in Virginia. The simulator has been adapted and is being tested for this purpose at the Texas Institute for Rehabilitation and Research (*Figure 8-5*).

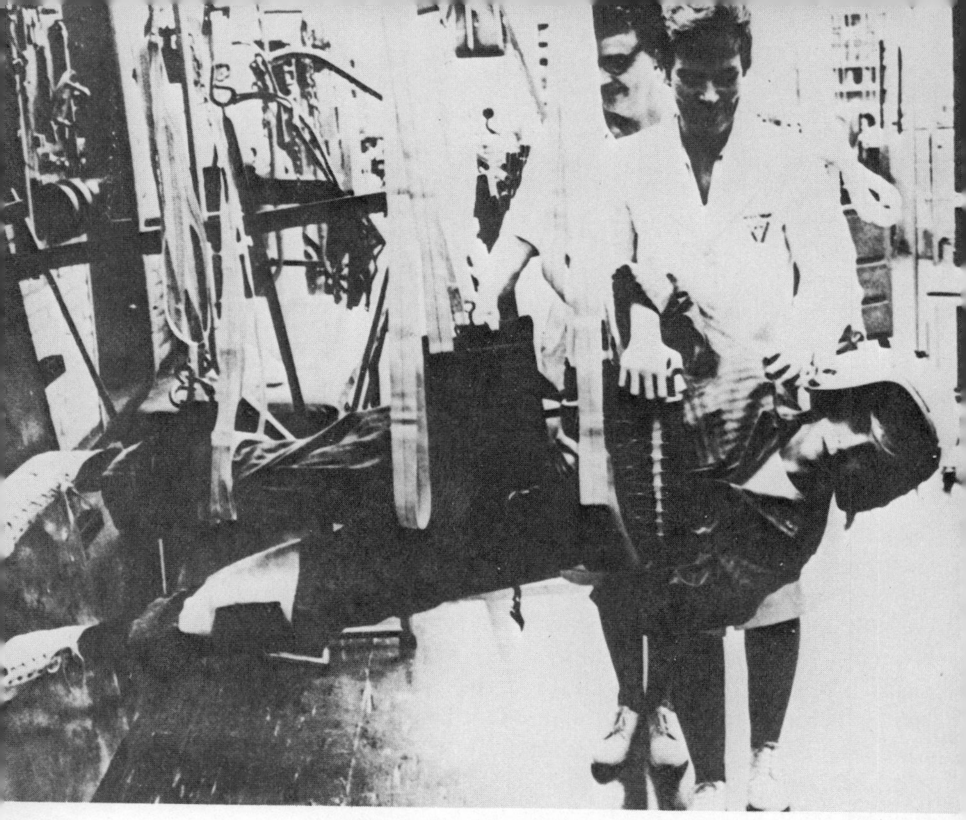

Figure 8-5. Lunar Gravity Simulator.

A paralyzed person can now operate a motor-driven wheel chair, turn the pages of a book, change a TV station or call for help — all with a switch which can be operated by the movement of his eyes (*Figure 8-6*). The switch operates on the principle of infrared reflection from the eyeball.

The "clean room" techniques developed by NASA for assembly of small components and those used at the Lunar Receiving Laboratory to prevent contamination are being adapted for use in hospitals.

A muscle motion measuring device has evolved from the equipment designed to measure the force of meteors. It is so sensitive that it can measure the heartbeat of an embryo chick inside the egg shell. It can detect the muscle tremor of Parkinson's disease long before it becomes visible (*Figure 8-7*).

For several years NASA's Jet Propulsion Laboratory has been

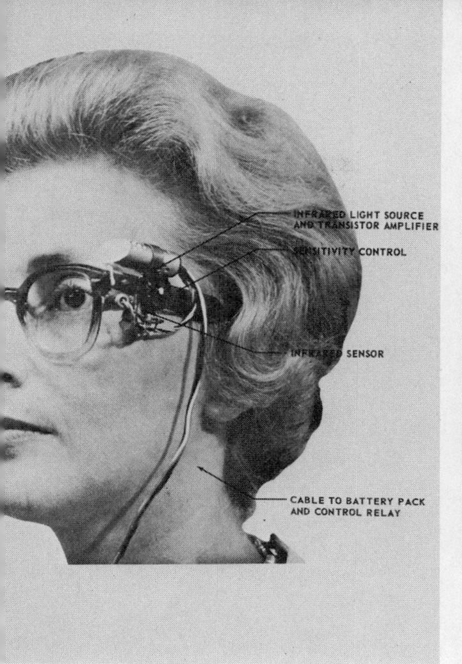

igure 8-6. "Sight Switch" mounted n eyeglass frame can be flicked by mere movement of an eyeball.

Figure 8-7. The Muscle Accelerometer, adapted from a momentum transducer developed at the NASA Ames Research Centre, California.

using digital computers to enhance the clarity of television pictures of the Moon and Mars transmitted from spacecraft. Although the original pictures were surprisingly sharp considering the circumstances, technical limitations of the camera and electrical distortion encountered in the transmission reduced the clarity desired. By processing the photos in a digital computer, details that were obscured in the original became apparent.

In 1966, this technique was applied to medical and biological X-rays, with promising results. Doctors were able to observe clarity of detail that would otherwise be lost or overlooked. Computer enhancement of X-ray photos currently continues under development (*Figure 8-8*).

Laser technology developed originally for defence and space use is being adapted for delicate and precise use in medicine. Aerospace engineers have been working since 1962 with medical research teams to find new ways to use lasers in "knifeless surgery" and as diagnostic tools.

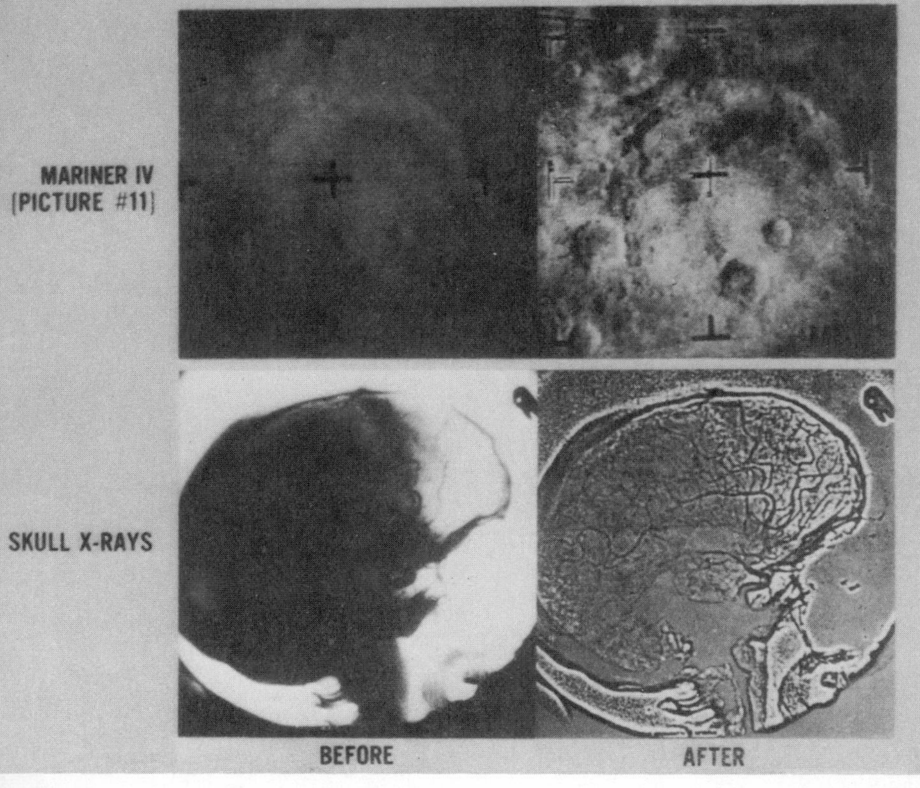

Figure 8-8. Photographs of the human skull and the surface of Mars before and after the application of digital filtering techniques.

Lasers are being used with outstanding success in eye surgery, and are being evaluated in dermatology, organ repair, amputation and in microbiological studies. Such surgery can be painless and, in many cases, practically bloodless. As a diagnostic tool, the microscopically small point of focus of a laser beam might be used to lift painlessly and instantaneously a minuscule particle of skin from a patient's arm. Or by passing the beam through a hypodermic needle, the laser can sample tissue inside the body.

A transducer developed for the Manned Spacecraft Centre to measure the impact of the Apollo spacecraft Command Module during water landings is being used in the fitting of artificial limbs. The transducer is smaller than a dime and weighs less than an ounce. The sensing diaphragm is stainless steel and the whole unit is waterproof. It will respond to changes in pressure, and is not affected by temperatures between freezing and 120°F.

Of course you have heard about the wheel chair that climbs stairs and walks in sand, and about the computerized clinics that take the health history of a patient and set it up for easy reference by the doctor. There is an interesting experiment now going on in the state of New Mexico where the medical backgrounds of a large but widely scattered rural population are being compiled and computerized so that a few doctors will be able to remotely treat a great number of patients.

All of these are rather like the tip of an iceberg in that they are only the most obvious developments. In this time of rapidly multiplying populations, the systems which have been invented for the maintenance of astronaut health are going to be applied to the health of people on the ground.

One of the most important of all aids to man will play an important role in public health aspects as well as many others — man's newest and most universally applicable tool, the computer (*Figure 8-9*). Computers have been so improved and developed to meet the demands of the space programme that our spacecraft can be directed on a perfect course through the vast and empty reaches of space, all the way to the Moon — all the way to Mars and the farther planets. This most capable servant has grown from a system which made one million calculations a minute only a few years ago for the Mercury flights to the Apollo computer complex which now produces 50 times that many — 50 million a minute, 80 billion calculations in a day. As computer capability has grown it has been adapted to serve most of industry and is now a $20 billion annual business in the United States alone, employing one in every 100 of the entire work force. NASA's investment in the development of computers amounts to almost three quarters of a billion dollars.

It is difficult to categorize such a useful tool, for it has application in almost everything we do. It will, of course, play a major role in the prevention and cure of pollution, as well as the management of health, education, slum clearance, traffic regulation, in short, wherever large numbers of units or more than a few kinds of considerations must be accommodated. While the computer is certainly not a direct product of our space activity, its development

Figure 8-9. Overall view of the Mission Operations Control Room, in the Mission Control Centre at Manned Spacecraft Centre, during television transmission from Apollo 13 mission.

has been so markedly accelerated by the requirements of Apollo and other space programmes that we might almost claim it as a step-child. There is no question that computer capability has been one of the principal building blocks of our space capability, or that it will continue to be central to most of man's activity in the future.

One might also say that engineering has also been so forcefully impacted by the needs of space that it is indeed a new technology. The requirement for the Saturn V for the Moon voyages demanded a kind of scale and perfection which had never existed before — not even in our most sophisticated aerospace products. The standards normally associated with commercial or military products had to be increased a hundredfold for the manufacture and testing of the five and a half million parts of the Saturn launch vehicle (*Figure 8-10*). And these systems worked with remarkable reliability. On the epochal flight of Apollo 11, only one

Figure 8-10. Saturn V first stage for the Apollo 8 manned mission being erected at the Vehicle Assembly Building.

mechanical part failed, the non-critical timer in the Lunar Module. That mission achieved a reliability of 99.99996 per cent — a truly unbelievable accomplishment.

Except for Apollo 6 and Apollo 13, on which there were mechanical problems, the Saturn vehicle performed with outstanding perfection. The engineering which created them had, therefore to be of a higher order than that ever practised before. That quality of workmanship has become a new way of working for thousands of people who spent years building these spacecraft. It can reasonably be expected that they will carry their work habits over to whatever other equipment they manufacture.

The chemically fuelled rocket engine, first demonstrated by Robert Goddard in 1927 and significantly improved during World War II by the Germans, has been brought under control, both as to performance and to stability, and man-rated.

The rocket engines used on the Saturn family of launch vehicles have performed with amazing perfection. Ten Saturn Is, each using eight LOX/Kerosene engines and six LOH/H_2 engines; five Saturn IBs with eight LOX/Kerosene 200,000 pound thrust engines and one 200,000 pound thrust LOX/H_2 engine and eight of the giant Saturn Vs, the world's largest flight equipment — each using five one million pound thrust engines and six two hundred thousand pound thrust LOX/H_2 engines — have all sent their payloads out of our atmosphere with almost perfect dependability. Including Mercury and Gemini flights, the total is 41 successful launches in 41 attempts.

These engines, therefore, are a new highly dependable source of power. The high pressure LOX/H_2 reusable engines now being developed will produce 400,000 pounds of thrust with a specific impulse of 450 seconds or more. This technology is expected to be used in the second generation of space vehicles, the space shuttles, which will provide economical reusable transportation between Earth and Earth orbit.

Requirements for construction of the Saturn vehicles advanced the state-of-the-art in welding and joining as well as in metallurgy. The need for high-strength, lightweight metals brought forth a new honeycomb structure. A new aluminium casting alloy called M-45 was created and is now being used in industry because of its superior strength and ductility at cryogenic temperatures.

Liquid hydrogen and liquid oxygen flow rates of 10,000 gallons per minute were desired for the rapid loading of the Saturn V. These rates were 10 times higher than any previously achieved. Liquid hydrogen is transferred through a 10-inch diameter vacuum jacketed line by pressurizing the 850,000 gallon storage tank to 60 pounds per square inch. Liquid oxygen, which is too heavy for efficient high speed transfer using only pressurization, required the development of a 10,000 gallon per minute pumping system. We do not know that these techniques have yet been used for any other purposes, but the know-how is developed and may well be applicable to irrigation, transportation or water pollution problems.

Innovation and invention characterized the Apollo Programme and created many outstanding examples of new technology. A

Figure 8-11.
Integrated Thermal Micrometeorite Garment.

good example is the astronauts' space suit (*Figure 8-11*). Because it contains all essential elements, it has been called the "world's smallest manned spacecraft". To build it, a new fabric has been developed — teflon coated fibreglass cloth which is non-flammable in a pure oxygen atmosphere at temperatures up to 1200°F. It is used for the exterior cover, as well as for other layers where strength and wear resistance are of importance.

This is only one of the developments which resulted from the need for fireproof materials in the Apollo Command Module. Over 3200 nonmetallic materials were tested and the information which was developed has been computerized so that any manufacturer can easily ascertain the flammability characteristics of materials which he contemplates using. Reference was made to this information during the preparation of cabin furnishings for the new Boeing 747. As this information becomes more widely used it can be expected to have a beneficial effect upon the furnishing of public places as well as upon consumed goods. One important manufacturer of bedding is already using for mattress filling the same material which was developed for the astronaut couches. Fire fighter suits are being manufactured of the same fabric which is used for the outer layer of the space suit.

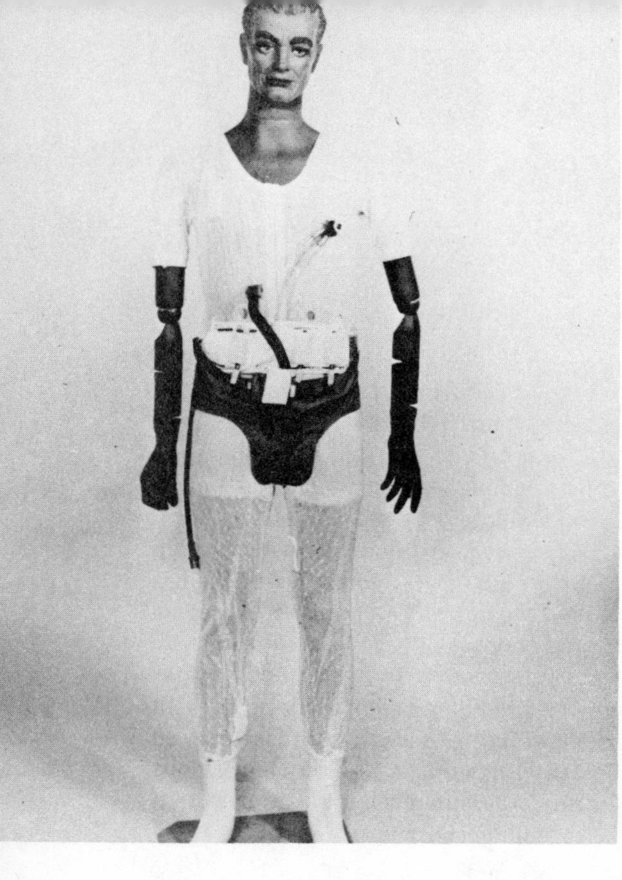

Figure 8-12.
Liquid Cooling Garment.

Fourteen layers compose the integrated thermal micrometeorite garment, which, as its name implies, protects the astronaut against the hazards of temperature as well as the possibility of micrometeorite penetration during extravehicular activity. The new materials which constitute these layers are alternately kapton, aluminized kapton and Beta cloth, under the teflon coated super Beta cloth outer layer.

The suit proper is composed of the Liquid Cooling Garment, the gas bladder for pressure control, and the restraint layer. The Liquid Cooling Undergarment and the insulation provided by the covering garment are designed for extravehicular activity.

The Liquid Cooling Garment is already being used in some hospitals for temperature control of patients with very high fevers (*Figure 8-12*). The pressurizing equipment in the space suit has also been used in hospitals as a medical aid.

The helmet with its extravehicular visor assembly completes the pressure vessel of the miniature spacecraft. It is composed of a non-breakable polycarbonate which is optically clear. The layers of the visor can attenuate visible radiation including the ultra-violet and infra-red ends of the spectrum to protect the eyes of the astronauts from damage.

Even before the refinements which were incorporated into the Apollo helmet were accomplished the Gemini helmet was adapted for an interesting and life-saving use (*Figure 8-13*). Children with respiratory ailments were difficult to treat since keeping them in bed prevented doctors from properly analyzing their response to activity. The Gemini helmet, fitted with the necessary medical equipment, not only solved that problem but also cheered the children, who are usually ardent space fans.

Pressurization and oxygen are supplied through connections on the front of the suit. During extravehicular activity, including exploration of the lunar surface, the back-pack carries expendables, communications equipment and all other units necessary to life support.

The Portable Life Support System, carrying all of the expendables needed by the astronauts, as well as their air conditioning system, can sustain the astronauts for four hours on a single filling. It can then be resupplied from the Lunar Module with an additional four hours of supplies.

The control unit for this Portable Life Support System is easily available to the astronauts as are its gas connectors. Equally accessible is the biosignal conditioner which transmits coded information from the biosensors to the medical team at the Manned Spacecraft Centre at Houston while the astronauts are in space.

The communications carrier, located in the helmet, provides radio contact through the Apollo spacecraft to the Flight Controllers at Houston, by means of the Apollo tracking system.

The oxygen purge system is able to maintain oxygen pressure in the event of a tear.

In this space suit we have brought to use many materials, devices and techniques which might otherwise have spent many years awaiting application. It is thus that space needs have forced

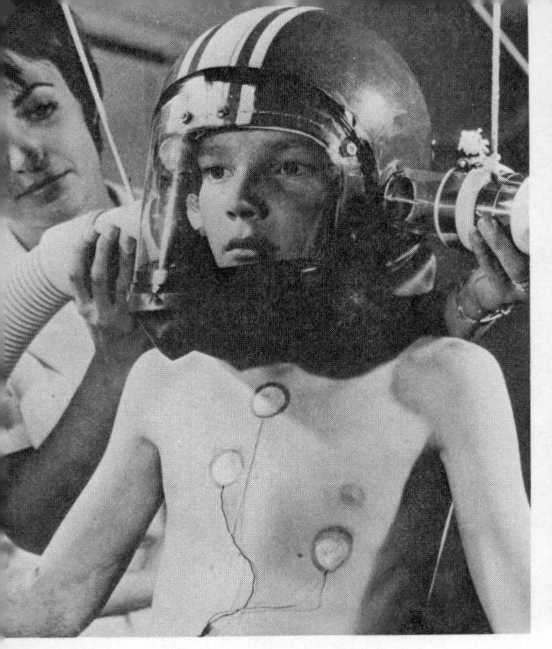

Figure 8-13. The Gemini helmet is being used in the treatment of children with respiratory ailments.

Figure 8-14. The crawler-transpo carrying Apollo 8's 363-foot-t Saturn V space vehicle to the lau pad.

the development and application of new technology. Most of the complex systems and fabrications contained in the suit have not yet been adapted for any other use, primarily because many are under continuing development.

Unprecedented feats of engineering were accomplished to support the great space adventure. Among them were the mobile launcher, a 12 million pound, 405 foot tall structure built so that it could be transported while carrying an Apollo Saturn V vehicle. The launch platform base is 135 feet wide, 160 feet long and 25 feet deep, divided internally into two levels which contain checkout equipment, a full computer and a telemetry terminal. Since the launcher is exposed to the rocket exhaust of the 7.5 million pound thrust first stage, special attention was required in the design to protect the equipment from the extreme thermal, vibration and acoustic environments.

The crawler-transporter (*Figure 8-14*), which carries the assembled launch vehicle from the Vehicle Assembly Building to the launch pad, was adapted from technology originally developed in the field of strip mining; however, adaptation of this type of vehicle to the task of transporting the Apollo Saturn V space vehicle imposed many new requirements, which presented challenging problems to the design engineers.

The crawler-transporter's load carrying capacity is 12 million pounds. With this load it can traverse a 500 foot horizontal curve, and climb a five per cent grade at one mile per hour in a 46 knot wind measured at the top of the 405 foot Mobile Launcher. The requirements were complicated even more by restricting the rate of turn per individual crawler track to 10° per minute, limiting value of 2.6 feet per second and maintaining the Mobile Launcher level in the horizontal plane within 1/12 of a degree at all times even while negotiating the 5° slope.

In 1966, the American Society of Civil Engineers presented the "1966 Outstanding Civil Engineering Achievement Award" for Kennedy Space Centre's Launch Complex 39.

The most prominent feature of Launch Complex 39 is the 525-foot tall Vehicle Assembly Building (*Figure 8-15*). This building was built to stringent requirements, which included a capability to withstand 125 miles per hour hurricane winds and a minimum deflection of the building under high winds. The particular modular truss structure employed also provides flexibility for accommodating internal structural changes that might be required for future space vehicle configurations.

Although the engineering award was a tribute to the design and construction engineering teams who overcame difficult and frustrating problems, the most challenging aspect of constructing the complex fell to the engineering management team, whose innovations in project time phasing resulted in an "on-time" completion of the facility from which man's first journey to the Moon would be made.

This concept of identifying and measuring progress towards critical milestones was a characteristic of the total Apollo Programme. These milestones were constantly and carefully monitored by project offices at the Centres and by the Headquarters Office in Washington. Time management and time saving were

Figure 8-15. The 525-foot-tall Vehicle Assembly Building is the most prominent feature of Launch Complex 39.

the essential elements involved in accomplishing the lunar landing at the lowest cost estimated for its completion.

Another of the critical engineering problems was thermal protection. The velocity of the spacecraft entering the Earth's atmosphere on its return from the Moon is 24,500 miles per hour. At that speed, friction created by contact with the air results in temperatures on the spacecraft of approximately 5000°F. Temperature environments during other parts of the flight vary from minus 150°F to plus 150°F. Therefore, heat shield technology was essential to the success of the lunar mission. The aerodynamic heating predictions involved both basic heating distribution on the blunt end of the spacecraft and the effects on windows, hatches and other protuberances.

Brazed stainless steel honeycomb core sandwich material was selected as the minimum weight support for the thermal protection system, the ablator. However, the conventional material was too sensitive at the minimum spacecraft temperature of minus 150°F, so a new alloy was created for which vacuum melting techniques were required. One ablative material met the requirements for total thermal protection but density reduction of that material as well as the nonporous nature of the structure made it necessary to develop new application techniques. Filled honeycomb core was found acceptable but new processes were again needed to fill the 350,000 cells of the comb. Thus, with the manufacture of the heat shield for the spacecraft, several new materials and processes were developed so that another of the "impossible" problems could be solved.

All of the research and development in thermal protection performed to date for Apollo will provide the fundamental knowledge for the development of protective materials and systems for the reusable space shuttle.

It became clear, early in the Apollo Programme, that an important element would be a reliable electrical power system of reasonable weight for the Command and Service Module. The fuel cell was an interesting candidate because it offered an attractive energy density ratio, but it had yet to be developed for practical applications.

The decision to proceed with development of the Apollo fuel cell power plant was made on the basis of a 250 watt demonstration power plant (*Figure 8-16*). An important consideration was the fact that fuel cells require less weight in the spacecraft than batteries. Development began early in 1962, and in August, 1965, the unit passed its qualification tests. A comparable system performed well in operational tests in Gemini flights and the decision for the development of this technology was vindicated.

The fuel cell is now under study by 27 natural gas companies co-operating in a $20 million research and development programme. It gives promise of supplying total home power for lighting, heating, cooling and air purification — and eventually for total waste disposal. Since the only product of the fuel cell is distilled water, it becomes an attractive candidate as a power source in the very near future when pollution control will be a pervading influence.

Certainly, when we are searching for ways to clean the air in our cities we may find that much of the basic research on air purification has already been done — in support of perfecting the Apollo spacecraft.

Figure 8-16. CSM fuel cell cryogenic storage.

Electronics and electrical engineering have been more fundamentally affected by the demands of space activity than almost any other regimes. The NASA Communications Network, which provides voice, television and telemetry over a quarter of a million miles to the Moon, consists of several systems of diversely routed communications channels including communications satellites, common carrier systems and high frequency radio facilities. Both narrow- and wide-band channels and some TV channels are used.

A primary switching centre, located at Greenbelt, Maryland, and intermediate switching and control points located at Canberra, Madrid, London, Honolulu, Guam and Kennedy Space Centre in Florida, provide centralized control of the message and switching operations of the more than 600 computers that relay commands to the spacecraft and bring telemetry back to Houston.

During the launch, the Kennedy Space Centre is connected directly to the Mission Control Centre, Houston, Texas, and to the Marshall Space Flight Centre, Huntsville, Alabama.

After launch, all network tracking and telemetry data centres at Greenbelt for transmission to Houston via two 50,000 bits-per-second circuits used for redundancy and in case of data overflow.

Two Intelsat communications satellites are used for Apollo. The Atlantic satellite services the Ascension Island Unified S-Band station, the Atlantic Ocean ship and the Canary Islands site.

The second Apollo Intelsat communications satellite over the mid-Pacific services the site at Carnarvon, Australia, and the recovery ships in mid-ocean. All these stations are able to transmit simultaneously through the satellite to Houston. The space programme requirement for Intelsat services has stimulated that commercial enterprise, the first major transfer from space to the private sector.

We have all seen initial results of the basic changes which have occurred in satellite relayed TV broadcasts from the Moon and from Mars. The Manned Spaceflight Network can really be considered as a pilot system for future communications networks which will operate throughout our solar system.

Microminiaturization, forced upon the space programme in its early years by the high cost of putting payloads into orbit, has turned out to be one of the first space products to be assimilated into our daily lives. Its use is evident in everything from medical instrumentation to electric razors, from transistor radios to tiny tape recorders.

Apollo is the largest and most complex programme that has ever been accomplished. It has been compared to the building of the Pyramids of Egypt, the harnessing of atomic power, and the building of Boulder Dam — all rolled into one. And because of its size and complexity, because it depended upon constant change and constant invention, it was undoubtedly the most difficult programme which will ever be undertaken. I say the most difficult, because what we have learned in Apollo we can now apply to any of the problems which will have to be solved in the future.

Figure 8-17. Management Council.

Apollo was done, as it had to be, by the development and employment of new methods and materials as well as the judicious use of the old. Government, science and industry were so blended that they worked as a single cohesive organization for the attainment of man's most ambitious goal — the landing on the Moon.

New management techniques had to be invented which would be strong enough to maintain the pace and the perfection which the programme demanded, and still be resilient enough to withstand accident and to accommodate change (*Figure 8-17*). It was as important to plan for opportunities afforded by success as for the setbacks which would be inevitable in connection with any research and development programme.

Hardware was being manufactured in almost every state in the nation. Millions of parts had to be factory tested, certified "man-rated" and then shipped to Cape Kennedy where they would be integrated into the total system.

Figure 8-18. Manned Spacecraft Centre, Site 1.

Figure 8-19. Manned Spaceflight Centre Headquarters in Huntsville, Alabama.

Figure 8-20. Members of "Apollo Executives' Group" visit Pad 37.

Three major centres and a score of smaller ones, each under the direction of strong and innovative people, needed to be carefully focused upon the goal — the lunar landing (*Figure 8-18*). A total systems approach was obviously necessary. A pattern of project management was instituted which *Fortune Magazine* recently called a "management revolution" (*Figure 8-19*).

Apollo offices were set up at each Centre as well as on the premises of the major contractors, much as airline technical offices are installed in the plants of major suppliers of equipment. These project offices monitored design, schedule, cost, performance and quality control and formed a separate network headed by the programme offices in headquarters in Washington. Communication lines between all offices remained open at all times. Visibility of all problems at the earliest possible moment was essential to prevent their escalation.

A most experienced group of individuals, the chief executives of the manufacturing companies which were producing the bulk of the hardware, served as members of the "Apollo Executives' Group" which acted both as an advisory board and, at the same time, became an effective communications link between Apollo management and their own companies (*Figure 8-20*). The dedication of these men was one of the greatest assets of the programme and a continuing source of real help to me.

In another area of equal importance, the Science and Technology Advisory Committee was formed. This group of 15 renowned scientists, led by Dr. C. H. Townes, included three Nobel Laureates as well as Dr. Lee DuBridge, now Science Adviser to President Nixon. Their constant interest and thoughtful watch over the scientific and new technological developments were responsible for a great part of the success of the programme, both in science and in execution.

The intimate association and involvement of all participants in the conduct of Apollo can be expected to set a useful precedent for other large programmes, both in the private and the public sectors.

It was the intent of the Apollo Programme not only to create the capability to land men on the Moon and return them safely to Earth, but to do it within the time and cost constraints originally set. Therefore tight scheduling was obligatory for getting the job done in a timely fashion and was the major factor in cost savings. Toward this objective a new concept was introduced — "all-up" testing. Instead of the incremental testing which had been used in previous research and development programmes, each part was thoroughly tested on the ground, and then the whole system was assembled and tested all together in the first flight of this totally new equipment.

Thus, the first Saturn V flight tested for the first time the first, second and third stages of the Saturn rocket as well as the guidance system, and in addition simulated the flight out to the Moon and back with the Command and Service Module. This simulation provided the first test at lunar re-entry velocities of the spacecraft heat shield as well as the first test of the Apollo recovery forces.

Figure 8-21. Apollo 8 astronauts on board U.S.S. Yorktown. From left: Frank Borman, James Lovell, William Anders.

This successful mission allowed the third Saturn V to be manned and the first lunar flight (Apollo 8) to be carried out on this first manned test of the Saturn V (*Figure 8-21*).

One of the consequences of this highly successful activity was the fact that after Apollo 11, additional equipment for many more trips to the Moon remained. And the programme goal, the first lunar landing, was carried out within the time specified.

Another vital result is the fact that we have learned those management techniques which will, in the future, facilitate the management of research and development projects. Accurately scheduled invention was as essential as accurately scheduled production. Every one of the millions of elements which comprised the whole of this vast enterprise had to be co-ordinated with every other. It was as if every single note played by every instrument in a symphony concert would have had to be written, the instrument invented, the artist trained and all blended together to create the perfect whole.

In our efforts to manage the already badly damaged ecology of our "spaceship Earth", we are going to draw heavily upon the lessons which we learned in Apollo. Solutions to the problems of air and water pollution and of solid waste disposal are of such magnitude that they will require the most sophisticated total systems management techniques as well as the adaptation or invention of equipment more complex than we have yet seen. The experience and many of the researches and developments conducted for Apollo will provide the foundation upon which to build.

Our present marked concern with the condition of our environment is undoubtedly attributable in some degree to the Apollo flights. Man's first view of the Blue Planet as a whole has stimulated universal realization that the Earth is indeed a closed ecology.

This is only one of the cultural and philosophical results of man's first venture beyond the pull of the Earth's gravity and outside the protective shield of its atmosphere. With the completion of the first lunar landing, the whole future of mankind was irrevocably changed. The limitations which have confined him throughout history are removed. Slowly, over the centuries, men have pushed their horizons outward. With each advance in transportation, communications and territory, men have moved their degree of civilization a little farther forward until now the whole planet is becoming the concern of every man — quite a contrasting concept to the tribal or local concerns which motivated most of our ancestors.

Such a revolutionary change must, of necessity, have repercussions of profound social nature. Breaking the bonds of Earth must increase man's confidence in his own ability, but this new condition may also increase his sense of insecurity as he finds no safe haven in which to isolate himself. This will, indeed, be the time for daring, for exploration, for new concepts and new thinking about man and his place in the universe. The landing on the Moon will, in my view, have more significance on the future course of history than any previous discovery or event.

Never again will celestial bodies be unattainable. That mythology which has been our heritage will now be replaced by real knowledge, constantly increasing, and far more wondrous than the imaginings of the ancients. All children will learn at the mother's knee that the silver disc in the sky has already been visited by man — and that the next flight is scheduled for some near date.

We can expect to find changes in our language and in our poetry as men eliminate those classical references which are no longer applicable. We are already seeing some cultural effects of the space adventure as theatre, cinema and painting begin to reflect the absence of barriers in our real world.

The manner in which the Apollo Programme was performed has special meaning as the world rapidly becomes more populous than ever before. Great explorations have in the past been carried out by individuals, or by small groups headed by one man — Columbus, Magellan, Peary — but with Apollo we have seen, for the first time, the full strength that resides in an organization of hundreds of thousands of people in all walks of life dedicated to a single peaceful goal. The realization that men can effectively operate together for difficult achievements is both timely and appropriate.

Attraction to the mysterious and the unattainable has always been man's deepest and most basic motivation, and properly so, for only through his understanding of natural forces and his ability to adapt them to his needs has man improved his Earthly condition. New knowledge has brought new freedoms which have punctuated the evolution of the race.

It took four billion years for life to venture onto the land from the sea. Four hundred million years passed before man first appeared on Earth. Only four thousand years ago man developed his first real transportation system — the sailing ship — and with this as an aid he has populated the Earth.

Now we have taken another great evolutionary step in entering

Figure 8-22. Recovery of Apollo 11 crew after splashdown in the Pacific Ocean.

into a new environment. We now have a transportation system which will take men to the planets. Soon we will bring our life forms throughout the solar system. I believe that we will press forward, and that we will take those steps in the days ahead that will permit us to aim for the stars in the future.

We have made a beginning . . . you will continue the work . . . (*Figure 8-22*).

CHAPTER NINE

Future Apollo Missions, Apollo 14 Through 19

By L. B. James

INTRODUCTION

Man has left the Earth, visited and explored another celestial body 400,000 kilometres away, and returned safely (*Figure 9-1*). This will stand as the greatest and most far-reaching achievement of our time. History will refer to our generation as the one that ushered in the Space Age, and, in the course of a single decade, not only opened up vast reaches of space to man's machines, but made them accessible to man himself. A movement has been started which will not stop, and which will in the course of time extend man's domain throughout the solar system.

Prior to the inception of the Apollo Lunar Exploration

Figure 9-1. The Moon.

Programme, considerable information regarding the nature of the Moon was obtained from telescopic observations, from unmanned satellites and from early Apollo missions.

The Ranger and Lunar Orbiter Programmes increased photographic resolution 46 to 152 metres on the near side of the Moon and 46 to 460 metres on the far side. Detailed views of the selected sites have resolution to three feet. Surveyor and Luna Programmes indicated a basalt-like composition for the mare. Data from Luna II indicated that any lunar magnetic field must be less than 1/10,000th that of the Earth. Data from Explorer 35 satellite suggests a surface magnetic field of even less strength.

By tracking spacecraft in lunar orbit, the lunar orientation and rates of change and the physical librations are now known to about 198 metres.

Tracking data from the Lunar Orbiter series and Apollo missions indicate the presence of mass concentrations in the Moon and variations in its shape. These anomalies arouse curiosity regarding their origin and, since their distribution is not well known, pose operational problems (trajectory perturbations) for low orbital altitudes.

The broad objectives of the Apollo Lunar Exploration Programme are: to understand the Moon in terms of its origin and evolution; to search its surface for evidence related to the origin of life; and to apply new data on the differences and similarities between the Earth and Moon to the reasonable prediction of dynamic processes that shape our planet.

The specific objectives supporting these broad objectives are as follows:

- Investigate (a) the mare and highland lunar surface features, (b) the impact, volcanic and mountain-building surface processes, and (c) the regional problems such as mare-highland relation, the major basins and valleys, the volcanic provinces, the major faults and the sinuous rilles.

- Collect and completely characterize lunar material samples by detailed analyses on Earth, including rock identification, chemical composition and rock dating.

- Determine the gross structure, processes and energy budget of the lunar interior by measuring seismic activity, heat flow, and disturbance in the Moon's axis of rotation with long-lived surface instrumentation.

- Survey and measure the lunar surface from orbit about the Moon with metric and high resolution photography and remote sensing, tying together local studies into a regional framework. Provide detailed information for science planning of surface missions, and lunar-wide control of surface position and profile.
- Investigate the near-Moon environment and the interaction of the Moon with the solar wind; map the gravitational field and any internally produced magnetic fields; and detect atmospheric components resulting from the neutralized solar wind and micrometeorite flux-impact effects by long-term monitoring with lunar orbiting satellites.
- Return uncontaminated samples to Earth for analysis of biologically related organics, such as prebiotic material, fossil life forms, and micro-organisms, and determine their origin; conduct in-situ analyses of lunar samples for biological material; and relate these data to a comprehensive theory on the origin of life by comparison with the Earth and planets.
- Determine how geological processes work on the Moon in the absence of an atmosphere, fully exposed to the solar wind and with one-sixth the force of Earth gravity, in order to gain a much deeper understanding of the dynamic processes that shape our terrestrial environment.

The early Apollo lunar landing missions have shown that man can readily adapt to the work effectively in the lunar environment. Samples of lunar material have been returned to Earth for analysis and passive seismic experiments and laser retroreflectors were emplaced on the Moon for monitoring from the Earth (*Figure 9-2*).

In order to achieve the goals and objectives of the lunar exploration programme, it is necessary to expand the limited landing site accessibility, staytime, payload and mobility capabilities developed in the early Apollo Programme. Eventually the entire lunar surface must be made accessible for landing sites; the staytime must be extended from days to weeks; astronaut mobility must be improved to the extent of at least 10 kilometres radius

Figure 9-2. ALSEP deployed on Moon.

with science payloads of several hundred pounds; increased payloads to support the longer staytime and provide adequate scientific equipment are necessary; the ability to return larger payloads of lunar material for laboratory examination on Earth is needed; and finally, the capability must be expanded to conduct a broader spectrum of experiments.

LUNAR EXPLORATION RATIONALE

While demonstrating a capability for landing and returning men safely from the Moon, we have obtained a wealth of scientific data from the first two manned lunar landing missions. The future Apollo missions will give much more emphasis to increasing scientific return from the remaining missions in the approved programme.

Existing Apollo hardware is available to carry out seven more missions to the Moon. In re-establishing the schedule under the Space Programme's reduced budget situation, we determined that the interval between missions will be about six months for the next five years. The longer intervals will be consistent with careful scientific analysis of the findings and feedback of the results into plans for the future missions. At the present time, Apollo 14 is scheduled for 1970; Apollo 15 and 16 in 1971; Apollo 17 in 1972. Then we plan a two-year interval in which the scientific community can develop plans for two additional missions, using Apollo 18 and 19, currently scheduled for 1974.

In this stretchout of the schedule, we are preserving maximum flexibility for the scientists developing plans for those missions. However, it is still questionable whether the total of nine manned lunar landing missions will provide an adequate sample of the diverse surface of the Moon, to allow an integrated view of basic lunar characteristics. With numerous improvements, the later missions will allow us to exercise the full capability of the Apollo system to get maximum scientific and technological return.

LUNAR SURFACE EXPERIMENTS

From the beginning of our space programme, the age of the Moon has stood high on our list of questions because we hoped to find material on the Moon older than the 3½-billion-year age of the oldest rock on Earth. That is, we hoped to uncover evidence of the lost history of the Earth. Theory resulting from telescopic observation and pictures taken by automated spacecraft and Apollo missions indicated that the lunar seas overlie the highlands. Thus, the highlands are thought to be older than the seas. Therefore, the scientists are eager to obtain samples from the highlands as soon as possible. The age of materials taken from various parts of the Moon can supply evidence of its origin and perhaps even the origin of the solar system.

The composition of the highlands, the seas and their interior can tell us about the environment in which the material was found. We plan to correlate the results of the chemical and mineralogical analysis of material, with the age-dating of samples from various places to understand the sequence of events in the history of the Moon.

To understand the Moon's major structural body properties, such as whether it is layered like the Earth, we will emplace additional seismometers to record moonquakes and other vibrations of the lunar surface. We need to obtain data from several sites simultaneously to trace the way seismic waves travel through the Moon's interior and thus establish its structure.

We need to understand the dynamic processes that have acted and continue to act upon the Moon. The principal source of understanding here is the examination of the Moon's geology, at first hand and by study of photographs. The seismometer, magnetometer, heat-flow experiments, and instruments measuring the solar wind have helped considerably in this area.

Apollo 11 and 12 missions visited and sampled the eastern and western mare regions. Apollo 11 landed in the Sea of Tranquillity in the eastern hemisphere, which is the left eye of the man in the Moon. Apollo 12 landed in the Ocean of Storms in the western hemisphere, which forms a part of the misshapen nose of the man in the Moon. Both sites were in relatively flat regions and had suitable landing sites near the Moon's equator.

The Moon has a surface area of about 37,955,000 square kilometres, about the same size as the North and South American continents. On Apollo 11, Neil Armstrong and Buzz Aldrin walked in an area about the size of a suburban backyard. On Apollo 12, Pete Conrad and Alan Bean explored something like two city blocks. Thus one can understand the need to be cautious about drawing firm conclusions from so limited an examination.

On Apollo 11, the astronauts collected about 20 kilograms of rock and soil samples, took several hundred high-quality photographs and emplaced a seismometer, a laser retroreflector and a solar wind collector on the lunar surface. On Apollo 12, the astronauts brought back 34 kilograms of lunar material and emplaced a much more sophisticated scientific station that included five instruments. They also brought back some parts of the Surveyor III spacecraft which landed on the Moon 31 months earlier.

It is important to emphasize that the information gathered from the previous flights has triggered many new ideas and avenues of thought. Many of the contending theories were advanced years ago. The difference is that today the new information has established important limiting factors that must now be fitted into the jigsaw puzzle of the origin and evolution of our solar system.

Table I — LUNAR SURFACE EXPERIMENTS

Experiment	Apollo Missions							
	11	12	14	15	16	17	18	19
PASSIVE SEISMIC	X	X	X	X	X	P	P	P
ACTIVE SEISMIC			X		X			
SEISMIC PROFILING						P		P
MAGNETOMETER		X		X	X		P	P
LUNAR HAND-HELD MAGNETOMETER			X		P		P	
SOLAR WIND, SPECTROMETER		X		X				
SUPRATHERMAL ION DETECTOR		X	X	X				
HEAT FLOW					X	P	P	
CHARGED PARTICLE LUNAR ENVIRONMENT			X					
COLD CATHODE IONIZATION		X	X	X				
LUNAR GEOLOGY INVESTIGATION	X	X	X	X	X	P	P	P
LASER RANGING RETROREFLECTOR	X		X		P			
SOLAR WIND COMPOSITION	X	X	X					
FAR UV CAMERA/SPECTROSCOPE						P		
DUST DETECTOR		X	X	X	X			

X — APPROVED.
P — PROPOSED.

Equally important is that many of the missing pieces of the puzzle can be found on the Moon.

Table I lists the lunar surface experiments with tentative mission assignments. The experiments listed for Apollo 16 through Apollo 19 are tentatively assigned for planning purposes only.

Apollo Lunar Surface Experiments Package

The Apollo Lunar Surface Experiments Package (ALSEP), placed on the Moon by the Apollo astronauts (*Figure 9-3*), is designed to provide the scientific community with unprecedented knowledge of the lunar environment — especially in the areas of geology, geophysics, geochemistry, particles and magnetic fields.

One of the most interesting questions to be explored with the ALSEP instruments will be whether or not the Moon evolved in the same pattern as is now believed for Earth. On Earth, rock formations include granite and basalt, both with an almost

Figure 9-3. Simulated ALSEP Deployment.

bewildering variety of mineral combinations. The lunar geophysical information we have thus far does not permit scientists to determine whether similar lunar differentiation exists.

The extent of layer exposure that exists on the Moon is also of scientific interest. On the Earth, this exposure results from erosion and man's excavation. Exposure may occur on the Moon in regions of faulting and may provide scientists an opportunity to study the layering of the rock as it occurs in depth. It is only with this exposure and the use of the ALSEP instruments (particularly the seismic instruments) that scientists are able to determine lunar subsurface characteristics.

The ALSEP seismic instruments will allow a study of the internal structure and present tectonic activity of the Moon. Two principal sources of natural seismic energy expected on the Moon are moonquakes and meteoroid impacts.

If there are moonquakes, the compressional and shear velocity structure of the Moon may be revealed with a precision dependent upon the number and type of recorded seismic events, and distribution of the quakes. Scientists may then be able to answer such basic questions as: (1) Is the internal structure of the Moon radially symmetrical as Earth, and, if so, is it differentiated? (2) Does the Moon have a core and a crust? and (3) Is the Moon's core fluid or solid?

The role of the unexpected must not be under-rated. A series of scientific experiment instruments successfully deployed and operating on the lunar surface may reveal heretofore unexpected and perhaps inexplicable information. Indeed, the course of extraterrestrial exploration and our understanding of the forces in the universe may change dramatically as the ALSEP experiments report their data.

Objectives will be achieved through the use of a number of scientific experiment instruments and their supporting subsystems. While in operation on the Moon, the ALSEP system will be self-sufficient and use a Radioisotope Thermoelectric Generator for electrical power.

The ALSEP system consists of two subpackages and a fuel cask assembly. The two subpackages are mounted within the scientific equipment bay of the lunar module for transit to the

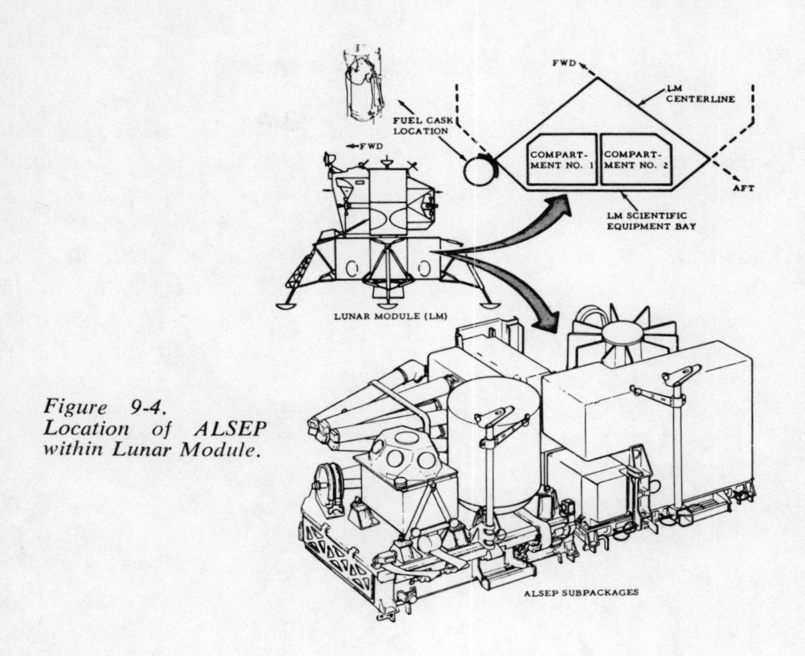

Figure 9-4. Location of ALSEP within Lunar Module.

Moon (*Figure 9-4*). The subpackages occupy a volume of approximately 425 cubic decimetres and, together with the fuel cask assembly and lunar hand tools, weigh approximately 127 kilograms. The fuel cask assembly is part of the electrical power subsystem.

Passive Seismic Experiment. The Passive Seismic Experiment (*Figure 9-5*) is designed to determine the natural seismicity of the Moon. Seismic energy is expected to be produced on the Moon by tectonic disturbances and meteoroid impacts. Knowledge of moonquakes is essential for definition of the strain regime of the Moon. It is also important to know the location of quake epicentres, thus permitting correlation of seismic events with surface features. In this way, insight into the origin of visible features on the Moon may be achieved. Analysis of the form and characteristics of seismic waves will provide data on the physical properties of the lunar interior. Subsurface materials will differ in compressibility, rigidity and temperature. These differences will cause variation in seismic wave velocities and character, from which the material characteristics may be inferred.

Figure 9-5. Passive Seismic Experiment.

Finally, this experiment will permit study of the free oscillations and tidal deformations of the Moon and provide data on the gross physical properties of the Moon. The passive seismic experiment is a portable nine-kilogram package which has a shape similar to a drum rounded on one end. The astronaut places the instrument on a small levelling stool 10 feet from the central station, manually levels the instrument and deploys its thermal shroud. The shroud (or radiation shield) minimizes temperature fluctuations within the instrument.

If the Moon should prove to be seismically dead, then active seismic surveys must be used to obtain information concerning characteristics of the lunar interior.

Active Seismic Experiment. The Active Seismic Experiment (*Figure 9-6*) provides information for determining the structure, thickness, physical properties and elasticity of surface and shallow depth materials of the Moon. The active seismic experiment uses

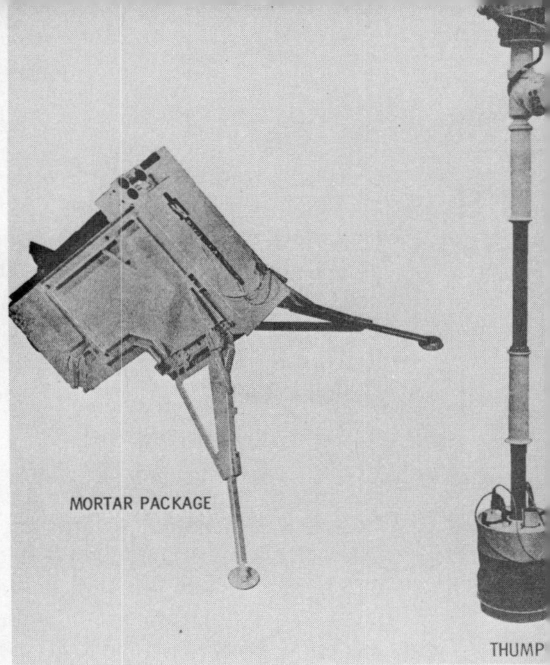

Figure 9-6.
Active Seismic Experiment.

explosive devices detonated at various distances to measure the elastic properties of lunar subsurface material to a depth of approximately 150 metres. Seismic energy will be transmitted through lunar subsurface material and detected by a geophone array.

The active seismic experiment contains the seismic energy sources and the detection system. Two energy sources will be employed: a mortar box assembly, from which four explosive grenades will be launched to detonate at various distances up to 1500 metres from the geophone detectors, and a "thumper" assembly containing 21 explosive Apollo standard initiator cartridges which will be activated by the astronaut at specified locations along the geophone line. The detection system is a linear array of three geophones together with amplifier systems and electronics.

The thumper assembly is used for investigation of material characteristics within a 23-metre depth of the lunar surface. The upper section contains electronics for the firing mechanism, the cartridge barrel and contact points. The lower section is a hollow cylinder containing a plate which couples the energy source to the lunar surface and imparts seismic waves to surface materials for detection by the geophones.

The mortar box contains four explosive grenades to be activated by Earth command near the end of the one-year operation on the lunar surface. It contains electronics and grenade-launching rockets and is designed to minimize the effects of recoil. Since it is necessary to know the distance from the geophone array at which the grenade is detonated as well as the time of detonation, the design provides for measurement of grenade launch angle, grenade launch velocity, and time of flight.

Refraction velocity surveys by the active seismic instrument will be used to study the subsurface relations between the maria and the highlands, possible internal layering within the maria and the existence and nature of isostatic lunar topographic features. On a smaller scale, data on the thickness, strength and the variation of physical properties with depth in a possible surface fragmental layer is pertinent to a full interpretation of the fine structure of the lunar surface. It is also possible that surface-bearing strength and the degree of hardening subsurface materials may be inferred from active seismic refraction data. A controlled active seismic survey will also be of particular importance in the search for water on the Moon. Local concentrations of ice may be present on the lunar surface — beneath the depth of penetration of the diurnal heat wave. A seismic velocity survey will be used to detect the

Figure 9-7. Magnetometer.

presence or absence of buried ice layers on the Moon, at landing sites some distance from the equator.

Magnetometer. The lunar surface magnetometer (*Figure 9-7*) measures the magnitude and direction of the surface magnetic field of the Moon and changes in the field direction up to a frequency of about one cycle per second. The placement of this ALSEP instrument on the Moon is such that the equatorial magnetic field is determined. Magnetic fields connected with interplanetary space should show periodic variations; fields associated with the Moon will be stationary during the lunar rotation.

As the solar wind sweeps the interplanetary magnetic field against the Moon some of this field should diffuse into the interior in a manner roughly analogous to heat flow. By studying the surface manifestations of this interior field during lunar day and night, it may be possible to infer the electrical conductivity and magnetic permeability of the lunar interior. These quantities must depend upon the composition of the Moon and its internal temperature, and therefore are related to the origin and thermal history of that body. If the Moon has a small core of iron-like material, magnetic field lines diffusing in from the solar wind should "hand up" on the core and impede the diffusion. It is possible, then, to imagine a lunar magnetic field streaming out through the Moon on its dark side, raising the possibility of utilizing the magnetometer for determining deep structure in the Moon. Other approaches to the problem of the interior composition are found by examining the propagation of electromagnetic disturbances which originate in the solar wind and are carried through the Moon. The response of the Moon should be that of a negative-gain conductor.

An additional purpose of this experiment is to monitor the passage of the Moon through the magnetic tail of Earth. It will obtain specific information of the interaction of the solar wind with the lunar surface and record whether the process results in the generation of plasma waves and produces some compression of the interplanetary field during the impacting of the solar plasma. Lastly, the site-surveying property of the magnetometer instrument allows detection of plasma currents and the presence of subsurface magnetic materials such as meteorites.

Figure 9-8.
Solar Wind Experiment.

Solar Wind Experiment. The Solar Wind Experiment (*Figure 9-8*) measures medium energy ranges of the solar wind particles. The solar wind is a flow of electrons, protons and other charged particles from the Sun. The nature of the interaction of the solar wind with the Moon is an intriguing problem in basic plasma physics. This interaction is different from that with Earth's magnetic field, and cannot be predicted theoretically with any certainty. Because of these uncertainties, the solar wind instrument is equipped to accept fluxes from all directions above the lunar horizon and has a wide range of sensitivities down to fluxes much smaller than an undisturbed interplanetary solar wind.

The structure and propagation velocity of the solar wind can be studied by measuring the time intervals between the observations of sudden changes in solar wind properties at the Moon and at Earth. The time intervals are expected to be as long as 15 minutes, depending on the relative positions of the Sun, Moon and Earth. The measurements of the solar wind experiment will permit knowledge to be gained about the length, breadth and structure of the magnetic turbulent wake of Earth.

The solar wind experiment will measure the number of charged particles impinging on it, and their energy up to 1330 electron

Figure 9-9.
Suprathermal Ion Detector.

volts for electrons and to 9780 electron volts for protons. The direction of these particles will be obtained by observing which of seven sensors (each sensitive to an overlapping portion of the lunar sky) indicates their flow.

Suprathermal Ion Detector. The Suprathermal Ion Detector Experiment (*Figure 9-9*) will measure the flux, number, density, velocity and energy per unit charge of positive ions in the vicinity of the lunar surface.

The experiment uses two curved plate analyzers to detect and count ions. The low-energy analyzer has a velocity filter of crossed electric and magnetic fields. The velocity filter passes ions with discrete velocities and the curved plate analyzer passes ions with discrete energy, permitting determination of mass as well as number density. The second curved plate analyzer, without a velocity filter, detects higher energy particles, as in the solar wind. The experiment is emplaced on a wire mesh ground screen on the lunar surface and a voltage is applied between the electronics and ground plane to monitor any electrical field effects.

Cold Cathode Gauge Experiment. The Cold Cathode Gauge Experiment provides data pertaining to the density of the lunar ambient atmosphere. Of particular interest will be any variations

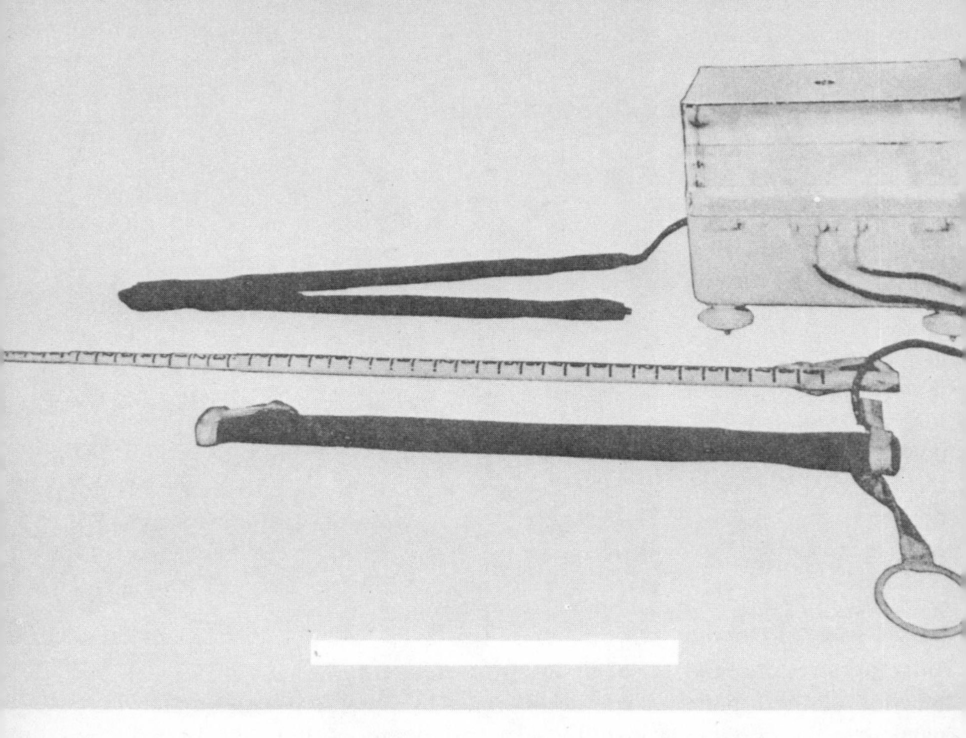

Figure 9-10. Heat Flow Experiment.

of the particle density associated with the lunar phase of solar activity. This instrument will study the effects of foreign material left by the LM and the astronauts, and rate of loss of contaminants.

When the astronaut deploys the Ion Detector package, he removes the Cold Cathode Gauge and emplaces it three to five feet away from the Ion Detector. An electrical cable connects the cold cathode gauge instrument to the Ion Detector.

Heat Flow Experiment. The Heat Flow Experiment (*Figure 9-10*) measures the lunar temperature profile at depths up to three metres and the value of the Moon's thermal conductivity over the same depth. From these measurements, information may be deduced regarding the net outward flux of heat from the Moon's interior and the radioactive content of the Moon's interior compared to that of the Earth's mantle. It will also provide data from which it is possible to reconstruct the temperature profile of the subsurface layers of the Moon and to determine whether the melting point may be approached towards its interior.

The Heat Flow Experiment consists of two sensor probes and a common electronics package. Two 2½-centimetre diameter, three-metre holes will be drilled into the lunar surface by the Apollo astronaut. This will be accomplished by a specially designed heat flow drill. A two-section probe approximately 115 centimetres long will be lowered into each of the two holes. The probes contain sensors to measure absolute temperature and temperature difference. Thermal conductivity is investigated by measuring absolute and differential temperatures while actuating small electric heaters in the probes.

The Apollo Lunar Surface Drill allows the astronaut to implant heat flow temperature probes below the lunar surface and to collect subsurface core material.

The Lunar Surface Drill is designed as a system which can be removed as a single package from the ALSEP pallet and carried to the drilling site. There it will be used to drill two holes. The holes are cased to prevent cave-in and to facilitate insertion of probes of the heat flow experiment. The subsurface core material from the second hole will be retained in the drill string and returned to Earth in a sample return container.

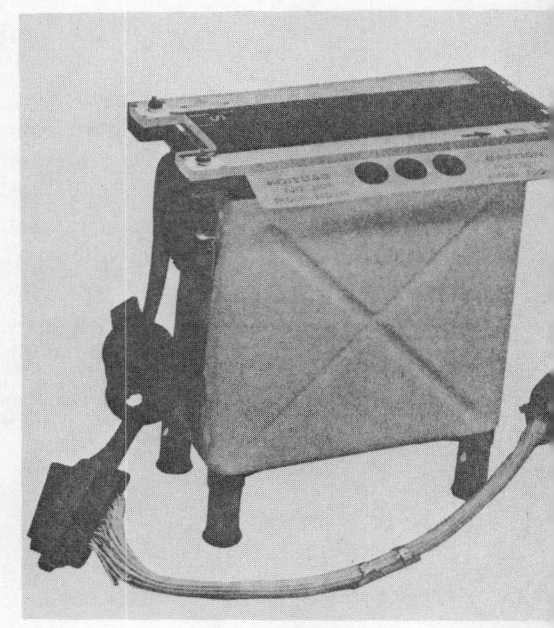

Figure 9-11. Charged Particle Lunar Environment.

Charged Particle Lunar Environment Experiment*. The Charged Particle Lunar Environment Experiment (*Figure 9-11*) will study the energy distribution and time variations of proton and electron fluxes in 18 energy intervals over the range of about 50 to 150,000 electron volts.

The lunar surface may be bombarded by electrons and protons of the solar wind. This wind is caused by the expansion of the outer gaseous envelope of the Sun into interplanetary space. Because the solar wind is supersonic and the Moon is a large body, it is possible that, at times, there may be a standing shock front of the solar wind between the Moon and the Sun. The detailed physical processes occurring at such a shock front are largely not understood, and they are of considerable interest in fundamental plasma research. If there is such a shock front near the Moon, the Charged Particle Experiment will detect the disordered or thermalized fluxes of electrons and protons on the downstream side of the shock front.

At times of the full Moon, when the Moon is in the "magnetic tail" of Earth, the Charged Particle Experiment will detect the accelerated electrons and protons that cause auroras when they plunge into the terrestrial atmosphere. These acceleration processes are not understood, and their simultaneous observation near Earth and the Moon is essential for detailed study. The Charged Particle Experiment will also measure the lower-energy solar cosmic rays occasionally produced in solar eruptions or flares.

Dust Detector*. The Dust Detector will measure the accumulation and effect of lunar dust accretion over the ALSEP central station. The package is mounted on top of the central station sunshield with the photocells facing the ecliptic path of the Sun. Each cell is protected by a blue filter to cut off ultraviolet wavelengths below 0.4 micron and a cover slide for protection against radiation damage. Attached to the rear of each photocell is a thermistor to

---

* Both these experiments were developed by Dr. Brian O'Brien of the School of Physics, University of Sydney, who, as principal investigator, will analyze the data in Sydney. This work will be supported in part by the Science Foundation for Physics within the University of Sydney.—Ed.

monitor the individual cell's temperature. The temperature of each photocell, compared to the anticipated value for exposure at a given solar angle, is a measure of dust accretion and insulating values.

Communications. The ALSEP telemetry system consists of two distinct links. The Earth-to-Moon link (the up-link) provides for remote control of ALSEP command functions such as experiment mode selection, transmitter selection, change of subsystems, data rates and subsystem operation flexibilities (turn-on, turn-off, etc.). The Moon-to-Earth link (the down-link) provides for the transmission of scientific and engineering data from the ALSEP subsystems to Earth receiving stations.

ALSEP communications are through a helical antenna attached to the central station. This type of antenna obtains high gain over a moderately narrow beam width. When deployed on the lunar surface, the antenna will be aimed at Earth using a sun compass and adjustment knobs. The data subsystem is located in the base of the central station.

Command data (up-link) are received by the helical antenna and go to the command receiver which demodulates the input carrier signal and provides a modulated subcarrier output to the command decoder. The different data transmission frequencies will be used to permit simultaneous operation of three separate ALSEP systems.

The data subsystem is the "nerve centre" of the ALSEP system. It accepts the experimental subsystem scientific and engineering data and encodes the data for phase modulated radio frequency signal transmissions to Earth. The data subsystem also receives command data from Earth and decodes and distributes the commands to the ALSEP subsystems. The data subsystem is capable of the simultaneous reception of commands and the transmission of data. As a backup measure to help ensure a one-year operation, redundant transmitters and data processor components are included in such a way that, by command, the output of either data processor may be connected to either transmitter.

Electrical Power System. The Electrical Power System provides all the electrical power for ALSEP system operation on the lunar

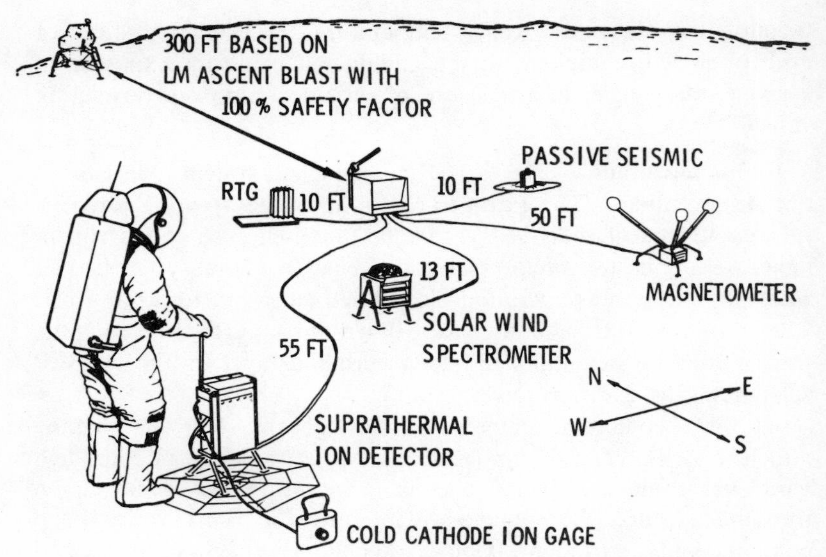

Figure 9-12. ALSEP Deployment.

surface for a period of at least one year. The major components include the Radioisotope Thermoelectric Generator, the Fuel Capsule Assembly and the Power Conditioning Unit which is located in the central station. The supporting components include the graphite LM fuel cask, the fuel cask mounting assembly, and the fuel transfer tool.

The Fuel Capsule Assembly uses the nuclear fuel Plutonium 238 and produces 1500 watts of thermal energy. When the fuel capsule assembly is combined with the generator assembly to form the SNAP 27 Radioisotope, Thermoelectric Generator at least 63 watts are converted by thermocouples from thermal energy to electrical energy and supplied (at + 16 VDC nominal) to the Power Conditioning Unit.

Deployment. (*Figure 9-12*) As one of the initial ALSEP deployment tasks, the astronaut transfers the fuel capsule from the cask to the generator. The fuel cask mounting assembly is tilted for access to the cask and the fuel transfer tool is used to effect a transfer. Thermal equilibrium (i.e., full power) of the Radioisotope Thermoelectric Generator is reached in approximately 1½ hours.

Figure 9-13.
ALSEP Subpackage.

The conditions of the lunar environment during ALSEP deployment by the Apollo astronauts (temperature extremes, vacuum, a one-sixth gravitational pull, and extreme light intensity) are moderated by the extravehicular mobility unit consisting of a pressure suit, thermal overgarment, a helmet with multiple visors, and a portable life support system. Development of the extravehicular mobility unit is independent of ALSEP. However, the extravehicular mobility unit characteristics influenced the design of ALSEP handling features.

ALSEP is inoperative during its trip to the Moon. After landing, it is deployed and activated by a series of astronaut tasks together with a series of Earth commands to the data subsystem from the manned space flight network. Deployment begins with ALSEP subpackages 1 and 2 (*Figure 9-13*) are separately removed from the storage equipment bay and lowered to the lunar surface. The astronaut opens the storage equipment bay door on the lunar module, removes the package restraints, and grasps a deployment lanyard which is attached to a boom and one subpackage. Pulling the lanyard extends the boom and allows the package to be withdrawn from the storage equipment bay and lowered to the surface in a continuous motion. The other subpackage is similarly unloaded.

The radioisotope fuel capsule is next transferred from the fuel cask to the generator mounted on subpackage 2. This is accomplished by rotating the fuel cask to a horizontal position, removing its dome and withdrawing the fuel capsule with the fuel transfer tool. Using the fuel transfer tool as a handle, the astronaut inserts the capsule into the generator locking it in place with a twisting motion which also frees the transfer tool.

The Apollo astronauts then carry the ALSEP some 91 metres from the LM to the final deployment site. The primary transport mode uses the antenna mast attached to the two subpackages to form a "barbell". A simple, slip-fit, trigger-actuated lock secures the mast to the subpackages. The alternative "suitcase" carry mode makes use of individual handles on the subpackages.

During the traverse, the astronauts determine the most desirable site, about 91 metres from the LM, to locate ALSEP. They look for a smooth area, large enough to accommodate the planned 30-metre separation between the magnetometer and the Suprathermal Ion Detector; a level site, free from rubble.

The 91-metre distance assures that there are no destructive LM ascent blast effects on ALSEP and also reflects the need to keep the astronaut at all times within a safe distance for return to the LM in case of failure in his oxygen supply.

Figure 9-14.
ALSEP Central Station.

At the end of the traverse, the astronaut deploys the Radioisotope Thermoelectric Generator by removing all other equipment from subpackage 2 and placing it in its upright position. The Radioisotope Thermoelectric Generator-to-central station interconnecting cable is connected to a receptacle located on subpackage 1.

Subpackage 1, which contains the central station (*Figure 9-14*), is deployed by placing it in an upright position three metres from subpackage 2, removing the experiments from the sunshield, raising the sunshield, and installing the antenna.

Experiments are removed from the sunshield by using a Universal Handling Tool to release the tie-down fasteners and to lift the experiments to other locations.

LUNAR ORBIT EXPERIMENTS

In the Command Module cameras and other small experiments are carried on each mission. On Apollo 16 through Apollo 19 a substantial orbital experiment payload will be carried in the Service Module.

Table II lists the experiments that have been approved for flight on the CSM in lunar orbit. Orbital experiments will complement and extend information gained from the lunar surface science. For example, compositional analysis can be accomplished from lunar orbit through radioactivity detection and spectral reflectivity measurements, radar sounders can probe the subsurface nature of volcanic features which the astronauts have examined; measurements will be obtained in the fields of geochemistry, imagery, geodesy, temperature, subsurface profile, particles and fields, and transient atmosphere.

ADDITIONAL EXPERIMENTS

During the past year (September, 1969 to 1970) NASA asked the scientific community to propose experiments to be carried on the later Apollo missions. This request stimulated the submission of 99 orbital experiment proposals and 42 surface experiment proposals. The number that can be selected exceeds, by a considerable amount, our ability to carry them. Thus, only the very best can be developed. However, with the schedule stretchout, we will be able to introduce some of these new experiments when this is indicated by the results coming in from the missions.

Table II — LUNAR ORBITAL EXPERIMENTS

	Experiment	14	15	16	17	18	19
S-160	GAMMA RAY SPECTROMETER			X	X		
S-161	X-RAY SPECTROMETER			X	X		
S-162	ALPHA PARTICLE SPECTROMETER			X	X		
S-163	24-inch PANORAMIC CAMERA			X	X	X	X
S-164	S-BAND TRANSPONDER			X	X	X	X
S-165	MASS SPECTROMETER			X	X		
S-166	3-inch MAPPING CAMERA			X	X	X	X
S-167	SOUNDING RADAR						
S-168	EM SOUNDER A						
S-169	ULTRA-VIOLET SPECTROMETER					X	X
S-170	DOWNLINK BISTATIC RADAR	X	X				
S-171	INFRA-RED SCANNING RADIOMETER					X	X
S-173	PARTICLE SHADOWS/BOUNDARY LAYER			X		X	
S-174	SUB-SATELLITE MAGNETOMETER			X		X	
S-175	LASER ALTIMETER			X	X	X	X

APOLLO LANDING SITES

Landing Site Selection

The difference between Apollo 11 and 12 results proves that the Moon is very different at different places on its surface. It is to a number of these different places that we want to go if we are to understand the age, composition processes and structure of this celestial body.

The criteria for selection of the lunar landing sites includes the following significant factors:

- Scientific interest and uniqueness—its own characteristics and how they relate to those of the rest of the Moon with regard to providing answers to scientific questions.
- Scientific variety—the number of varied features which could be explored in the vicinity of the landing site.
- Instrument networks—the selenographic location as a part of a viable network of scientific instruments (primarily seismic).
- Photography, i.e., data availability—sufficient high resolution photography of the site and its landing approach to determine landing feasibility capabilities and to verify scientific interest.
- Match to capability—the ability of the Apollo system and programme of missions to achieve a landing site and of the surface exploration to achieve scientific objectives.

The placement of seismic detectors in such a way as to form networks for locating and measuring lunar seismic activity is a very significant part of the scientific exploration programme. The ground rules used for the development of seismic networks for this programme are that the ALSEP design lifetime goal is one year with a maximum lifetime of two years. Acceptable seismic networks require three concurrently operating stations where the included angle between any two of the three legs of the triangle is greater than five degrees (*Figure 9-15*).

Spacecraft orbital tracks over the lunar surface must be considered in mission planning because of the requirements to perform photography of subsequent landing sites and to plan the orbital science programme. The extent of nominal mission

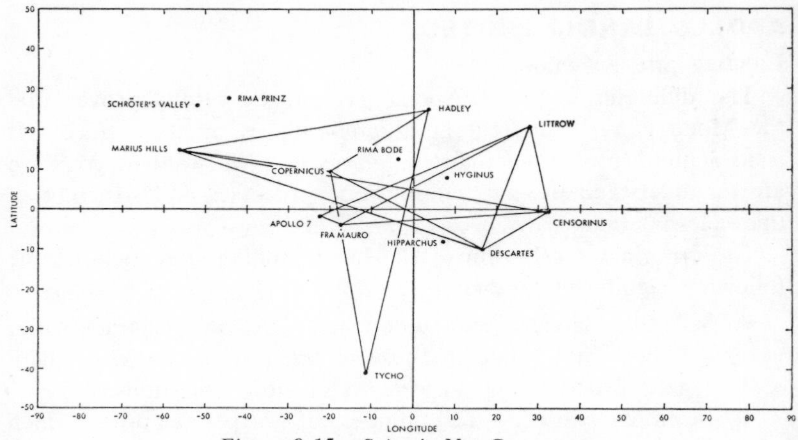

Figure 9-15. Seismic Net Coverage.

coverage is determined by the landing site location, the launch date, dependent trajectory characteristics, and the nominal mission timeline. After LM/CSM docking, subsequent coverage is limited by the SPS$\triangle$V which can be provided for plane change if LM rescue manoeuvres are not required and by the amount of remaining consumables.

The manoeuvre is executed no earlier than 17 hours after LM ascent and no later than four hours before trans-Earth injection.

- The $\triangle$V for Lunar Orbit Plane Change 2 (LOPC 2) is 152 metres per second. This manoeuvre can be performed at any point in the parking orbit and results in a plane change of about 5.5 degrees.
- Trans-Earth Injection (TEI) can be targeted to the originally planned flight time and inclination. Preliminary studies indicate that the maximum TEI$\triangle$V penalty for high latitude landing sites is about 76 metres per second. It is usually less than 21 metres per second for equatorial sites.
- For plane changes of less than 5.5 degrees, the area covered, the $\triangle$V required for plane change and the maximum TEI$\triangle$V penalty are all reduced proportionally.

The surface area of the Moon covered on each mission is determined by the landing site location and launch date dependent

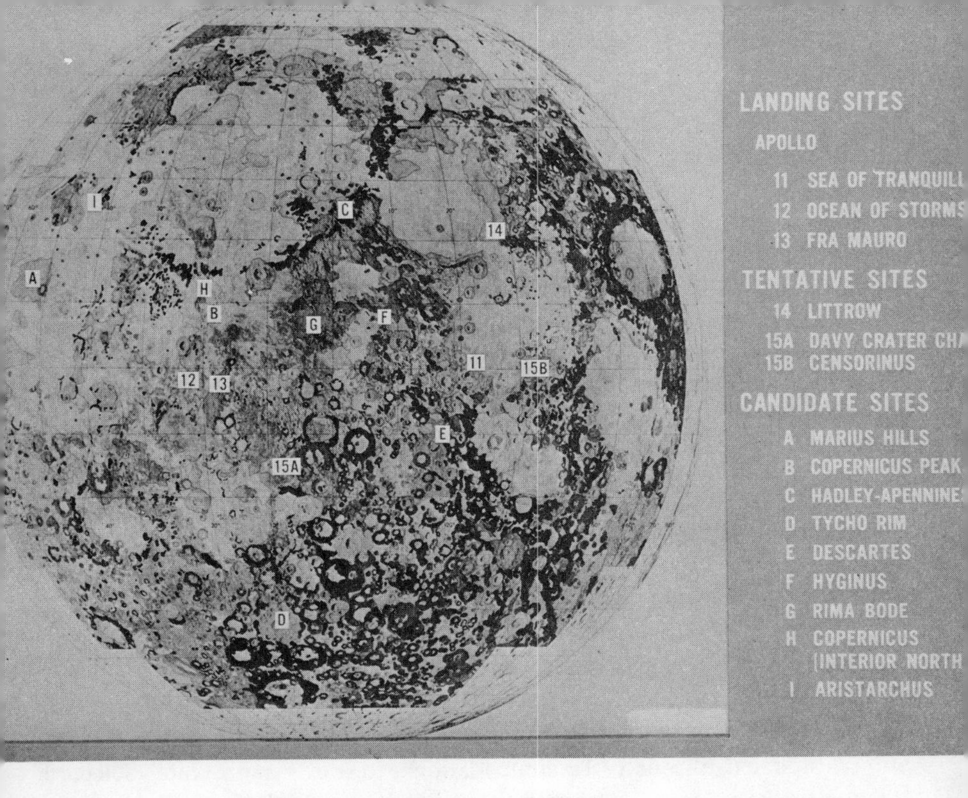

Figure 9-16. Future Landing Sites.

trajectory characteristics. Since the landing site and mission sequence assignments are preliminary, the precise area of surface coverage cannot be defined at this time. The objectives of orbital surveys are best satisfied by maximizing surface coverage and/or overflying targets of specific interest. The flexibility to select one of these options on each mission is available with the LM rescue budget after a nominal rendezvous sequence is completed.

Site selection (*Figure 9-16*) must meet both the geological and geochemical objectives of the lunar exploration. The geophysical objectives require a specific mission assignment plan, particularly for the construction of seismic networks. Other scientific objectives do not call for much that would contradict this rationalization.

With these considerations in mind, prime landing sites (Table III) have been selected by the Group for Lunar Exploration Planning. The list of prime landing sites will be reviewed

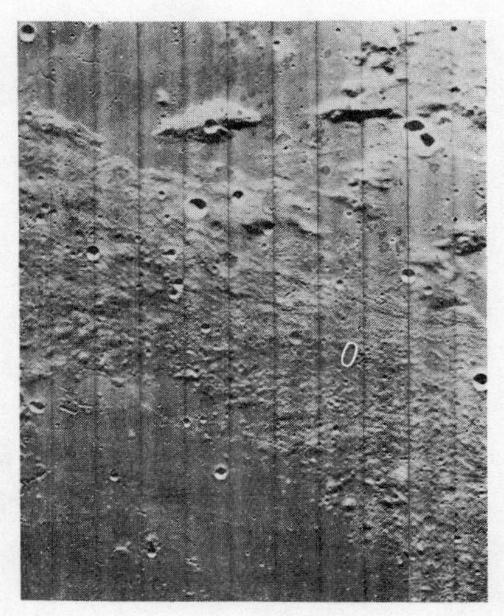

Figure 9-17. Fra Mauro Formation.

periodically by the Group for Lunar Exploration Planning and updated through recommendations to the Apollo Site Selection Board.

Table IV summarizes the salient geological, geophysical, and geochemical characteristics of these landing sites.

Landing Site Description

Apollo 13—Fra Mauro Formation. The Fra Mauro Formation (*Figures 9-17* and *9-18*), an extensive geologic unit covering large portions of the lunar surface around Mare Imbrium, has been interpreted as the ejecta blanket deposited during the formation of the Imbrium basin. Sampling of the Fra Mauro Formation may provide information on ejecta blanket formation and modification, and yield samples of deep-seated crustal material, giving information on the composition of the lunar interior and processes active in its formation. Age dating the returned samples should establish the age of premare deep-seated material and the time of formation of the Imbrium basin, thus providing important points on the geologic time scale leading to an understanding of the early history of the Moon.

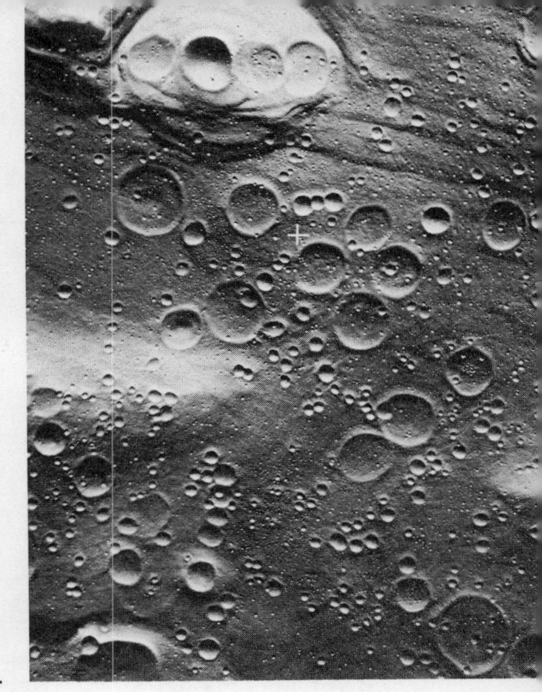

Figure 9-18. Fra Mauro Formation.

Table III

RECOMMENDED APOLLO LANDING SITES

Flight	Mission	Landing Site
Apollo 13	H-2	Fra Mauro
Apollo 14	H-3	Littrow
Apollo 15*	H-4	Davy Crater Chain or Censorinus
Apollo 16	J-1	Copernicus
Apollo 17*	J-2	Descartes or Tycho
Apollo 18	J-3	Marius Hills
Apollo 19	J-4	Hadley Apennines

* A decision on the landing site for Apollo 15 and 17 will be made after photography from previous Apollo mission has been analyzed.

Table IV

CANDIDATE SITES	GEOLOGY, GEOCHEMISTRY AND GEOPHYSICS																			
	CHRONOLOGY				COMPOSTION					PROCESSES										
										CRATERING			TRANSPORT			VOLCANIC-TECTONIC				
	"ORIGINAL CRUST"	ORIGIN OF MAJOR MARE BASINS	MARE FLOODING	POST-MARE TIME SCALE	PRIMITIVE ROCKS	DEEP-SEATED ROCKS	DIFFERENTIATED ROCKS	TRANSIENT EVENTS	ATMOSPHERE	IMPACT	VOLCANIC	CHAIN	RILLE	EJECTA	GRAVITY FLOW AND SLUMP	FAULT	DOME	FLOWS	RIDGES	SEISMICITY
FRA MAURO FORMATION	?	×			×	×				×				×						
LITTROW			×				?	?	?							×		×	×	×
CENSORINUS	?	?		×	×					×				×						
DAVY CRATER CHAIN	?			?	?	×	?		?		×	×		×	?			?		?
MARIUS HILLS			?	×		×	×	?	×		×		×				×	×	×	×
DESCARTES				?	?	?	×				×					?	×	×	×	×
TYCHO	?			×	?	×	?	?		×				×				×		?
COPERNICUS CENTRAL PEAKS	?			×	?	×	?			×	?				?		×	?		?
HADLEY-APENNINES	?	×	×	×	×	×	?	?	?	×	?		×		×		×	×	×	?
HYGINUS -	?		?	?	?	×	?	?	?		×	×		?		×	?	?		?

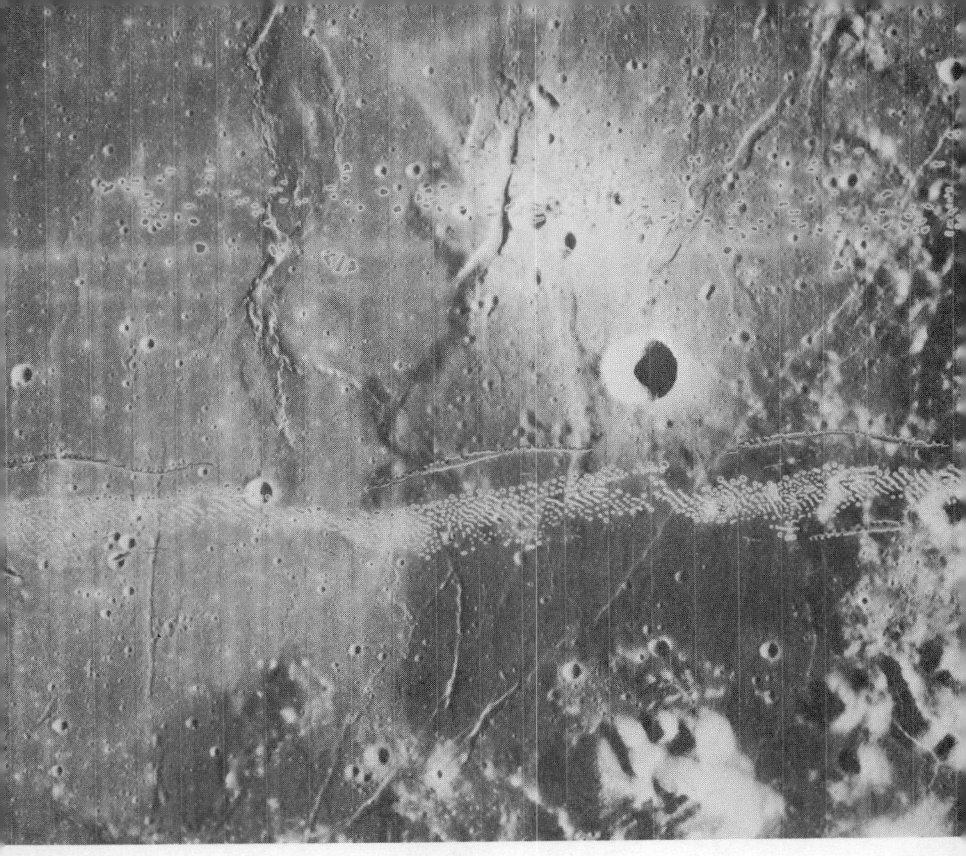

Figure 9-19. Littrow.

Apollo 14—Littrow. The Littrow area landing site (*Figure 9-19*) lies on the eastern edge of Mare Serenitatis in the vicinity of a series of straight rilles and wrinkled ridges oriented parallel and sub-parallel to the edge of the basin. A mantling material of very low albedo as well as a topographic bench lie in the vicinity of the landing site. Analysis of material from Mare Serenitatis will provide geochemical and age data which can be related to results from Apollo 11 and 12 to show compositional and age differences from different maria. Investigation of the wrinkled ridges should provide an understanding of the composition, origin and significance of these widespread mare features. The dark mantling material appears to be younger than most other features in the site. It is probably among the youngest of lunar surface materials and may record the latest stages in the process of basin filling in Mare Serenitatis.

Figure 9-20. Censorinus

Apollo 15—Censorinus. Censorinus (*Figure 9-20*) is a small (3.5 kilometre diameter) bright crater of probable impact origin located in a segment of the highlands just southeast of Mare Tranquillitatis. The composition and age of highland materials and mechanics of crater formation at a young crater are among the primary objectives of a mission to the edge of the continuous ejecta blanket of Censorinus. Study of the distribution, structure and morphology of the ejecta material should provide information relating to the mechanics of crater formation. Sampling of the landing area will not only provide data on the composition of the highland surface material, but should also provide information about shallow highland material excavated by the event which produced Censorinus. Since the crater Censorinus is relatively very young, possible age dating of the event producing the crater will provide an important point on the lunar time scale. Age

Figure 9-21. Davy Rille.

dating of the highland material sampled at this site will serve to clarify an understanding of the relationship of this area to the extensive mare regions.

Apollo 15—Davy Crater Chain. The Davy Crater Chain (*Figure 9-21*) is a probable volcanic crater chain crossing the highland-mare boundary slightly northwest of the crater Alphonsus. The chain of craters, several of which are thermally anomalous, stretches some 60 kilometres from Davy C, located in plains material, to Davy G on an upland plateau. Since the craters forming the Davy Crater Chain are analogous to terrestrial mare-type volcanic craters which often bring up deep mantle material, the primary objective of this landing site concerns the acquisition of material from deep within the lunar interior. A landing near the point where the crater chain crosses into the highlands should

Figure 9-22. Marius Hills.

Figure 9-23. Marius Hills.

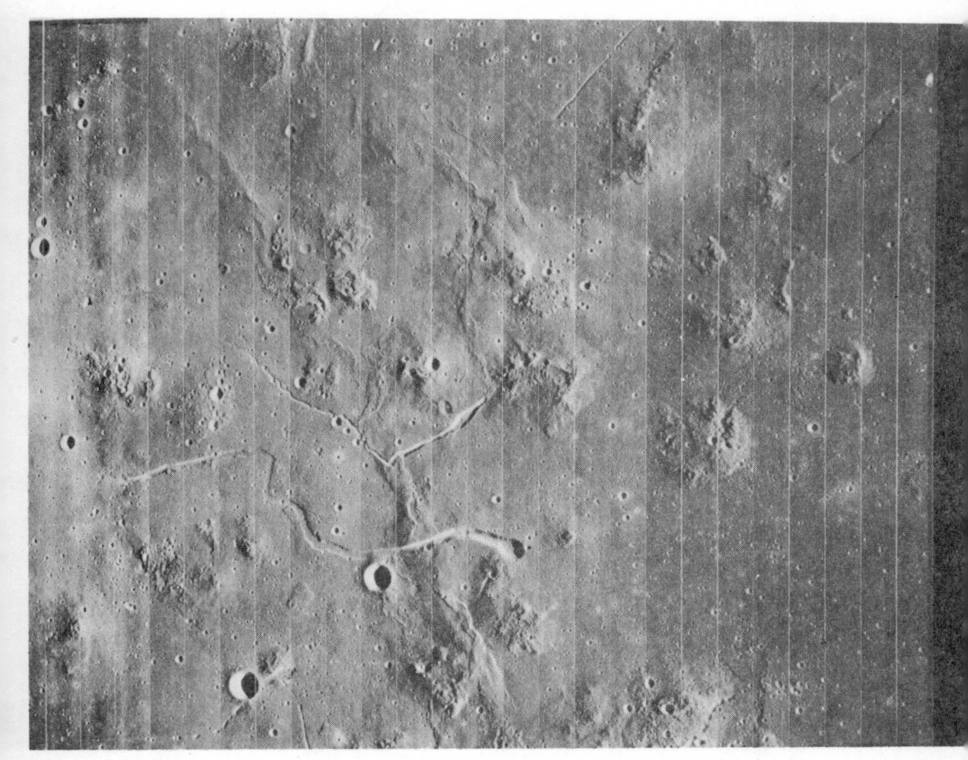

also provide samples of the plains material on the floor of Davy Y, a widespread unit in highland basins and highland material. Acquisition of these materials will provide data on the physical properties of the lunar interior as well as on the characteristics and age of several widespread geologic units.

Apollo 16—Marius Hills. The Marius Hills (*Figures 9-22* and *9-23*) are a series of domes and cones located northwest of the crater Marius near the centre of Oceanus Procellarium. The morphologic units which comprise these hills are analogous in form and sequence to terrestrial volcanic complexes which display a spectrum of rock compositions and ages. The various geologic units suggest that a prolonged period of volcanic activity has occurred in the Marius Hills area and that magmatic differentiation has produced a spectrum of rock types and a series of volcanic landforms displaying characteristic structural relationships.

Figure 9-24. Descartes.

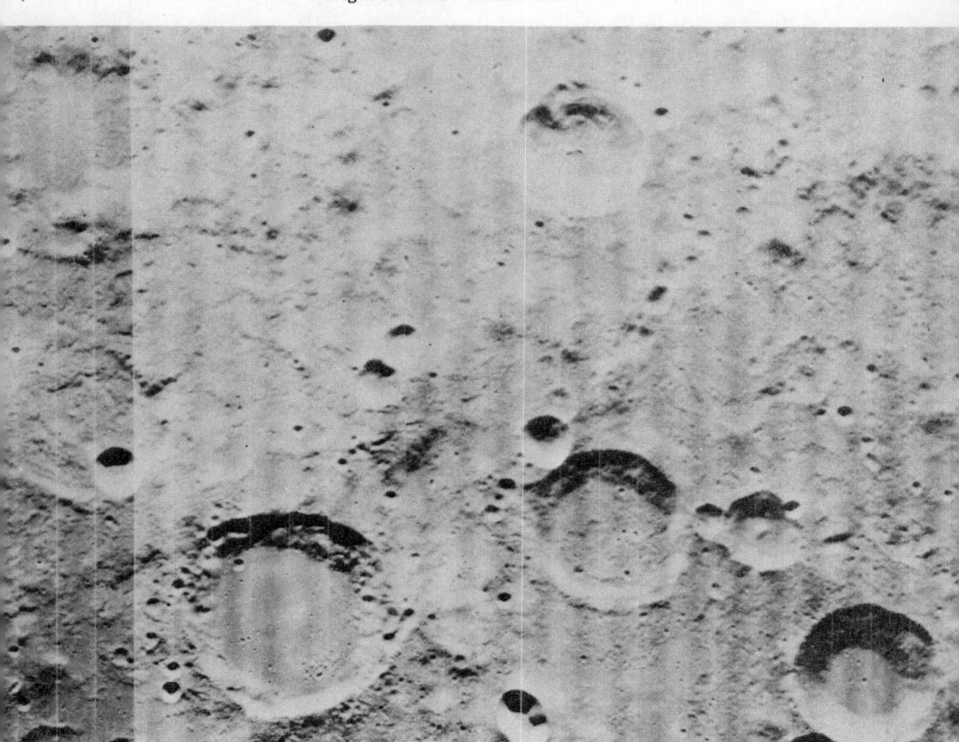

Therefore, the primary objectives of a mission to the Marius Hills are to study the spectrum of geologic units in order to establish the extent and age of possible magmatic differentiation and to determine the structural relationships of volcanic landforms in the maria.

Apollo 17—Descartes. The Descartes landing site (*Figure 9-24*) lies in the central lunar highlands several hundred kilometres west of Mare Nectaris, and is the site of hilly, grooved and furrowed terrain which is morphologically similar to many terrestrial areas of volcanism.

The Descartes area is also the site of extensive development of highland plains material, a geologic unit of widespread occurrence in the lunar highlands. The primary objectives of a mission to this site would be the examination and sampling of a highland volcanic complex and of the plains material. Knowledge of the composition, age and extent of magmatic differentiation in a highland volcanic complex will be particularly important in understanding lunar volcanism and its contribution to the evolution of the lunar highlands. Comparison of this highland volcanic complex to mare volcanic complexes such as Marius Hills will provide a sample of a wide spectrum of lunar volcanic activity. An understanding of the composition and age of the highland plains material will add to our knowledge of the processes which modify large areas of the lunar highlands.

Apollo 17—Tycho. The crater Tycho (*Figure 9-25*) is an 85 kilometre diameter very young crater of probable impact origin located in the southern lunar highlands. Bright rays from Tycho spread across the near side of the Moon. A mission to the northern crater rim of Tycho would land in the vicinity of the Surveyor VII spacecraft. Among the principal objectives would be the investigation of the composition of the highlands and of features associated with a young large impact crater. The origin and nature of the ejecta, flows and associated volcanism located on the crater rim are of interest in this regard. Since Tycho is approximately four kilometres deep, the ejecta material should provide samples from deep within the highlands. The composition and age of this material will provide important information about the formation

Figure 9-25. Tycho.

and evolution of the lunar highlands. Establishment of the age of the relatively young event which produced Tycho will add an important point to the lunar time scale.

Apollo 18—Copernicus Peaks. Copernicus (*Figures 9-26* and *9-27*) is a relatively young, very large bright-rayed probable impact crater approximately 95 kilometres in diameter and located just south of Mare Imbrium. A mission to the floor of the crater Copernicus, four kilometres below the crater rim, would have as its primary objectives the examination of the central peaks and the crater floor material. The central peaks, which rise up to 800 metres from the crater floor, probably represent deep-seated material, which is of importance in determining the internal characteristics of the Moon. Examination of the domes and textured material of the crater floor will provide an understanding

N

Figure 9-26. Copernicus.

Figure 9-27. Copernicus (Artist Concept).

Figure 9-28. Hadley.

Figure 9-29. Hadley (Artist Concept).

of the process of crater floor filling and help clarify the role of volcanism in post-event crater modification. Age determinations of the central peak material, the cratering event, and the subsequent crater fill material will provide a time scale of importance in understanding the origin and modification of large impact craters.

Apollo 19—Hadley-Apennines. Rima Hadley (*Figures 9-28* and *9-29*) is a V-shaped lunar sinuous rille which parallels the Apennine Mountain front along the eastern boundary of Mare Imbrium. The rille, an elongated depression, originated in an area of associated volcanic domes and generally maintains a width of about one kilometre and a depth of 200 metres until it merges with a second rille approximately 100 kilometres to the north. The origin of sinuous rilles such as Rima Hadley is enigmatic but is probably due to some type of fluid flow. The Apennine Mountains rise up to two kilometres from the area of Rima Hadley and contain ancient material exposed during the excavation of the Imbrium basin. The determination of the nature and origin of a sinuous rille and its associated elongated depression and deposits will provide information on an important lunar surface process and

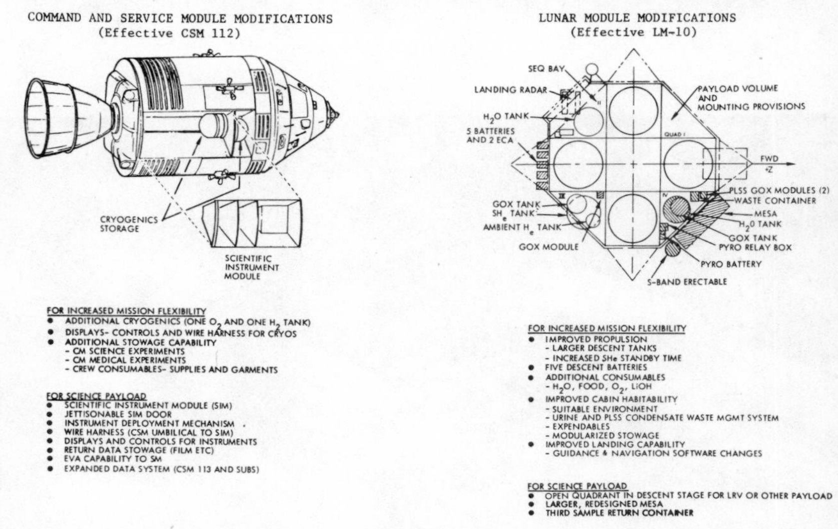

Figure 9-30. Command Service Module and Lunar Module Modifications.

MODIFIED (-7) PORTABLE LIFE SUPPORT SYSTEM (PLSS)

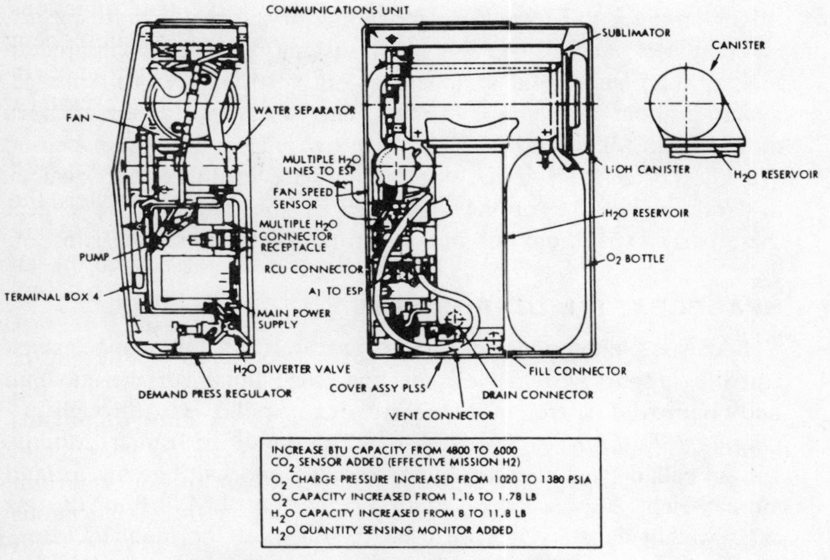

SECONDARY LIFE SUPPORT SYSTEM (SLSS)

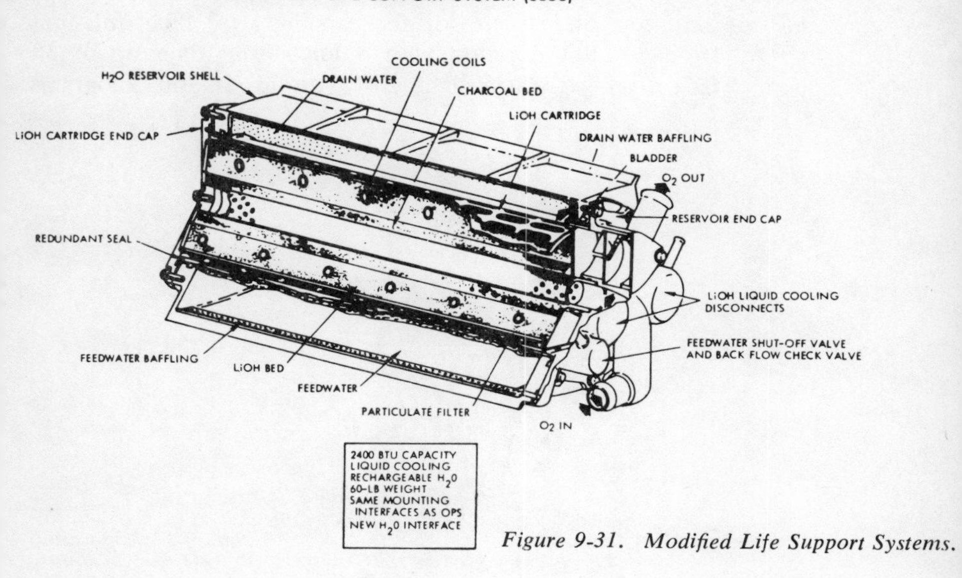

Figure 9-31. Modified Life Support Systems.

may yield data on the history of lunar volatiles. Sampling of Apenninian material should provide very ancient rocks whose origin predates the formation and filling of the major mare basins.

These few landings on the lunar surface might be compared to making nine landings at various sites on North and South America, which is about the same land mass. One would not expect to have more than a token knowledge of that area under the circumstances. We know that these initial explorations are insufficient to give us in-depth knowledge of the vast area of the Moon. We expect these early expeditions to guide us to the most interesting sites.

SPACECRAFT MODIFICATIONS

NASA is planning to modify the spacecraft, astronaut space suits, and life support systems to allow increased lunar surface staytime and improved astronaut mobility for Apollo 16 through 19 (*Figures 9-30, 9-31* and *9-32*). The modified Lunar Module will be called the Extended Lunar Module and will be able to land more weight on the Moon than the current LM. Part of this extra weight capacity will be used to transport the manned Lunar Rover Vehicle. The modifications are required to satisfy the mission requirements; to provide the capability for flight missions up to 16 days total duration with a lunar stay time up to 54 hours; to deliver a heavyweight LM (up to 16,760 kilograms

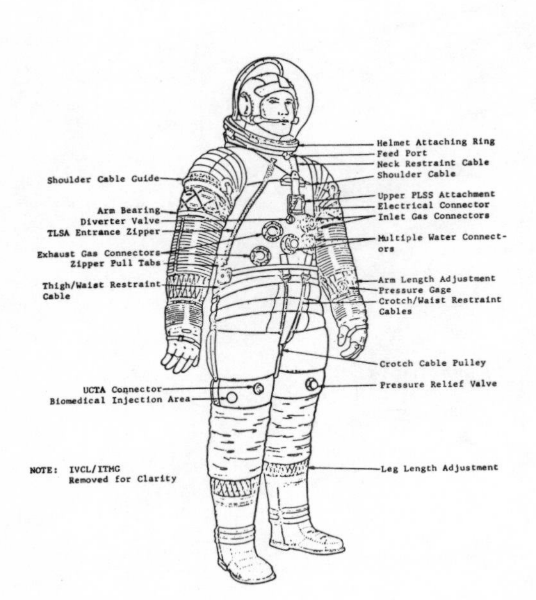

Figure 9-32. Spacesuit.

manned) to a 111 × 5 kilometre lunar orbit; and to provide the capability to carry and operate a scientific instrument payload through the addition of a general purpose scientific instrument module with interfacing subsystem modification, including Extra-vehicular Activity recovery of stored scientific instrument module data.

LUNAR ROVER VEHICLE

NASA is developing a Lunar Rover Vehicle (LRV) (*Figure 9-33*) to transport two astronauts as they explore the lunar surface. Four operational vehicles are being readied, the first of which is scheduled to be flown on the Apollo 16 lunar mission in late 1971.

The four-wheel vehicle will provide transportation for the astronauts and their tools, scientific equipment and lunar samples collected during traverses. It will be manually operated by one of the astronauts, and will not be capable of automatic control.

Figure 9-33. Lunar Rover Vehicle.

The driver will operate the vehicle much as he would on Earth, using a pistol grip hand control rather than a steering wheel. The vehicle will be able to travel at variable speeds up to 16 kilometres per hour on a relatively smooth surface.

Simplicity, in both design and operation, is one of the most important features being emphasized in the development and construction of the vehicle.

The Lunar Rover will be about 3.2 metres long, 1.8 metres wide and have a 2.2 metre wheel base. The four wheels will be individually powered by electric motors. The Rover's power source will be two primary (non-rechargeable) batteries.

The LRV will weigh no more than 181 kilograms, including tie-down and unloading systems. It will carry a total weight of 438 kilograms, which will include the two astronauts and their life support systems (181 kilograms for each man and his equipment) plus 45 kilograms of scientific experiments. It must also be able to carry up to 32 kilograms of lunar soil and rock samples.

The delivery date of the first operational vehicle is compatible with the completion of improvements to the Apollo spacecraft and astronaut equipment, which, with the LRV, will extend the lunar staytime and give exploring astronauts much greater mobility — two basic requirements for increasing the scientific value of the Apollo Lunar Exploration Programme.

The operational lifetime of the LRV on the Moon, according to current plans, will be 54 hours during the lunar day. It will be able to make any number of short trips up to a cumulative distance of 120 kilometres. Because of the limitations of the life support systems, the vehicle's range will be restricted to a radius of about five kilometres from the Lunar Module.

The LRV will be delivered to the Moon in the "cargo section" (descent stage) of the Extended Lunar Module (*Figure 9-34*). The LRV will be stowed in the stage's quadrant number one with its four wheels folded over its chassis. Deployment will be semi-automated; one astronaut must be able to quickly and easily deploy, activate, check and operate the vehicle.

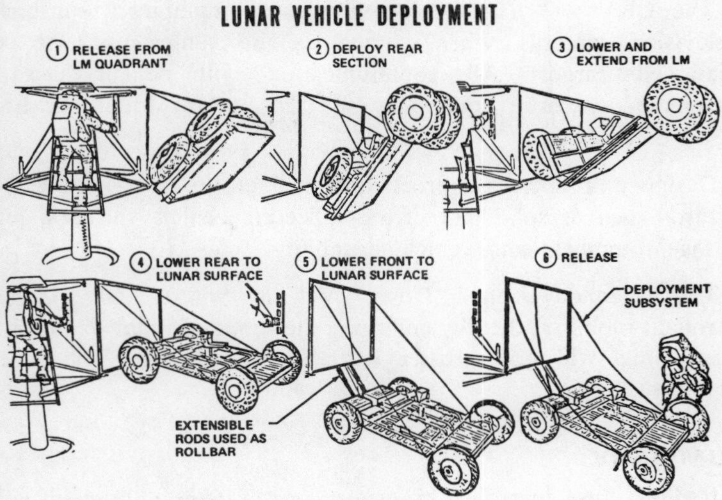

LUNAR VEHICLE DEPLOYMENT

1. RELEASE FROM LM QUADRANT
2. DEPLOY REAR SECTION
3. LOWER AND EXTEND FROM LM
4. LOWER REAR TO LUNAR SURFACE
5. LOWER FRONT TO LUNAR SURFACE
6. RELEASE

DEPLOYMENT SUBSYSTEM

EXTENSIBLE RODS USED AS ROLLBAR

Figure 9-34. Lunar Rover Deployment.

The LRV will be designed to negotiate step-like obstacles 30 centimetres high with two wheels in contact at zero velocity and be able to cross crevasses 70 centimetres wide with two wheels at zero velocity. The fully loaded vehicle will be able to climb and descend 25 degree slopes. A parking brake will be provided that can stop and hold the LRV on slopes up to 35 degrees.

The vehicle will have ground clearance of at least 35 centimetres on a flat surface. Pitch and roll stability angles will be at least 45 degrees with a full load. The turn radius will be no more than one vehicle length (less than 3.3 metres).

The driver of the LRV will be seated so that both front wheels are visible during normal driving. Mirrors will be used for rear vision. The driver will navigate through a simple (dead reckoning) navigation system that will determine the direction and distance between the LRV and the Lunar Module, and the total distance travelled at any point during a traverse.

The LRV will have no communication equipment on board. Television cameras, when carried on the vehicle, will be self-contained cargo. All communication will be between the astronauts' spacesuit equipment and the Lunar Module or Earth based controllers.

Design and operating procedures will ensure crew safety from hazards such as solar glare from reflecting vehicle surfaces, lunar surface roughness and vehicle instability.

The Manned Lunar Rover Vehicle will greatly increase astronaut mobility, thereby enlarging the scope of lunar exploration. The results will give greater scientific returns from the Apollo missions.

SUMMARY

In the future improved transportation systems will make going to the Moon relatively inexpensive. We will want to put a laboratory into lunar orbit which will be the base of operation for crews of scientists who will visit the lunar surface by means of a shuttle system, bring their samples back to the lunar orbiting laboratory for examination, and return again for samples from another site.

When a reasonably good evaluation of various kinds of surfaces has been obtained, it may be advisable to establish a base on the Moon to continue exploration or exploitation, depending upon what we find.

Within the next decades we should be able to make three vehicles operational. A space shuttle to operate between the Earth and Earth orbit, a nuclear powered shuttle to operate between an Earth orbiting space station and the lunar orbiting laboratory, and another vehicle to fly between the lunar orbiting laboratory and the lunar surface.

Transportation has always been the key to new dimensions in man's progress. With the rocket engined launch vehicles we have taken the first step into space. With the tri-shuttle system, we may be able to explore the entire solar system.

CHAPTER TEN

The Skylab Programme

By L. B. James

The next developmental step in the exploration of space is the establishment of a place for men to live and work in this new territory. Proceeding in an orderly manner from the development in the Apollo Programme of the capability to operate outside the Earth's atmosphere, the Skylab Programme is composed, primarily, of equipment which was developed for the lunar landing programme. Because the testing methods used in Apollo were eminently successful, back-up equipment was not required for many of the planned tests. Therefore, it is now available for the conduct of this follow-on project which will be the first experimental space station (*Figure 10-1*).

Apollo Programme planning and management provided not only against unanticipated setbacks, but also for unexpected opportunities to move forward. It became evident some time before the successful flight of Apollo 11 that all of the launch equipment and spacecraft might not be required to make the lunar landing. Looking toward this eventuality, plans were begun to take full advantage of such a circumstance, should it arise, and the Skylab (then called Apollo Applications) Programme was described. Work has continued and today we are preparing for the first of four Skylab missions in the latter part of 1972.

As we have ventured into space we have begun to learn of some of the marvels which are available there, and to design the ways to make use of them. Some conditions are unique while others can only be duplicated on Earth at great expense and with considerable difficulty. Hard vacuum is attainable on Earth but it is a costly condition to imitate and does not approach that to be found 200 miles above our planet.

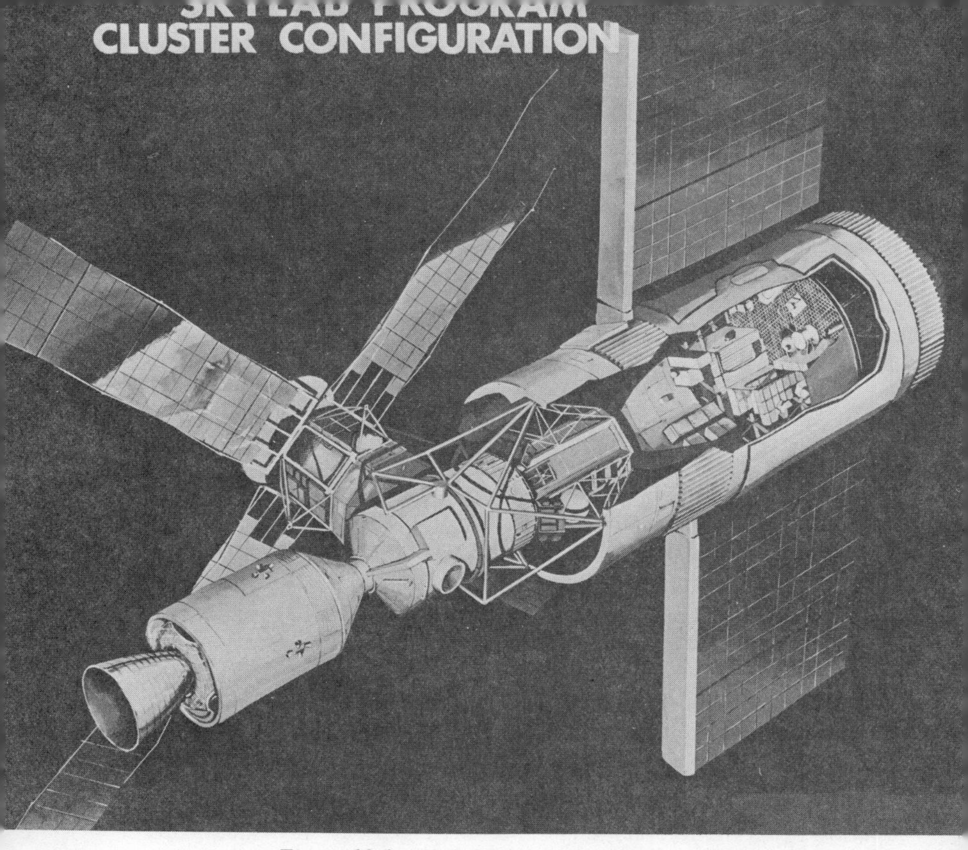

Figure 10-1. Skylab Programme Cluster Configuration.

Weightlessness can be created here, but for only a few seconds in an aircraft performing a Keplerian trajectory. It is also created for a few seconds by means of what the American colonists knew as a "shot-tower". The sterility of space is a goal of laboratories and pharmaceutical manufacturers, but it is an expensive condition to create.

The one unique quality of near-Earth space is its role as a vantage point for looking at this planet. This is a condition which we can not even try to imitate (*Figure 10-2*).

This particular characteristic has already begun to add to our knowledge. For the first time the true shape of the Earth is known—and the science of geodesy has accommodated to this new fact. For some time it has been known that different conditions on Earth, and different substances, give off differing amounts of

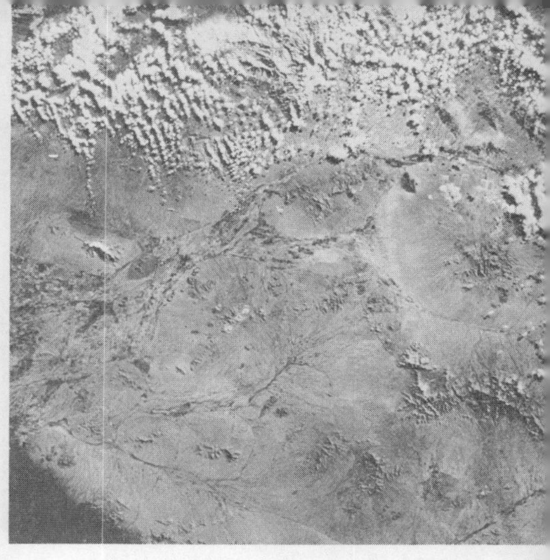

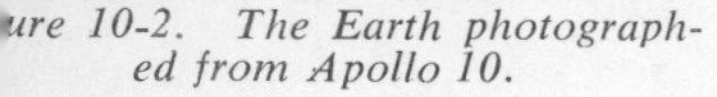

ure 10-2. The Earth photographed from Apollo 10.

Figure 10-3. The Earth.

heat and reflect different parts of light. From an aircraft, using filtered photography, different crops growing in adjacent fields can be identified. The chemical and mineral content of the soil is also detectable from altitude.

Soon after Gemini astronauts began to qualify as the world's greatest photographers, scientists began to look closely at the beautiful pictures which were being returned from space. Not only were these some of the most colourful and exquisitely composed photographs ever taken, they were also revealing facts about the Earth which had not been known before (*Figure 10-3*).

A remarkable finding was that instead of details on Earth being blurred from a distance of hundreds of miles, they were indeed sharper than photos taken from high flying aircraft. The "remarkable eyesight" attributed to some of the astronauts was also a characteristic of the cameras they carried. From a photograph taken from 200 miles above the surface of the Gulf of Mexico, a shrimp fisherman was able to pick out the beds from which his crop came. And he could tell from looking at one space photo why the shrimp had moved from their accustomed place to a new location — water pollution, which was clearly evident in the picture.

Figure 10-4. Photograph of the Western desert of Egypt taken from Gemini.

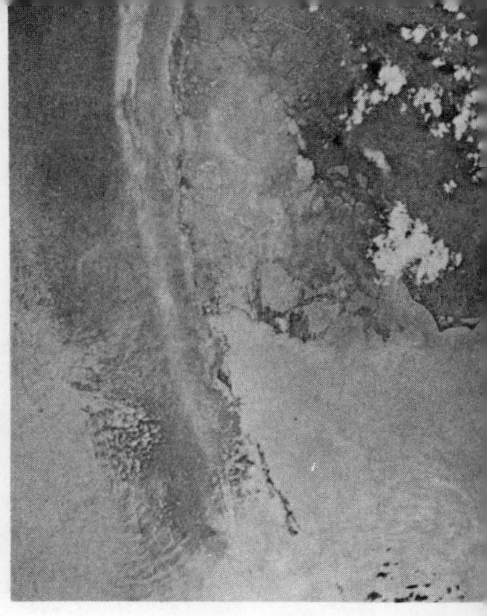

Figure 10-5. The Ocean Floor.

An Egyptian geologist, working at the University of California, saw a picture of the Western desert of Egypt which was taken from Gemini. He knew the area well, having worked there before coming to the U.S. It caught his attention that earth formations indicated three times as large an area of mineral interests as the size of the land already being mined (*Figure 10-4*).

Hydrologists and oceanographers were excited by the space photographs. They were amazed at the clarity which showed the ocean floor as it had never been seen previously. They rightly reasoned that thermal sensors, reporting from space, would reveal the varying temperatures of the waters of the world. Their theories were confirmed on subsequent flights, and thermal maps of the oceans are in preparation. These will not only be of great scientific value, but fishermen will be interested, for fish follow plankton, and plankton follow warm water (*Figure 10-5*).

An earth fault has been seen on a photograph taken from Apollo that shows the same characteristics as those faults which are coincident with rich oil bearing lands. But this fault is in the northeast of Africa, where oil has never been found.

Mineralogists have realized that different minerals give off different amounts of heat, and that if these measures can be identified, prospecting by means of space-carried heat sensors would

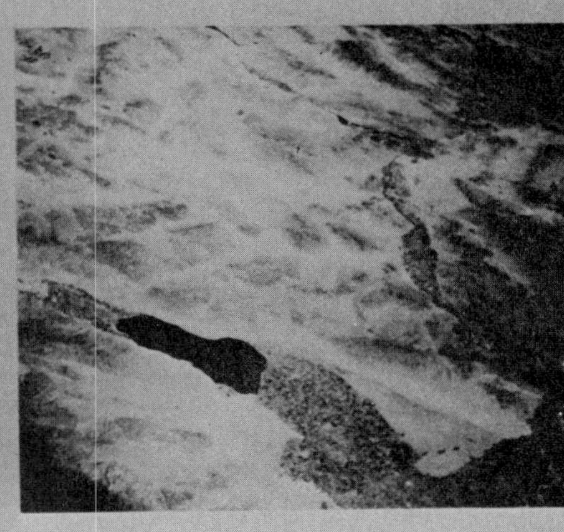

Figure 10-6. Earth Resources Survey.

be a practical possibility. The scientific development of such instruments is under way, and it is hoped that initial equipment will be ready for use on the first Space Workshop in 1972.

It has been estimated by a research organization that a few satellites in synchronous orbit, using multispectral photography, could assess the world's crop of food grains, oats, corn, wheat and rice. This essential information could be used by planners to manage food resources. For the first time it would be possible to prevent famine by the proper management of the world's food (*Figure 10-6*).

Of course, we are all already accustomed to the cloud-cover reports which are relayed from Earth orbiting satellites every day. Most of the countries of the world are making good use of these facilities which are easily accessible to them by means of inexpensive receiving stations. With more sophisticated meteorological equipment now under construction, we can expect to be appraised of forthcoming weather as much as two weeks in advance. Even today, Great Britain reports an annual saving in agriculture alone of $20 million by using cloud-cover predictions.

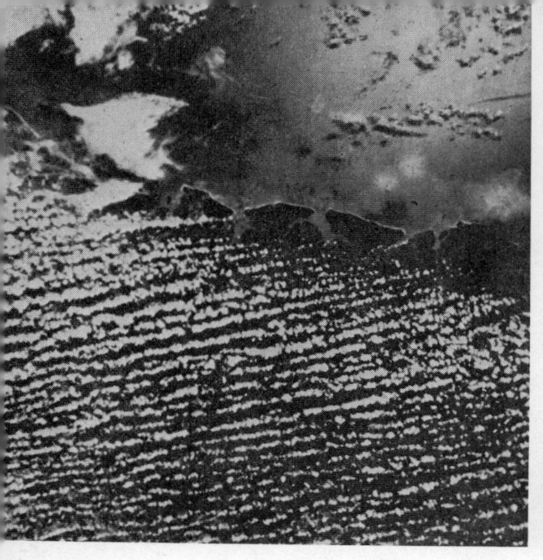

Figure 10-7. Cloud-Cover.

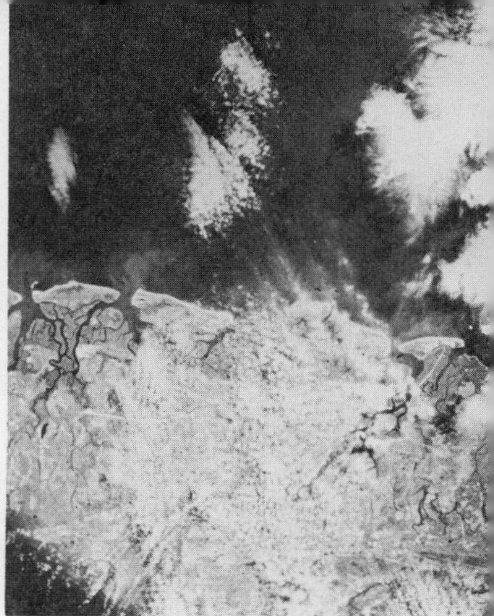

Figure 10-8. Urban Area.

With two weeks in advance weather information, savings would be accrued by farmers, the construction industry, all forms of transportation, resorts, expositions, and the retail business — and school holidays could be arranged for bright sunny days (*Figure 10-7*).

Understanding and predicting the weather are the first steps toward doing something about it. For the first time in man's history this is going to be possible. Throughout the millennia he has provided against the weather — now for the first time he will be able to use knowledge of the weather to his advantage — and, eventually, to control it for his benefit.

Satellites have been making comprehensive maps of clouds, day after day for years, and they have also been making superb maps of the terrain. Some boundaries in South America have been definitized for the first time by means of mosaics of photographs taken from space. Cartographers have estimated huge savings in map making by the use of satellites, for these pictures from space, covering hundreds of thousands of square miles, are considerably cheaper than multiple aircraft flights, and orders of magnitude less costly than ground based measuring techniques. Space photography has, indeed, revolutionized the map making business.

Not only the borders and natural formations of Earth can be clearly seen from space, but most of the man-made characteristics are equally discernible. Smog, for instance, and polluted water are easily distinguished. And urban areas give off more heat than rural ones, as well as showing man's distinctive straight lines of construction. Planners can easily identify urban growth from space photographs (*Figure 10-8*).

Apollo 9 carried the world's first controlled multispectral photography experiment into space. With hand-held cameras, the astronauts took simultaneous pictures with photographs being taken of the same terrain with the same lighting conditions from high and low flying aircraft. In addition, multispectral photographs were obtained through four different filters from synchronized cameras mounted on the spacecraft. Immensely valuable work was done which will be incorporated into the Skylab Programme, and into the unmanned Earth Resources Satellite which is also in preparation.

Using aircraft, it is possible to survey a considerable amount of land on a single flight. However, a good many flights would be required to survey a whole state — or a whole country. From space, it is possible to see almost all of Australia at one time. An early Gemini flight actually took a picture which encompassed most of your vast country. So you can readily see that taking filtered photographs from space can have some significant value (*Figure 10-9*).

Figure 10-9. Australia's Northern Territory as photographed from Gemini.

Figure 10-10. Six Camera Multispectral Photography Experiment.

Research into the equipment and the systems for using our platform in space to measure the crops of the world, to ascertain areas of blight in time to take corrective action, to measure snow cover in order to predict floods, to find the schools of fish which follow warm currents, to ascertain what lies beneath certain kinds of geological faults, to map the inaccessible areas of the Earth — these are among the principal purposes of Skylab. We expect to learn enough about multispectral photography from space and the rapid transmission of information to the Earth so that automated satellites can be put to work on a long-time basis, to orbit the planet every 90 minutes and report their findings back to Earth automatically.

The Skylab Multispectral Photographic experiment will add to our knowledge of how multi-band photography may be effectively applied to Earth sciences. Photographs will be made of selected ground sites using six Itek electric cameras with synchronized shutters. Each camera will use a different film and/or filter combination to allow photographs to be obtained in selected spectral bands of visible and near IR portions of the electromagnetic spectrum. The multispectral photographs will be analyzed by experts studying oceanography, water management, agriculture, forestry, geology, geography and ecology. The orbital path of the Saturn Workshop I will permit Earth resources survey coverage of the entire contiguous United States (*Figure 10-10*).

The use of Skylab to initiate the development of these capabilities can be expected to materially hasten their utility. Testing such systems by the trial and error method of putting unmanned satellites into orbit, waiting for return of information and then producing new equipment, correcting and improving the old, is a time and money consuming process. Skylab will have a set of equipment with some options and modifications for a test programme, perhaps in the breadboard stage so that the crew can carry out some of the essential developmental steps on sensors. In a single mission they can be expected to accomplish what would have required several years and many launches of automated equipment. Thus this programme can not only lessen the cost of equipment development but can also bring Earth sensing activity into practical use at a much earlier date.

The Sun will also be a prime target for observation in the Skylab Programme. The experiments to be conducted by astronaut-astronomers are the most exciting of the more than 50 experiments to be performed. Approximately 32,000 times as much energy as the human race is now using bombards the Earth each year from the Sun. Therefore, it can be to our great advantage to learn as much as possible about the Sun, not only for the scientific value of such fundamental research, but also in the hope that eventually some portion of the Sun's energy may be harnessed for man's use — either on Earth or in space. It was, after all, the fundamental research of scientists at the turn of the century which eventuated in the harnessing of nuclear energy in the service of man.

The Apollo Telescope Mount is being developed in the Skylab Programme to be used by man for performing high resolution studies of solar phenomena. The Apollo Telescope Mount consists of a structural rack, an experiment package containing a number of large solar telescopes and auxiliary equipment and subsystems for providing power, stabilization, communications and other supporting functions. The Mount employs a three-axis stabilization and control system utilizing control moment gyros to maintain the entire Workshop in a solar inertial mode and a vernier gimbal arrangement to maintain experiment orientation to within ± 2.5 arc seconds of any point within a 40 arc minute square centred on the solar disc.

The Apollo Telescope Mount experiment package accommodates five separate experiments for studying solar phenomena (*Figure 10-11*). These astronomical instruments are the largest, most advanced devices ever designed for performing solar research from an orbiting spacecraft. They permit measurements of the Sun's radiation with a combination of high spatial, spectral and time resolution never before achieved. Special emphasis is placed on observation of those portions of the Sun's emissions which are invisible to ground observatories because of apsorption in the

APOLLO TELESCOPE MOUNT SCIENTIFIC EXPERIMENTS

EXPERIMENT NUMBERS	ORGANIZATION	PRINCIPAL INVESTIGATOR	INSTRUMENT	PURPOSE
S052	HIGH ALTITUDE OBSERVATORY	DR. G. NEWKIRK, JR.	WHITE LIGHT CORONAGRAPH	MONITOR THE BRIGHTNESS, FORM AND POLARIZATION OF THE SOLAR CORONA IN WHITE LIGHT.
S082	NAVAL RESEARCH LABORATORY	MR. J. D. PURCELL	CORONAL SPECTROHELIOGRAPH	MAKE HIGH-SPATIAL RESOLUTION MONOCHROMETRIC SOLAR IMAGES IN THE 160-650 ANGSTROM RANGE
			CHROMOSPHERIC SPECTROGRAPH	RECORD SOLAR SPECTRA IN THE 800-3000 ANGSTROM RANGE WITH HIGH SPECTRAL RESOLUTION
S054	AMERICAN SCIENCE AND ENGINEERING CO.	DR. R. GIACCONI	X-RAY SPECTROGRAPHIC TELESCOPE	STUDY SOLAR FLARE EMISSIONS IN THE SOFT X-RAY WAVELENGTHS (2-60 ANGSTROMS)
S055	HARVARD COLLEGE OBSERVATORY	DR. L. GOLDBERG	UV SCANNING POLYCHROMATOR SPECTROHELIOMETER	PHOTOELECTRICALLY RECORD HIGH RESOLUTION SOLAR IMAGES AND STUDY EMISSION SPECTRA OF SELECTED FEATURES OF SOLAR DISC.
S056	GODDARD SPACE FLIGHT CENTER	MR. J. E. MILLIGAN	HI-RESOLUTION X-RAY TELESCOPES	OBTAIN TIME-HISTORIES OF THE DYNAMICS OF THE SOLAR ATMOSPHERE IN X-RAYS IN THE 3-100 ANGSTROM RANGE

Figure 10-11. Apollo Telescope Mount Scientific Experiments.

Earth's atmosphere. With the control, power, thermal and communications subsystems necessary to exploit the full capabilities of these powerful research instruments, this equipment becomes a true solar observatory.

The Apollo Telescope Mount utilizes man on board to perform, in space, activities characteristic of those used in carrying out solar research programmes at ground-based observatories.

On-board displays will permit astronauts to visually scan the Sun to locate targets of high scientific interest. They will assist in the alignment and calibration of the instruments, point them to the appropriate targets, using an on-board television monitor, make judgements of operating modes, and generally conduct a comprehensive programme of solar investigation geared to maximize the scientific return of the mission. The crew will be prepared to perform any possible failure circumvention activities to preserve the Apollo Telescope Mount's scientific success, and will be directly responsible for retrieval of film from cameras by means of extravehicular activity. Thus the primary solar data will be returned to Earth for postflight analysis by the scientists who have designed the experiments.

The majority of the data from the Apollo Telescope Mount experiments will be recorded on photographic film, enabling the collection of large quantities of precision information. The advantages attendant to the utilization of film exploit the full high resolution potential of the large instruments (*Figure 10-12*).

To complement the photographic mode, the Apollo Telescope Mount communication systems include high transmission rate telemetry channels which provide important data to the ground in real time. A television system permits ground observers to monitor periodically the same displays of solar information available to the astronauts. Voice channels will insure good communication between the Apollo Telescope Mount crew and the scientists on the ground, and a teleprinter will provide an additional communications link from the ground up to the Skylab (*Figure 10-13*).

The Apollo Telescope Mount experiment payload consists of two X-ray telescopes, an extreme ultraviolet spectroheliograph, an extreme ultraviolet spectograph, an extreme ultraviolet scanning spectrometer, and a white light coronagraph.

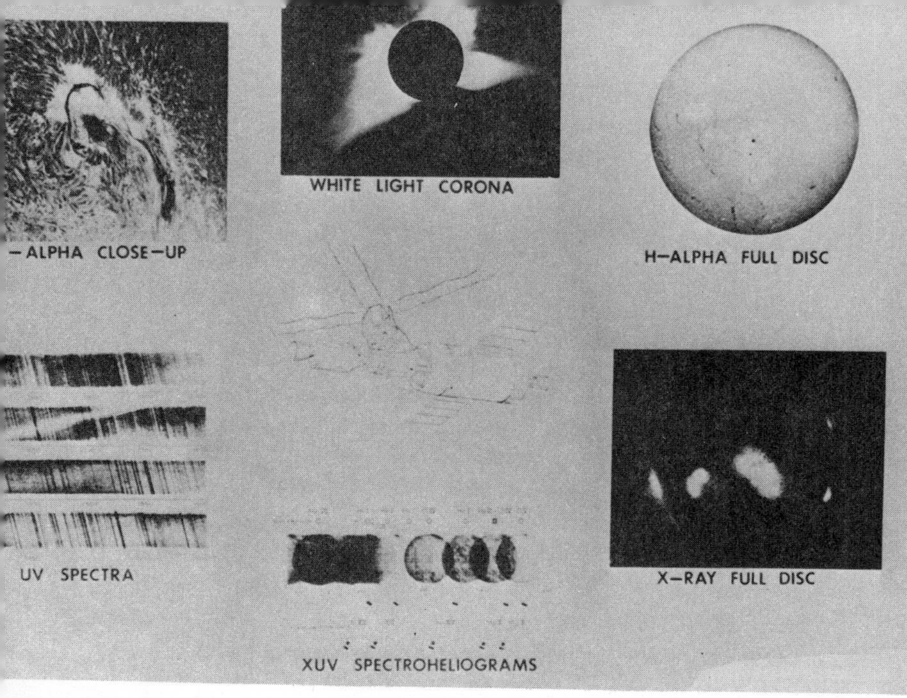

Figure 10-12. Data from the Apollo Telescope Mount Experiments.

Figure 10-13. Apollo Telescope Mount. Neutral Buoyancy Testing.

The spectrographic grazing incidence X-ray telescope will obtain photographs of the time variation of solar activity-related X-ray emissions in the 2 to 8 and 44 to 60 angstrom range with 2 arc-second spatial resolution and spectral resolution of a fraction of an angstrom.

X-ray telescopes, using similar focusing techniques, will photograph the solar X-ray distribution over the complete solar disc and near corona in the 0 to 33 angstrom region. Both X-ray experiments provide a complementary and comprehensive investigation of the results of extremely high temperature ionization of the elements present at considerable heights above the Sun's photosphere.

An XUV spectroheliograph in the 150-650 angstrom range is designed to photograph images of the solar disc with good spatial and spectral resolution. Another instrument, an XUV spectrograph, will photographically record line spectrograms of the solar radiation between 900 and 3900 angstroms from small selected areas on the disc and across the limb into the corona.

A scanning UV spectrometer will obtain photoelectric scans in the 300-1300 angstrom range of 5 $\times$ 5 arc minute selected areas on the disc with good resolution. Though overlapping in spectral range, some instruments perform complementary measurements with one instrument yielding precise quantitative intensity information with only moderate spectral resolution and another providing qualitative intensity measurements with very high spectral resolution.

The white light coronagraph will photograph the brightness, form and polarization of the solar corona between 1.5 and 6 solar radii from the Sun's centre. Measurements somewhat comparable to these can be made from the ground only during the rare and extremely brief total solar eclipse transits.

In addition to the five basic experiments, an XUV disc monitor, an X-ray monitor, a corona display, and two H-Alpha telescopes provide a display of the Sun to the astronauts, permitting them to select the most interesting area for study and to point the appropriate instruments accordingly.

Two related unmanned scientific instruments were launched on Aerobee Rockets from the White Sands Missile Range, November 4, 1969, as part of the Apollo Telescope Mount Programme to verify design and evaluate performance characteristics. One rocket carried a one-half scale model of the spectroheliograph intended for flight on the Apollo Telescope Mount, and the other carried an instrument similar in many respects to the solar X-ray telescope. These instruments were launched during a Class 1 bright flare with excellent operational and scientific results. The information obtained on solar emission intensity, filter performance, film response, exposure time, etc., has been fed directly back into the development of the Apollo Telescope Mount instruments, which are now in the prototype phase. The Apollo Telescope Mount programme was able to achieve highly successful results from the simultaneously launched rocket flights. The solar X-ray telescope, for the first time, obtained a cinemagraphic record of the early phase of a solar flare. The extreme ultraviolet spectroheliograph, for the first time, obtained high quality filmed XUV spectroheliograms of the same event. A TV camera carried aboard one rocket provided ground observers with the first real-time views ever obtained of the Sun in the XUV wavelength band. These development flights are just the prelude to the complete and detailed investigation of the Sun and its physical processes which the Apollo Telescope Mount programme will pursue, beginning in 1972.

Among the 50-odd experiments to be performed on board Skylab are some which are of very special interest. Included are three examinations of circadian rhythm. One is the study of the parameters of the circadian rhythm of a sprouting potato tuber in Earth orbit. These measurements will be compared with identical measurements made on similar samples in laboratories on Earth. In order to arrive at these facts the potato tuber will be maintained in a controlled environment and the oxygen consumption and temperature changes will be monitored. Pressure and temperature will be sampled periodically over the 28-day period and results will be returned for comparison with the control sample on Earth.

Also in this basic study is one concerning the "temperature compensation" in the circadian rhythm of an insect. For this purpose, the rate of hatching of the vinegar gnat will be measured at different temperatures in space. The pupae of the gnats will be hermetically sealed. Measurements will be compared with those made at sea level.

Pocket mice will also be used in corresponding study of the same phenomena. The stability of the circadian rhythm of a mammalian system under space flight conditions will be studied by observing the body temperature, heart rate, and animal movement of six pocket mice, housed separately in individual cages in total darkness. The atmosphere will be supplied and controlled to simulate an Earth atmosphere.

The conduct of these three experiments in the Skylab Programme is expected to shed some light on this rather mysterious phenomenon of inhabitants of the planet Earth, which makes them subject to a 24-hour cycle.

Naturally, the most sophisticated study of this subject will result from the astronauts' observations of themselves and of each other. Biomedical data, either accumulated in the Workshop or telemetered to the Manned Spacecraft Centre at Houston will add to the information.

As each flight in the Apollo Programme was designed to be more complex than the last, and to perform experiments which would be important to the conduct of later missions, so, each of the missions in the Skylab Programme will perform tests of systems and equipment which may be useful on a later flight. In addition to the study of the physiological and psychological reactions of the astronauts, and the evaluation of present equipment, some experiments will check out experimental equipment which is designed for later use.

A foot-controlled manoeuvring unit which may be a valuable tool for extravehicular activity will be examined in the weightlessness of the Workshop. The principal advantage of such a mechanism is that it would allow the crews to move about in space with their hands free to perform work tasks. Another similar test will evaluate the usefulness of a hand-held manoeuvring unit similar to the backpack type used in Gemini. Both of these pieces

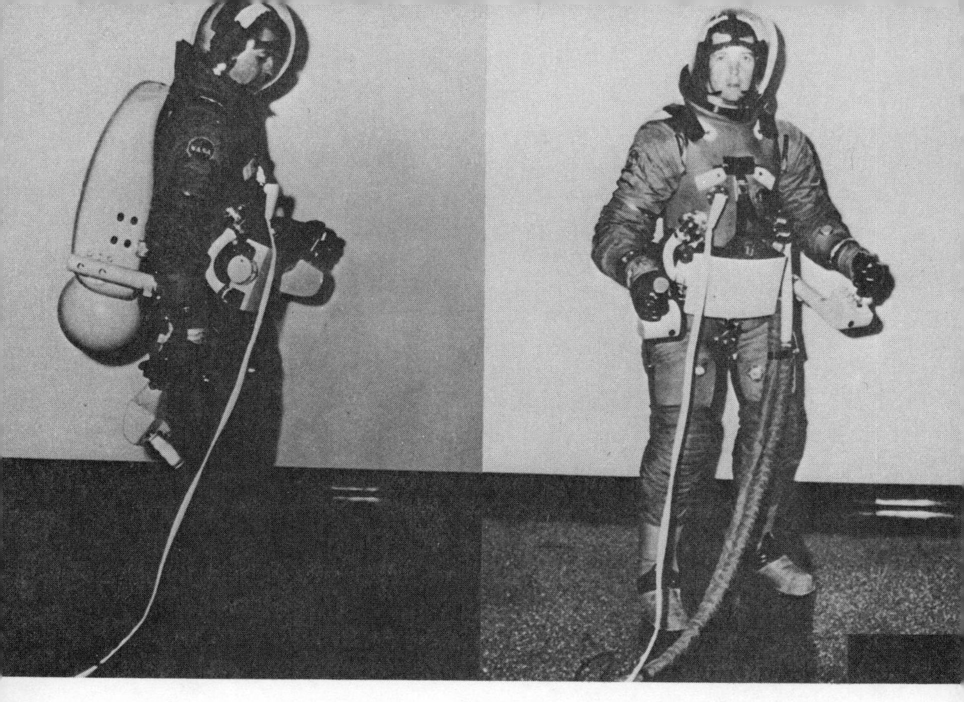

Figure 10-14. Astronaut Manoeuvring Equipment Experiment.

of equipment will be thoroughly checked out inside the Workshop which has adequate volume for this type of testing. It either is successful, or can be modified, it may well be used on later space missions (*Figure 10-14*).

One of the fundamental researches which will be conducted concerns the effects of zero gravity on the flammability of non-metallic materials in the spacecraft environment. Under closely controlled conditions, within an enclosed chamber with a vent to space, information will be obtained on the ignition of various non-metallic materials, flame propagation and extinguishment characteristics in order to build a basic understanding of the behaviour of fire in weightlessness.

To evaluate some of the exciting theories concerning the behaviour of materials in the space environment, experiments will be carried out which may eventuate in perfect ball bearings, large perfect crystals and exotic material compositions. Learning how

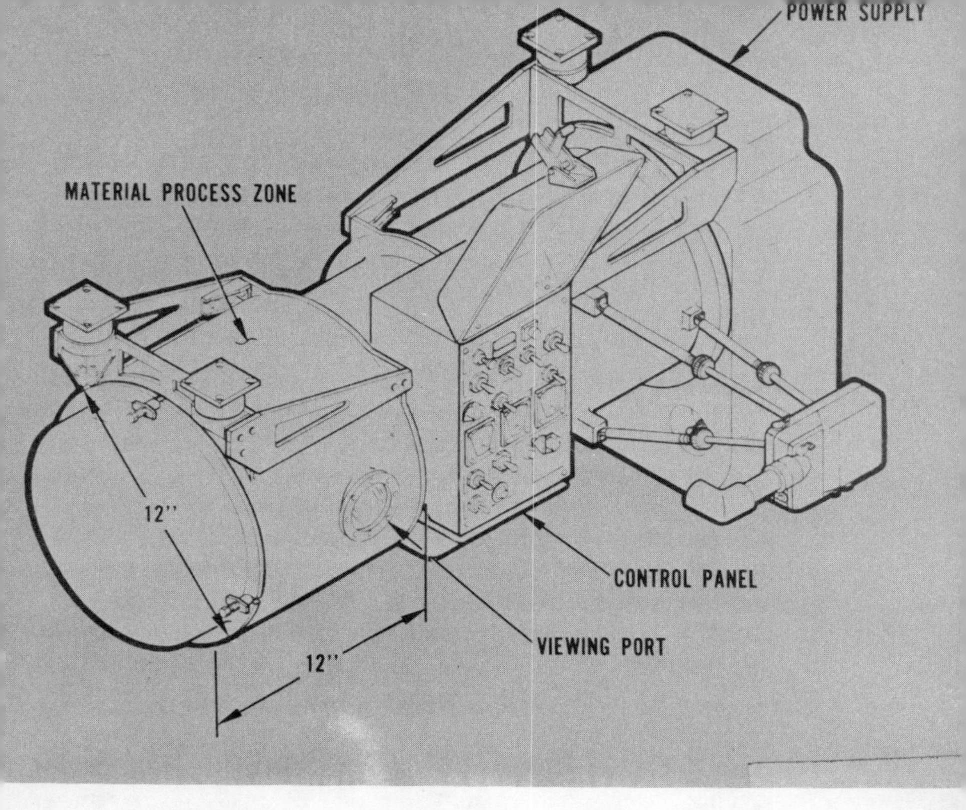

Figure 10-15. Manufacturing Process Chamber.

to produce such things will involve five separate tasks (*Figure 10-15*).

1. **Metals Melting.** Metal samples of various alloys of varying thickness will be melted and returned to Earth for examination.
2. **Spherical Casting.** Several specimens of a representative material will be melted and allowed to solidify.
3. **Exothermic Heating.** Several specimens of stainless steel tubing will be joined using a braze alloy. The heat source for the process will be exothermic material.
4. **Fibre Whisker Reinforced Composites.** Several specimens of a powdered composite material with fibre whiskers implaced in the material will be melted and allowed to solidify.
5. **Growth of Single Crystal.** A sample representative material will be melted, seeded and allowed to grow into a single crystal.

All of the products of these experiments will be returned to Earth for examination.

All of these tasks will be performed in a materials melting facility which will include an electron beam welding gun, a power supply and a control panel. Provisions will be made for astronaut access, for removing and replacing the experiment specimens, for viewing ports to permit visual observation and data recording by cameras, and for connecting the chamber to a vacuum port.

The heat source for the metals melting and spherical casting tasks will be the electron beam gun. The heat source for the exothermic heating and the fibre whisker reinforced composites will be exothermic material. The heat source for the growth of single crystals will be electric heaters.

The Skylab Programme consists of three missions, the first crew of three will spend 28 days in the Workshop, the second three-man crew will be scheduled for 56 days as will the third crew. On these missions we hope to establish man's long duration tolerance to the space environment by means of the extensive medical and habitability experiments which are planned.

The first step in the programme is placing the Skylab in Earth orbit. The largest module of this cluster is the Orbital Workshop, containing living quarters and experiment areas. This central structure is derived from a third stage of the Saturn V launch vehicle completely modified on the ground. Within its 10,000 cubic-foot fuel tank, a three-man crew of scientists-astronauts will have a bedroom with three separate sleeping compartments, a food preparation and recreation area, a bathroom, and a large area in which to conduct more than 50 experiments. The total available living and working space will be equivalent to a small three-bedroom house—vastly larger than the restricted cabins of present-day spacecraft (*Figure 10-16*).

The nerve centre of the orbital cluster is the cylinderically shaped Airlock Module attached to the forward end of the Workshop. It contains the communications, environmental control and electrical power systems required to operate the entire cluster and to provide a shirt-sleeve environment for the crew. The airlock also provides the hatch for extravehicular activities, such as retrieving film from the ATM, which precludes the need to depressurize the entire cluster. Though it is a new structure, the

-IVB-212 STAGE BEING REMOVED FROM STORAGE PRIOR
O MODIFICATION TO NEW WORKSHOP CONFIGURATION

Figure 10-16. Saturn Workshop 1.

Figure 10-17. Multiple Docking Adaptor.

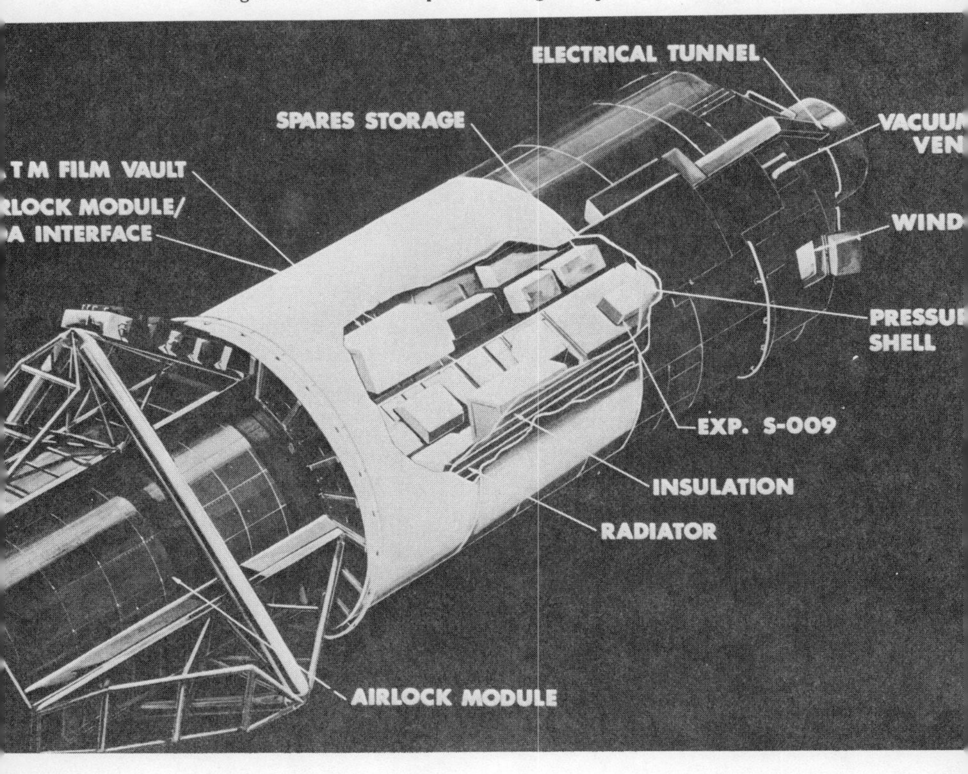

Airlock Module utiiizes many components developed in the Gemini Programme.

Attached to the Airlock is the larger cylindrically shaped Multiple Docking Adaptor. This unit, as its name implies, provides for leakproof docking of the spacecraft to the Cluster, so that the crew can pass back and forth between the two without any need for depressurization. The Multiple Docking Adaptor also houses other vital equipment such the control and display panel for the Apollo Telescope Mount (*Figure 10-17*).

The structure of the Apollo Telescope Mount was described in some detail earlier.

It is presently planned to launch Skylab (Orbital Workshop, Airlock Module, Multiple Docking Adaptor and Apollo Telescope Mount) into a 235 nautical mile orbit of 50° inclination to the equator using the first two stages of the Saturn V as the launch vehicle. This will permit viewing all of the United States and will also cover most of the heaviest areas of population of the world. Because of the inclination, Skylab will also have a view of Australia.

After the Workshop is inserted into orbit, the solar panels, which will provide on-board power, will automatically deploy. The Telescope Mount will position itself for operation which requires that it swing around to face the Sun. The solar panels to power the Telescope will then deploy and the system will become stabilized in the solar inertial mode, so that its instruments will face the Sun intermittently throughout the life of the Skylab missions.

Making use of a Command and Service Module prepared for the Apollo Programme, the crew will be launched from Cape Kennedy by a Saturn IB launch vehicle a day after the Workshop has been put into orbit. In the Command Module, the three-man crew will rendezvous and dock with the Workshop, using the multiple docking adaptor for this purpose. After all equipment and materials have been checked for readiness, the crew will enter the Workshop for their 28-day stay. During this period the Command and Service Module will be powered-down to await the entrance of the crew and return them to Earth. The Apollo

Figure 10-18. Airlock Structural Test Article.

Command Module is now undergoing some minor modifications to fit it for its role in the Skylab Programme. Principal among these is the addition of a back-up deorbit capability for the return to the surface of the Earth (*Figure 10-18*).

The investigations of man's reactions to long-duration space flight are a primary objective of the Skylab Programme. There will be a number of medical questions to be answered, including the measurement of physiological responses to measured amounts of work, examination of eating and sleeping habits and cycles, and the assessment of mineral balance. The mineral investigation will be done by carefully measuring the mineral intake of the astronauts, and collecting all waste material from them for return to Earth for analysis. In this manner we expect to evaluate the effect of zero gravity on such things as the calcium content of the bone structure.

A cardiovascular study involves the use of a lower body negative pressure, pre-flight, in-flight, and post-flight. A device tests the

Figure 10-19. In-Flight Lower Body Negative Pressure Experiment.

cardiovascular system reflexes which normally operate to regulate regional perfusion pressure in distribution of blood throughout the body as man changes his posture on Earth. This is a vitally important measurement of cardiovascular system response. The in-flight measurement will allow us, for the first time, to establish the onset, the rate of progression, and the severity of adverse functional changes in the protective reflex responses. This procedure requires a medically trained observer as a member of the flight crew. The cardiovascular investigation also includes obtaining a vectorcardiogram during given workloads on a bicycle ergometer (*Figure 10-19*).

Investigations in the area of haematology and immunology are directed toward determining the effects of space flight on the physiology of the formed blood elements, the fluid compartments of the body, the haemostatic mechanism, selected aspects of humoral and cellular immunity with reference to microfloral alterations, and

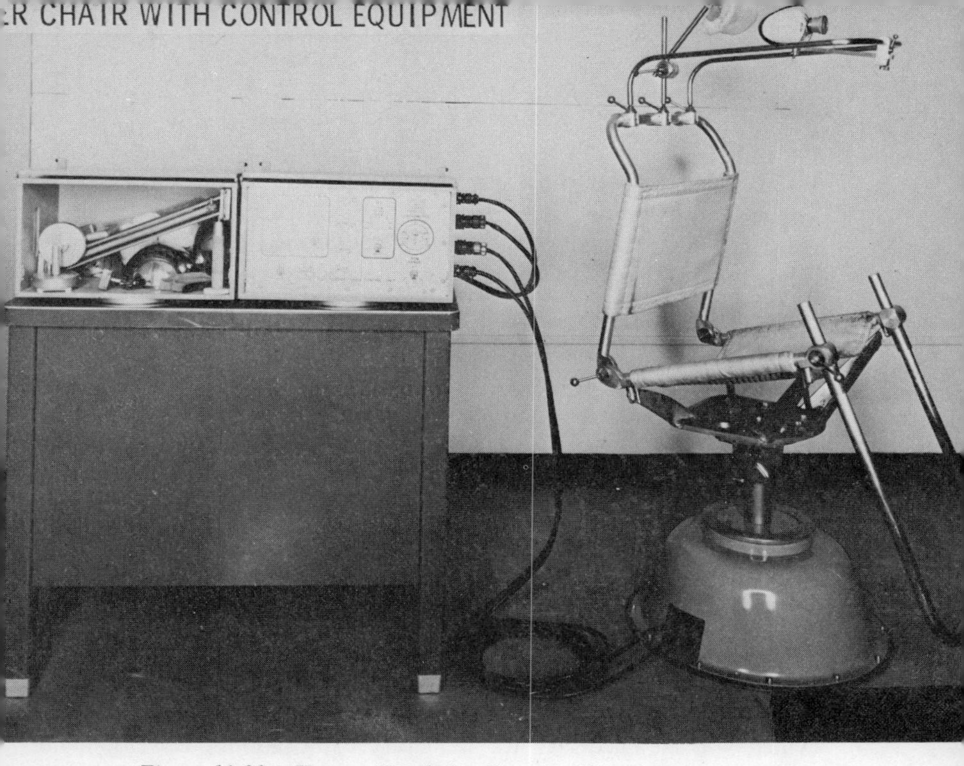

Figure 10-20. Human Vestibular Function Experiment.

the frequency of leukocytic chromosomal aberrations. Four pre- and post-flight experiments will be conducted in this general area. Blood will not be drawn from astronauts in-flight.

In the area of neurophysiology, two investigations are planned to evaluate central nervous system responses, and changes, if any, as a function of space flight. The first, a human vestibular experiment, is an extension of studies initiated during the Gemini Programme. The objectives are to investigate the effects of weightlessness and sub-gravity states on the perception and the organization of personal and extrapersonal space and to establish the integrity of the vestibular apparatus during prolonged weightlessness (*Figure 10-20*).

For the first time in space, energy expenditure will be measured by comparing metabolic rate of the astronauts during rest, during calibrated exercise using a bicycle ergometer, and while performing operational-type tasks.

To provide supporting and ancillary information to these experiments, we plan on being able to "weigh" men and materials in zero g. This "weighing" will be accomplished by the use of two mass measurement devices with ranges appropriate for small masses and for man.

An experiment support system is being developed to provide electronic, fluid, and electrical, and in specific instances, mechanical interfaces with the medical experiments.

Through the minute observation and constant surveillance of three well men, working under highly unusual circumstances and in a strange environment for one period of 28 days and two periods of 56 days each, we expect to learn a great deal about the actions and reactions of well people.

Habitability studies will be of major importance on all of these missions and are expected to develop the techniques and the equipment necessary for long duration stay in the large space stations which we envision for the future, and for the long journeys which will be necessary for planetary exploration. Developments now under way will result in equipment and procedures for bathing, for preparing foods, for exercising and for sleeping.

The large working area of the Workshop is divided by a floor, making two stories. Because of the lack of gravity there is no up or down, so astronauts will move about upon the "floor" or the "ceiling" with equal ease. Many of the experiments, as well as all of those concerned with intravehicular activity, will be conducted in these areas.

At the conclusion of the 28-day period, the first crew will enter the Command and Service Module, which has been in a quiescent state, docked to the Workshop. They will have prepared the Workshop to remain in orbit with systems powered-down, ready to be reactivated by the next crew. The Command and Service Module will then separate from the docking adaptor and return to Earth for recovery in the Atlantic.

The second in the series of Skylab flights is currently scheduled to be launched approximately three months after the launch of the first mission. Three crewmen, again using a modified Apollo

Command and Service Module and a Saturn IB booster, will be launched into orbit to rendezvous and dock with the Workshop which will have remained in a standby condition.

This crew may consist of one astronaut who is an astronomer, and one who is a medical doctor as well as a third astronaut who may have some other scientific speciality.

Their mission will be planned for a period of 56 days. However, we must bear in mind that all of these missions are open-ended, in that the crews can return at any time. It is possible that discoveries will be made of such overwhelming importance that it will be decided to bring the crew back early—or in the event of some anomaly, the same decision may be taken. However, it is hoped that the 56-day missions will go far toward establishing man's tolerance for the space environment as well as producing valuable scientific results.

Experiments begun on the previous mission will be continued, and others, more complex, will be started. Two film retrievals from the Telescope, using extravehicular activity, will be made, and a longer and more detailed examination of the Sun will be undertaken. Preliminary results from the photographs and measurements made on the previous mission will contribute to the work which will be scheduled for this flight, not only on the Apollo Telescope Mount, but also in medical, habitability and other scientific studies.

One month after the return and recovery of the crew of this flight, a third crew will enter the modified Apollo Command and Service Module and be launched aboard a Saturn IB into orbit to rendezvous and dock with the quiescent Workshop. Again, the medical, habitability and solar observation work will take place, building upon what will have been learned in the two previous missions.

Another experiment of considerable interest is the examination of the feasibility of a "gravity substitute" bench. In space there is no position reference, no down or up, and, while in theory materials would stay in one position, practically they do not. Therefore we are studying two different methods for holding

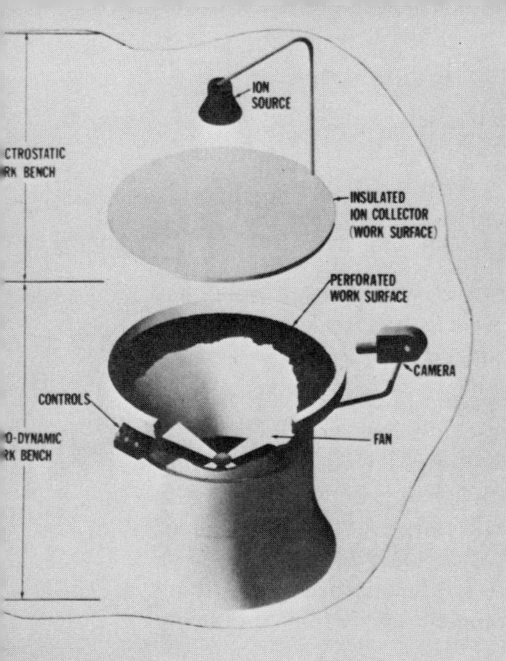

PURPOSE

TO EXPLORE AND ASSESS THE MERITS OF ALTERN
SOLUTIONS AND AIDS FOR GRAVITY SUBSTITUTION
APPLIED TO A WORK STATION.

SIGNIFICANCE

MAKE A POSITIVE CONTRIBUTION TO MAN'S BASIC
ABILITY IN SOLVING PROBLEMS ENCOUNTERED IN
FLIGHT; THE DEGRADATION OF PERFORMANCES B
EFFECTS OF ZERO GRAVITY.

DESCRIPTION

TWO MODES OF GRAVITY SUBSTITUTION ARE TO BE
UATED AND DATA FROM EACH MODE WILL BE IN T
OF FILM AND TAPED COMMENTS OF THE ASTRONA
DISASSEMBLES SELECTED COMPONENTS FROM THE
TASK BOARD.

DURING THE AERO–DYNAMIC MODE, THE FAN WILL
AN AERO–DYNAMIC FORCE FIELD AGAINST THE "
WORK TABLE. THE INSULATED ION COLLECTOR W
INSTALLED IN THE SAME FRAME AS THE AERO–DY
WORK BENCH. AN ELECTROSTATIC FORCE FIELD
CREATED AGAINST THE ION COLLECTOR BY THE I
SOURCE.

P. I. – Mr. John B. Rendall, MSFC

Figure 10-21. Gravity Substitute Workbench Experiment.

materials to a bench or work surface. One method holds materials to a surface by means of an airflow which uses suction. The other uses an electrostatic charge to maintain attraction between materials, thus holding an object in a stationary position (*Figure 10-21*).

The Skylab Programme is the precursor of large space stations where men will live and work in space. We know that we will learn enough from the Skylab flights to build a larger and more practical space station which, in the future, may allow us to have a continuously manned station in Earth orbit. Later missions will begin with crews which will remain in orbit for 120 days. They will be joined by the replacement crew, then returned to Earth. When sufficient medical information is in hand, we look forward to having crews in orbit for extended periods, possibly for more than a year.

Design, development and ground testing of subsystems and complete assemblies of the Workshop, Airlock Module, Multiple Docking Adaptor and Apollo Telescope Mount are all well advanced. Fabrication of the first flight article, Workshop I, has recently begun. The flight systems are scheduled to be delivered to Cape Kennedy in the latter part of 1971 and the first launch will take place toward the end of 1972.

While more than 50 experiments have been assigned to this programme, they represent only a fraction of those which scientists in many disciplines would like to perform in space. However, engineering, in this case, is the hand-maiden of science and the vital scientific investigations which will take place in space before the end of this century will have to await the establishment of a large space station which will be served by the space shuttle.

Many of the careers which you choose will involve work in the space stations of the future, for which the Skylab Programme is an initial test platform.

CHAPTER ELEVEN

The Highroad to Space

By L. B. James

Every new territory, each new area for exploration, has been opened by some new form of transportation. Just as the western territories of America and Australia were developed after rail transportation was available and Antarctica is yielding its mysteries to airborne investigators, the new environments beyond Earth's atmosphere will prove their usefulness to man when economical round-trip transportation is operational.

Our exploration and use of space has been, so far, constrained by the high cost of putting equipment into orbit, and its inaccessibility once it has been launched. Today, however, as we move into the 1970s, we are developing multiple programmes with emphasis on economy and on additional uses of space technology for the benefit of man.

One of these programmes now under study is the space shuttle, a reusable vehicle that will offer many economies and benefits. The shuttle will be available for use by more than one government agency or one space programme. Its development will stimulate both space and aeronautical technology. It will reduce the cost of payloads by allowing retrieval or repair of satellites in orbit and the transportation of cargo and passengers to and from orbit. It will have a quick response time and significant space rescue capability. Its design for reusability will provide for 100 or more flights (*Figure 11-1*).

Another programme also under study is the Modular Space Station. In Earth orbit, a space station will provide additional economic gains and practical benefits. The space station will reduce operating costs by its long life and its flexibility, combining

Figure 11-1.
Two Stage Fully
Reusable Concept.

many operations such as research, applications and support of space flight operations. It will be designed so people on board will be able to carry out their technical tasks without special flight training. The space station modules may be used in various Earth orbits and, ultimately, in lunar orbit or on a planetary mission.

The operation of the space shuttle and the space station will permit a considerable expansion in the scope of space activities, and a steady increase in the number of visitors into space. The expanded, more economical flight activities made possible by these advanced systems will conceivably open space to a broad range of public and private interests.

The space shuttle which we envision will be extrapolated from many concepts, combined with our latest knowledge of rocket engines, materials and technology. Our nine years of operation in the space environment will contribute significantly to our ability to produce an operational vehicle in a reasonably short period of time. Actually, we know a great deal more about how to build the space shuttle than we knew about how to build the Saturn V in 1961. Studies have been progressing for several years.

The characteristics of this machine will blend aircraft and rockets to achieve the characteristics which will make space operations economically practical.

The other essential element in space exploration and applications is a place to work. To fill this need, studies are under way for the definition of a space station. Because of the economics inherent in the space shuttle, the initiation of operation of the space station may depend to some degree upon the readiness of the shuttle. However, work is in progress on both of these new designs so that advantage may be taken of those breakthroughs in technology which can be expected to occur in research and development programmes. For we have learned from our Apollo experience that planning must provide not only for the unexpected mishap, but also for the unanticipated success which permits a flexible programme to make rapid progress.

An operational space shuttle will permit the inexpensive logistic servicing of a space station in orbit. Regular trips of the shuttle to the station would not only transport crews and supplies, but would also be used to return processed experiments to Earth and to replace or repair space station equipment.

Weather satellites, already in general use, have been costly to build, as have all unmanned satellites, since they have had to operate dependably without maintenance throughout their lifetimes. With a space shuttle in operation, it would be possible to perform on-orbit repairs, or, if the configuration and orbit were compatible, to carry the satellite back to Earth for repair or refurbishment. Thus two areas of high cost in unmanned satellites will be attacked by the shuttle: (1) the original high cost of construction, and (2) the waste occasioned by the failure of an unmanned space installation.

The space shuttle is expected to reduce the cost of all space activity, at the same time it will greatly increase our ability to operate effectively in space.

Those elements in the basic design philosophy of the space shuttle to which we are looking for cost reductions are aircraft manufacturing techniques, aircraft development test procedures, maximum flexibility for multiple use and volume production, long life components for repetitive reuse and airline type maintenance and handling procedures for economy of operation.

A major point of difference between the space shuttle and aircraft will be in the manner of launching this craft, for it will be powered by rocket engines and take off vertically from a launching pad. We expect that some day these could be located at many major airports since the cleared distance around the vehicle will need only to be in the order of a mile in radius. Many of the present jet ports could be used, especially those which have water on one or more sides (*Figure 11-2*).

A flight crew will fly the shuttle into Earth orbit to dock with

Figure 11-2. Space Shuttle Launch.

the space station. It will deliver and receive payload, perform its other functions which may include satellite maintenance and repair or space rescue and return through the atmosphere for a horizontal landing at its take-off point on a runway which would not need to be over 10,000 feet long. This flight pattern creates a different noise situation than that of present aircraft. There is, of course, the huge sound of the blast-off which lasts for a few seconds but is localized to the airport itself. The next sound is generated as the shuttle comes back into the atmosphere but this is hardly heard on Earth as the transition to subsonic speed occurs at an altitude of about 100,000 feet. As the craft approaches Earth, its speed is subsonic, and its landing sounds will probably be less than those of large jets.

As presently conceived, the space shuttle will consist of an orbiter vehicle containing crew, passenger and cargo accommoda-tions as well as power and fuel for orbital and landing phases, and one or more booster elements which will carry the bulk of the fuel necessary to achieve orbit. The boosters will also be manned and powered for return to Earth and horizontal airport landing Thus, the total vehicle is reusable. The vehicle is being designed to carry a flight crew and cargo and/or passengers into and out of orbit, with a 10,000 cubic foot internal payload volume. The operational altitude will range from 100 nautical miles up to approximately 800 nautical miles.

The design performance and operational characteristics of the shuttle will permit its use in orbits ranging from low inclination to polar and perhaps even Sun-synchronous (*Figure 11-3*).

The design parameters require operational flexibility to enable this craft to take off in any direction.

Designing the internal compartment to accommodate con-tainerized cargo will assure that a wide variety of payloads can be carried.

The vehicle configuration will provide crew and passenger safet comparable to that of current commercial jets. Extremely reliabl systems will operate redundantly where feasible; non-redundan systems will be operated well within established operational limits

SPACE SHUTTLE
DESIGN REFERENCE CHARACTERISTICS*

SYSTEMS

- TWO STAGE FULLY REUSABLE
- VERTICAL TAKEOFF/HORIZONTAL LANDING
- 3.5M LBS GROSS LIFT-OFF WEIGHT
- CARGO BAY SIZE 15' DIA X 60' LG
- MAXIMUM ATTAINABLE PAYLOAD

ENGINES

- HIGH PRESSURE BELL TYPE HYDROGEN/OXYGEN
- THROTTLEABLE
- 400K LBS THRUST PER ENGINE
- HYDROGEN FOR JET FUEL

OPERATIONAL

- REENTRY FOR INITIAL FLIGHTS AT LOW L/D FOR MINIMUM HEAT LOAD
- ON-ORBIT ΔV OF 1500 FPS
- SEQUENTIAL IGNITION OF STAGES
- NOMINAL 270 NMI ORBIT AT 55° INCLINATION
- TWO MAN CREW
- INTACT ABORT
- ORBITER SELF SUSTAINING FOR 7 DAYS

* APPLICABLE TO HIGH AND LOW CROSSRANGE DESIGNS

Figure 11-3. Space Shuttle Design Reference Characteristics.

This approach should allow for the graceful degradation of any system, precluding catastrophic failure and allowing time for a safe return to a landing site of passengers, crew and vehicle.

Two representative artist's concepts are similar in terms of size, performance and ascent and on-orbit operational modes. The booster element is "staged" at altitudes on the order of 200,000 feet and velocities of approximately 10,000 feet per second. The booster then descends and, at the proper conditions, jet engines are started for the approximately 200 nautical miles subsonic cruise back to the launch/landing site. The orbiter stage continues on to orbit to complete the mission. The fundamental difference between the two concepts is the design of the orbiter elements.

The first concept has fixed wings and is designed for re-entry at a high angle of attack. This mode of re-entry results in a significant deceleration at higher altitudes and thereby shortens the duration of the heat pulse. Therefore, by experiencing reduced heat inputs the thermal protection system design requirements may be reduced for the fixed wing configuration. On the other hand re-entry at a high angle of attack does not take advantage of the capability for achieving a large crossrange by manoeuvring at hypersonic speeds (*Figure 11-4*).

Figure 11-4. Fixed Wing Space Shuttle.

The second concept is a delta shaped configuration with a high sweep angle. This configuration may re-enter at lower angles of attack, and has a higher hypersonic lift to drag ratio and therefore a significantly higher crossrange capability than the previously mentioned fixed wing concept. Configurations re-entering at low angles of attack experience a somewhat more severe thermal environment which affects the design requirements for the thermal protection system (*Figure 11-5*).

These two concepts and their associated development and operational advantages and disadvantages are representative of the kinds of trade-offs that are being made in our continuing study efforts. There are a number of problems of a similar type that are currently being evaluated.

As is the case with most launch systems for space vehicles the pacing item is the new engine. The rocket engines for both the booster and orbiter elements will be throttleable high performance

Figure 11-5. Delta Shaped Wing Space Shuttle.

Figure 11-6. Space Shuttle Typical Mission Profile.

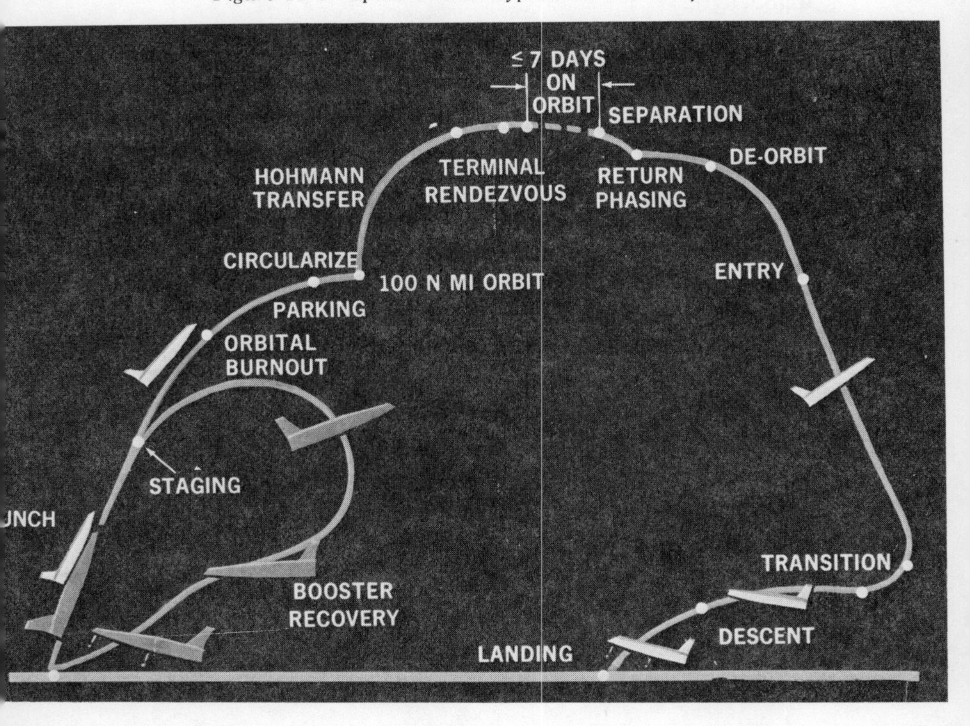

hydrogen/oxygen engines. Each engine will have a thrust of approximately 400,000 pounds. Depending on the design, the orbiter may use two or three engines while the booster may require 10 or more.

A typical mission profile for a space shuttle performing a space station logistics mission is shown in Figure 7. The first stage or booster accelerates the second stage or orbiter to a velocity of eight to 10 thousand feet per second and an altitude of approximately 200,000 feet. Staging occurs at this point in the trajectory. The booster coasts to a maximum altitude of about 230,000 feet and then begins descending. At approximately 250 nautical miles downrange and an altitude of about 50,000 feet the booster completes a 180 degree turn and cruises back to the launch site on jet engines (*Figure 11-6*).

The orbiter stage ignites at staging and continues on to the initial insertion orbit. Following a brief coast the orbiter circularizes in a nominal 100 nautical mile parking orbit. At the proper time a transfer is made to the space station altitude with a subsequent rendezvous manoeuvre. Following a period on-orbit, during which time passengers and cargo are transferred, the space shuttle separates from the station and prepares for the de-orbit manoeuvre.

After waiting for the phasing conditions to allow the proper relationship of the orbit with the landing site, the de-orbit manoeuvre is performed. The shuttle re-enters the atmosphere at a high angle of attack to achieve maximum deceleration at high altitudes and minimize the exposed surfaces to high temperature. Following the re-entry pullout and glide manoeuvres are performed. Transition occurs at an altitude of about 40,000 feet. The shuttle then lines up for the approach and lands at a speed of about 150 to 180 knots. Landings will be conducted under a power-on condition with the power being supplied by jet engines.

The mission profile for a logistics mission is typical of the various types of missions that would be flown by the shuttle. While the on-orbit condition would vary from mission to mission the ascent and descent phases would be essentially the same. The shuttle will provide round-trip transportation for passengers and cargo from Earth to low Earth orbits (*Figure 11-7*).

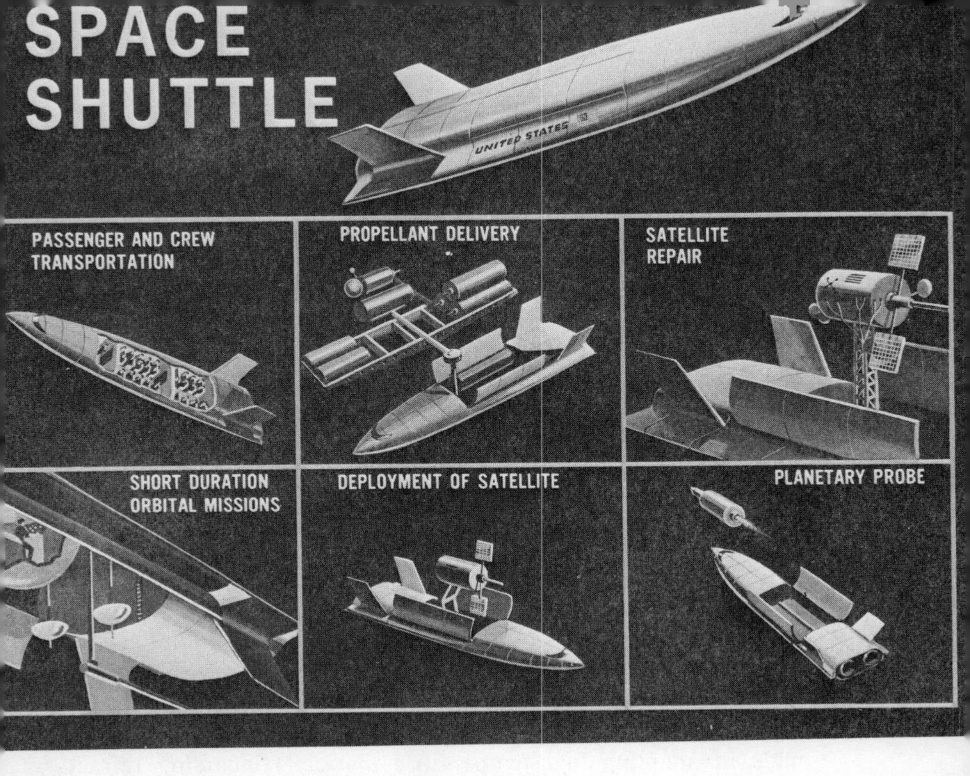

Figure 11-7. Space Shuttle.

The technology of all weather terminal flight control extending from the transition to subsonic flight at 100,000 feet through the final approach and landing of the several configurations under study is expected to be well in hand within the next several years. Extensive flight data from three lifting body configurations in regimes ranging from low supersonic speeds to approach and landing, show the feasibility of these designs. Related work on regular aircraft is continuing to investigate powered landings, automated control systems and operational problems. Results from this work will apply directly to the space shuttle.

The space shuttle will require additions to current technology in such areas as (1) efficient lightweight structures; (2) stage separation methods; (3) reusable, long-lived rocket propulsion; (4) lightweight airbreathing propulsion for terminal flight and landing; (5) automatic all-weather landing systems. In addition, detailed configuration analysis and wind-tunnel testing are needed to determine vehicle elements that can fly satisfactorily in all flight modes, considering different mission modes, variable cargo loads, and the large weights and areas associated with rocket engines at the base of the vehicles. Pertinent research is being expanded with a view to being ready to initiate a preliminary design a little over a year from now.

The technology for thermal protection during entry into Earth atmosphere for blunt vehicles is in hand as a result of the work done on Mercury, Gemini and Apollo. Velocities up to 36,000 feet per second, which produce temperatures in the order of 18,000 degrees Fahrenheit, are successfully withstood by ablative materials formed by any of three or four compounds and reinforced with glass fibres.

For re-usable entry vehicles such as the space shuttle, studies are being made of re-radiative heat shield materials and design concepts that can withstand a number of entries without replacement or refurbishment of the structure. Considerable progress has been made in adapting high-temperature materials including refractory metals such as columbium, as well as more conventional materials such as titanium and some nickel alloys, to this use. Composites of graphite and carbides are also being examined. The combined use of re-radiation with heat sink or convective cooling techniques for hot spots seems to promise an indefinitely reusable vehicle.

The high pressure, staged combustion rocket engines for the space shuttle will differ from the expendable engines used in our previous launch vehicles since they will be designed to have many of the characteristics of the engines in use in modern jet aircraft, including stability over a wide range of operating conditions, variable thrust to permit vehicle control, time between overhauls measured in hours of operations, high performance and operational dependability.

The choice of liquid oxygen and liquid hydrogen as propellants is based on our experience in Apollo, particularly on our ability to handle the propellants safely and relatively easily on the ground during loading and transferring. We feel that it will be possible to load these vehicles directly from tank trucks at the launch pad, thus greatly decreasing both facility and handling costs.

Another great advantage is that there are no toxic products or components of these chemicals and this propulsion system will not add to the pollution of the atmosphere. For simplicity and ease of ground handling we plan to use these propellants throughout the vehicle system, including using gaseous hydrogen for the subsonic engines for landing.

Onboard systems will provide the crew with necessary indications to make proper flight decisions. Today, there are some 20,000 people at Cape Kennedy who are directly involved with the checkout and launch of the Saturn V. Obviously, the space shuttle checkout and launch systems are going to have to be quite different if we are to radically lower operating costs. The approaches we are studying employ automation with the option of crew override and lead to airline type operations.

Some recent breakthroughs in electronics can supply the tools we need to achieve these goals. The large-scale integrated circuits which can provide as many as 1250 bits of memory in less than half a cubic centimetre of space are one of the tools which will be employed. Another is thin film memories. These advances will make possible the self-checking subsystems which we will need in the space shuttle. With immense capacity for logic, memory and multiple redundancy in a small volume, the self-checking black-box is going to be a practical reality.

A continuing goal in electronics is to perform more functions with less power, volume, and weight. Microelectronic circuits having a thousand active elements on a chip 150 mils (.001 inch) square are now available. It is estimated that through continued R&D, a million active elements can be placed on a chip 500 mils square. This will require new processes, such as laser or electron beam techniques, to replace photographic etching and the development of automated design, fabrication, and testing methods.

The necessity for new design for such electronic equipment creates the opportunity for an important advance toward simplicity. At the present time each of our subsystems, be it guidance, communications or engines, is interconnected by cables having literally thousands of wires. Not only does that represent one of the major components of weight, but it is by far the largest contributor to unreliability. We are ready for a breakthrough and I believe that there can and should be no more than six wires in one connector, going into or out of any black-box. These six wires might be allocated in the following manner: One wire would connect to a small computer inside the black-box which evaluates all the information from the circuits inside and reports its condition to the pilot. The second wire will carry all signals into the black-box. The third will carry all of the output signals or responses. The fourth and the fifth will be used for a standard power supply — all electrical equipment will be designed to operate off a 110 volt 400 cycle bus; any different power required will be generated by conversion within the black-box. The sixth wire is a spare.

The philosophy which will permit this concept is relatively simple — to assign the responsibility for the welfare and checkout of each subsystem to that subsystem. The microminiaturization which has been so notably advanced by the demands of space, now makes it possible and practical to build a small general purpose computer in a volume of about 10 cubic centimetres with enough logic and enough multiple redundancy to make it practical to use it to (1) self-check the internal circuits and (2) to provide time division multiplexing with proper addressing so that six wires can indeed effectively provide all the communications and power connections required.

Another ideal use for a computer on board the space shuttle is to throw switches. As a programme switching network a computer could not only flip hundreds of switches according to preset programmes but could also assure that they were actually in proper position. Some of the switching functions in Apollo are controlled by computers on the ground, but that task as well as the manual task of throwing and monitoring the 1000 switches in the cockpit can easily and effectively be taken over by a computer on board.

Using a computer inside the cockpit can also greatly simplify the control panel, which in spacecraft now have five times as many switches and instruments as a 727 or a DC8. For the future, I envisage a spacecraft command panel which contains three cathode ray displays, one digital input-output circuit which we call a DSKY, and an on-off switch for the computer. Two of the cathode ray displays would be used for control information, one for navigation and one for attitude control. These would operate interchangeably so that if one failed all information would be available from the other. These would replace the conventional "8 ball" attitude display systems which, being mechanical, are subject to breakdown. The cathode ray tube, on the other hand, has turned out to be one of our most reliable and long-lasting electronic products. The third cathode ray display would give the pilot information about any part of the total system or his computer. He can ask for any information by punching in the proper code for any system or subsystem, or internal programme. Stored in the computer memory is his checklist for each part of the flight, alternatives available to him in the event of an equipment malfunction, and such special information as propellant reserve, plot of flight path angle, and operational configuration of subsystems.

The electronic advances which are going to permit on-board checkout and control should make the space shuttle much simpler to operate than present commercial jets. Further, our design criteria require that all subsystems should be designed to continue to operate after the failure of any part except for the structure and to gracefully degrade in performance with subsequent failures. Electronic systems will be designed to give adequate warning of potential failures, to continue operation after the failure of two critical components, and to fail safe after any failure.

Present inertial guidance systems are gimballed, platform mounted gyroscopes and accelerometers. In the Lunar Module we have introduced a strapdown guidance system which replaces the mechanical gimbals with a somewhat more complex electronic system. It has demonstrated more than enough accuracy and computer capability to carry out the space shuttle missions. These strapdown systems with advanced reference devices, the next generation of gyroscopes and accelerometers with improved life and accuracy, should be capable of meeting all our requirements with increased reliability and life and a significant reduction in cost.

Star, planet, and horizons are basic references for navigation and present systems use gimballed trackers for attitude, reference position alignment, and velocity measurement. I expect that we will continue to use this system in our shuttle and space station. However, research is under way in several laboratories on holographic pattern recognition techniques for star patterns. This new reference system would have no moving parts and give three axis reference on any spacecraft orientation. This would also be true of aircraft navigation techniques. In another area, the technology of laser radars for rendezvous and docking, which is of course also applicable in clear weather to pin-point landing, has been demonstrated in the laboratory for a range of 65 miles and a docking accuracy of one inch. Efforts in the next five years should increase its range to over 200 miles, and make it an effective competitor with our present radio frequency radars.

Crew and passenger compartments will be maintained in "shirt-sleeve" environment. Spacesuits will not be required. As a matter of fact they are not now used in the Apollo programme except for extravehicular activity and in the event of certain emergencies, which, happily, have not yet been experienced. The atmosphere of oxygen and nitrogen in the space shuttle at a nominal pressure of 10 psia and a maximum flight load of three times Earth gravity, will allow any reasonably healthy individual to travel in the space shuttle without prior conditioning.

Communications with the ground can be similar to those now provided by FAA Air Traffic Control, but for orbital operations will be carried out through the use of communications satellites, thus greatly simplifying the ground network requirements.

Planning projections forecast that the space shuttle will have a useful life of 100 missions or more. As is characteristic of aircraft maintenance today, subsystems, particularly engines and thrusters, would be replaced on a progressive basis.

Launch operations will be simplified by the on-board system diagnostic instrumentation so that only vehicle erection, propellant loading and final boarding of payload and passengers will be necessary. Pre-launch checkout will be carried out on board by the pilots.

There are several areas which are now being studied. One is whether the shuttle should be designed with self-ferrying capability, so that if it lands at an airport other than its home base, it will be able to return home in a subsonic flight mode under its own power.

Another is whether the re-usable boost elements should have the capability to carry out their missions in an automatic mode, including a completely automatic landing system.

A third is whether the shuttle would have a "go around" capability in order to improve its ability to land under all conditions. Certainly included will be a powered landing capability which appears to be desirable to increase the ease of control and to improve performance in winds and gusts. This capability would be compatible with both powered landing and self-ferry ability.

Designs are providing landing visibility at least as good as that of high performance jet and SST aircraft and landing characteristics and handling qualities not more demanding than those of commercial jets.

One other feature that is being incorporated in the design is that wherever possible modules be used which can be replaced when the technology advances, without redesigning the total vehicle. Standard mountings and interconnections will be incorporated so that systems and subsystems can be replaced without basic configuration changes, as the design matures and improvements are available (*Figure 11-8*). Some unique development and test problems lie ahead of us since the shuttle is a combination of a rocket space vehicle and an airplane. It poses unique ground and flight test requirements.

The shuttle test programme will result from the combination of extensive experience which exists today in the development and test of high-speed aircraft, large rocket vehicles and manned spacecraft. Test philosophy which has evolved through Mercury, Gemini, and Apollo must be combined with the applicable experience of aircraft programmes such as the B-58, XB-70, X-15, SR-71, and the SST. We will be flight testing the shuttle two to three years after initial flights of the SST prototype and close liaison with that programme will be maintained. The flight regime of the shuttle is, of course, much more extensive than experienced in any other programme. Nearly 200 X-15 flights have provided valuable data for altitudes up to 67 miles and speeds in excess of Mach 6.

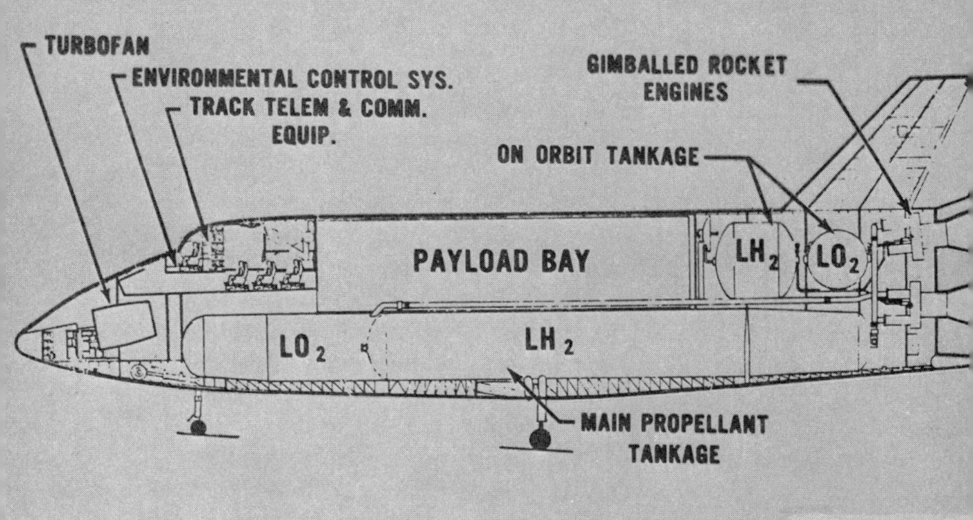

Figure 11-8. Typical Space Shuttle. Inboard Profile-Orbiter.

Figure 11-9. Size Comparison of Space Shuttle and Existing Flight Systems.

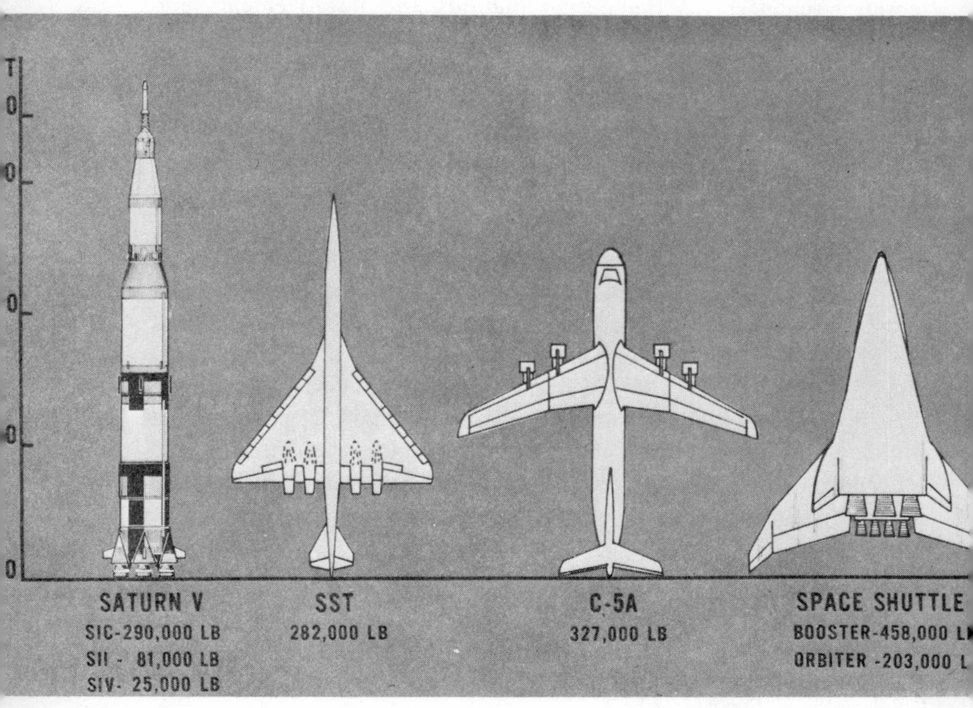

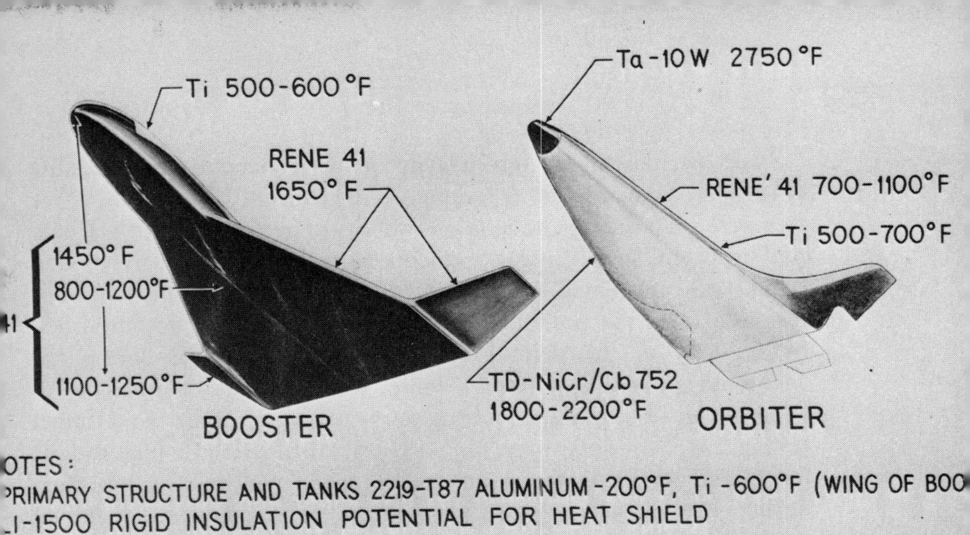

Figure 11-10. Summary of Materials and Predicted Temperature.

In terms of size, the shuttle is comparable to other systems. *Figure 11-9* shows the relative size of a typical delta wing configuration with the Saturn V, the SST, and the C-5A. Dry weights are indicated for comparison. The comparison of landing weights of both the booster and orbiter with several other aircraft are shown in *Figure 11-10.* The booster landing weight will be comparable to the maximum landing weight of the SST and about 100,000 pounds less than the maximum landing weight of the 747B. These weights will be representative of the orbiter and booster weights during the horizontal flight testing. Landing speeds of the booster of 145-155 knots will be only slightly higher than the 747. Orbiter landing speeds will be in the range of 155 to 170 knots, comparable to the SST. In terms of gross lift-off weights, that is quite a different situation. Fully fuelled the orbiter would be about 100,000 pounds less than the SST but the booster, of course, would be approximately 2.8 million pounds.

Plans for the shuttle rely heavily on the concept of early testing of critical components — the "proof-of-concept" approach. We therefore plan during the next 18 months to perform design verification tests on such critical concepts as lightweight structures and radiative heat shields. These will be large or full scale build-ups resulting from the preliminary designs which will be accomplished by industry in the definition contract phase. Also, during the 11-month definition contract there will be heavy emphasis on wind tunnel testing of the various configurations. Some wind tunnel testing has been done over the past year on possible configurations for the shuttle. As the point designs which will be accomplished during the definition phase are sharpened, considerable tunnel testing will be needed to identify the expected aerodynamics, stability and control, and thermal characteristics. Aerothermodynamics will be the highest percentage of wind tunnel testing. This is not surprising when one realizes the magnitude of the re-entry heating problem, coupled with the requirement for re-usable heat shields. *Figure 11-10* shows typical temperatures on the booster and orbiter during re-entry with candidate materials noted. As shown, Rene '41, TD-Nickel Chrome, and coated columbium are candidate materials for the large percentage of the vehicles for testing which are refurbished later for full operational use. This would be similar to the approach used for the C5A and the 747. The Saturn V was also a case where all vehicles were basically the same except for instrumentation and progressive engineering improvements.

Because the initial number of shuttle vehicles produced will be small, probably less than five, those vehicles allotted to flight test will represent a significant portion of the "fleet" cost. Hence, our desire not to plan for prototypes which can have no operational use. Of course, a recognition of the need for changes resulting from the flight test programme will mitigate against a strict "production approach" on the first flight test vehicles.

Secondly, a progressively more difficult flight test programme is planned which will parallel to some extent the ground testing. Rather than committing to sub-orbital or orbital flight, as in the case of past space vehicles, the shuttle test programme calls for an aircraft development approach with modifications and corrections for malfunctions being made following each test phase. Typically, in a spacecraft test programme the flight test programme cannot begin until the near-maximum level of design maturity has been reached. This must be improved on in the case of the shuttle to minimize development time and cost and provide an early initial operational capability.

The flight test concept presently envisioned is shown in *Figure 11-11.* The first phase of testing would involve horizontal subsonic flight of the orbiter and booster separately under jet power. The possibility of extending this phase of testing to higher altitudes and supersonic speeds by employing rocket propulsion will be studied. The second phase would involve vertical launches of the orbiter and booster. These would employ both jet and rocket propulsion. Initial studies have indicated that the booster flight envelope could be explored in this manner. Vertical launch of the orbiter by itself (fully fuelled without any payload) could explore the flight regime up to approximately Mach No. 8-10. In phase III vertical launch of the total vehicle would provide sub-orbital and orbital flight (*Figure 11-11*).

In addition to flights involving the basic booster and orbiter, there will be considerable testing of systems in support aircraft. For example, tests involving the avionics may be carried out in a military or commercial aircraft.

I believe that by the time we can produce the space shuttle we will be receiving so many benefits from our exploration of space and of the Moon, that many nations will want to use these economical space shuttles either by direct purchase or through chartering or leasing them from other nations.

If, as we expect, this shuttle will be the progenitor of a global transport, it will surely be necessary that multi-company and probably multi-national use of the vehicle be facilitated.

The exploration of space will of course be an international activity. As space engineering capability increases many more nations will take more active parts. It is interesting here to draw an analogy between space and aviation. Although only a few countries of the world manufacture the bulk of air transportation

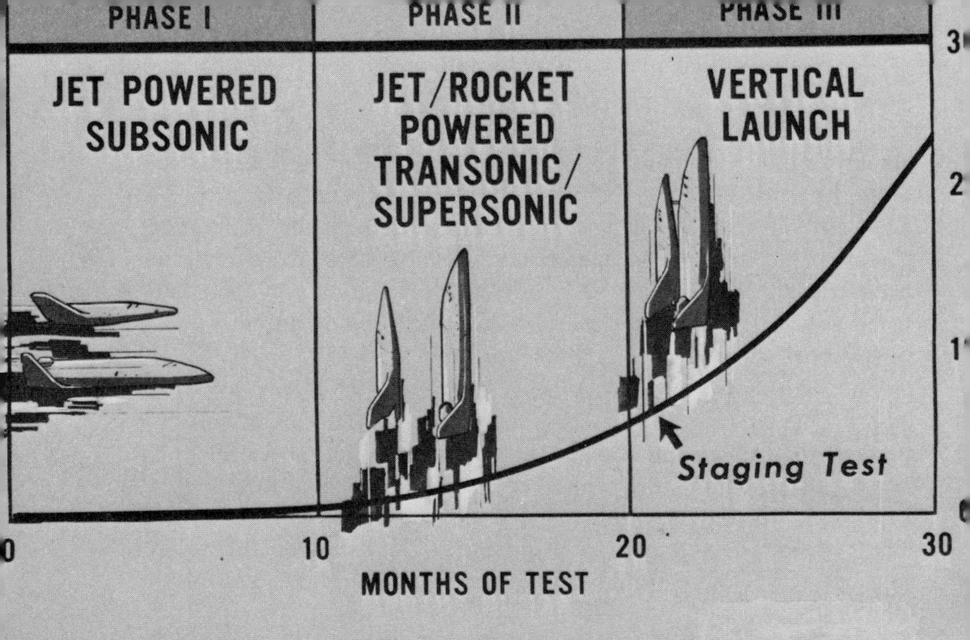

Figure 11-11. Flight Test Concept.

equipment, it is used by almost every nation for its own individual purposes. Within the next decade, no matter which nations provide the mechanical equipment for getting into space, that equipment will be used, by means of one or another kind of arrangement, to satisfy the space ambitions of all nations.

No really meaningful estimate of the price of an operational vehicle can yet be made, since the number of vehicles that will be needed is not yet clear. That number is a function, not only of the various jobs it will be called upon to perform, but also of the existence of the system itself. After all, no one needed a telephone, a computer or an airplane before they existed.

I fully expect that by the end of the next decade a number of the people in this room will have flown into and out of space. Consider with me for a moment the possibility of travelling in the space shuttle from, say, London to Sydney sometime in the late 1980s. You will board the regularly scheduled vertical take-off jitney in downtown London to fly quickly to Heathrow Airport. There, in the centre of the field, the Sydney-bound shuttle will be erected, being fuelled with liquid oxygen and hydrogen from tank trucks alongside. As the fuelling operation is completed, your luggage will be put into the baggage pod, and with your flight companions, you will enter the passenger module. The passenger compartment will be swung into place in the orbiter section of the shuttle. The flight crew will be conducting their final check-out of the vehicle, querying their on-board digital computer about the condition and configuration of each system and subsystem.

Passenger seats will swivel so that you will be sitting erect until power is applied, at which time the seats will recline so that the gravity load, not more than three times that of Earth, will be more easily accommodated. Your seat will be cushioned and provided with its own shock absorbers to lessen the load.

The shuttle will fly south-east by south on a sub-orbital course, taking advantage of the Earth's rotation to save considerable fuel. You will attain maximum speed of approximately 12,000 miles per hour at 100 miles altitude. Seven or eight minutes after lift-off from London, the booster element will peel off and return for a regular runway landing.

Although there was a huge explosive sound at the airport as the rockets ignited, it lasted only a few seconds, and you will not have heard it since it was far below your vehicle. As a matter of fact your whole trip will be practically noiseless. You will not hear the sound of entry into the atmosphere, but neither will people on the ground as it will occur at about 400,000 feet above the Earth.

While the passenger compartment will have no windows, television cameras mounted at strategic positions on the exterior of the orbiter vehicle will bring unobstructed panoramic views of the earth and of the sky into the cabin. As you leave the pad you will see the great thrusts of fire which propel your craft, and the land masses of Europe and Asia, the glistening clouds will occupy the television screens as you fly against the Sun, into the night.

While you are in flight you will, of course, be weightless. A cabin attendant will assist you to float about the cabin, or you may choose to remain in your seat which can now stay in the reclined position, or not, just as you like. Probably you will want to experiment with this new sensation. Certainly our astronauts seem to enjoy it. However, as descent begins you will have to return to your seat and belt yourself in — just as in other aircraft. Your views of the sunlit surface will be memorable as the earth seems to come up to meet you at 11,500 miles per hour. For this flight has not only been noiseless and weightless, but there has been no vibration and no feeling of motion after the gravity pull at the launching. Upon entry into the atmosphere, however, you will again encounter the forces of the Earth and your shock absorbing seats will return you to a reclining position.

Your next sensation will occur when the subsonic jet engines take over for your landing on the runway at Kingsford Smith Airport. Air traffic control will be programmed, using precise fixes on altitude and position verified by communications satellites, so you will not encounter any traffic delays. The sound of your craft coming in for its landing will actually be less than that of current jets.

Your landing at Kingsford Smith will be a subsonic jet runway landing and you will board the scheduled vertical take-off jitney there for its quick flight to downtown Sydney. Assuming that you took off from London at eight o'clock Monday morning, you would arrive in Sydney at eight o'clock Monday evening, in about an hour's flight time and 11 hours of Sun time. On the return, however, leaving Sydney on a Monday at eight in the morning, you would arrive in London at nine o'clock on Sunday evening, having out-distanced the sunshine.

We see no technological barriers to this kind of development. One lesson we have learned from the lunar landing, "impossible" as it seemed to most people less than a decade ago, is that no technological development can really be called "impossible" any more.

Figure 11-12. Space Station Concept.

But long before you and I travel in the space shuttle it will be carrying scientists, investigators, technicians and photographers from Earth to a space station in near-Earth orbit. Studies for the space station now in progress will draw upon our experience in the skylab workshop which will be operating in space in 1972 (*Figure 11-12*).

The three missions planned for the skylab workshop will contribute to the fund of design and operational data needed for the space station. Assuming successful accomplishment of the skylab workshop missions, the space station will obtain data from the experiments of the workshop on (1) physiological effects of zero gravity on the crew for periods up to 56 days, (2) demonstrated crew task performance data in station-type activity, (3) flight data on the habitability aspects of the workshop, (4) in-flight qualification and demonstration of several important new components including large solar arrays, control moment gyros and molecular sieves, and (5) general experience in logistic and orbital operations including major in-flight experiments. Other data required from skylab will be identified as the space station definition effort progresses.

The space station is now in the Phase B definition of the overall programme. This programme will achieve major advances in the nation's space endeavours by providing a centralized and general purpose laboratory in Earth orbit for the conduct and support of scientific and technological experiments, for beneficial applications, and for the further development of space exploration capability. This programme will fully use the experience gained in all previous flight programmes, both manned and unmanned, and those programmes now being initiated, and from appropriate ground-based research and development activities both within and outside of government.

The space station will be a centralized facility in Earth orbit supporting a wide variety of space activities. It will be similar to a highly flexible multi-disciplinary research, development and operations centre on Earth. The space station will utilize and exploit the unique features provided by its location in low Earth orbit — weightlessness, unlimited vacuum, wide scale Earth viewing, and unobstructed celestial viewing — with the direct presence of skilled scientists and engineers to pursue a wide variety of research and applications activities. Like its Earth-based counterpart, the space station will be configured for support and conduct of activities in many identified areas, and will have the flexibility to support others which may not be defined in detail as yet.

The nature of the space station programme can be established by examining the categories of activity which will be conducted. The combination of the environment of space, the facilities which will be available and the capabilities of the crews will provide unique research opportunities for investigations into many disciplines, especially astronomy, the life sciences, physics and chemistry.

Global surveys in Earth resources and meteorology will provide new knowledge for use on Earth as well as basic research into the systems and methods of collecting data of this kind.

Research into the possibility of producting unique material in the zero gravity and hard vacuum of space, for example the growth of large perfect crystals, precision casting, formation of composites and of perfect ball bearings, may lead to production of one or more of these candidate products. In that event it could be expected that the private sector of our economy would become interested.

Of course, the continual updating of a major space facility and its equipment is intrinsically a forcing function upon our technology in many regimes.

The space station will also be a prime source of information necessary to the continued exploration of the Moon and the solar system.

International co-operation, not yet possible on a day-to-day basis because of the constraints of present equipment, can be expected to expand and develop when it will be possible for the scientists of other nations to conduct their experiments on board the space station.

In all space activity, the space station will have an important role. It will act as a servicing and maintenance station for manned and unmanned equipment in space. Spacecraft of other nations will be able to call upon the space station for needed assistance. It will also act as a supply depot for the expendables necessary for expeditions to the Moon or to very high orbits, or to the planets.

These are the objectives which have already been established for the space station. Many more will undoubtedly be added as we move closer to its operation:

(a) Conduct beneficial space applications programmes, scientific investigations and technological and engineering experiments.

(b) Demonstrate the practicality of establishing, operating and maintaining long-duration manned orbital stations.

(c) Utilize Earth-orbital manned flights for test and development of equipment and operational techniques applicable to lunar and planetary exploration.

(d) Extend technology and develop space systems and subsystems required to increase useful life by at least several orders of magnitude.

(e) Develop new operational techniques and equipment which can demonstrate substantial reductions in unit operating costs.

(f) Extend the present knowledge of the long term biomedical and behavioural characteristics by man in space.

The in-flight operations of the space station will be more autonomous than present systems. On-board command and control capability will be incorporated and life support systems will be provided for extended periods without resupply (*Figure 11-13*). As a result, round-the-clock mission control activities on the ground will be minimal. While the space station will be reasonably independent of ground support except for routine resupply, it will use communications with the ground to augment the capabilities of the on-board research teams. Continuous communications capability may be implemented through the use of a data relay satellite system. Television and multi-channel voice links may allow principal investigators on board to consult with colleagues on the ground on a real-time basis (*Figure 11-14*).

Orbital research activities will ultimately be staffed with specialists with a minimum of astronaut-type training or physical conditioning. Therefore, particular attention will be paid to assuring comfortable, attractive and effective working and living conditions. The provisions may include individual quarters, kitchen and dining facilities, recreation areas, showers and greatly improved toilet facilities. Housekeeping functions will be highly automated to free the crew as much as possible for more productive work (*Figure 11-15*).

While the station will normally operate in a zero gravity mode, it will carry out an engineering and operational assessment of artificial gravity in the early weeks of its mission. A very extensive research and applications programme is being planned in the following years of space station operations.

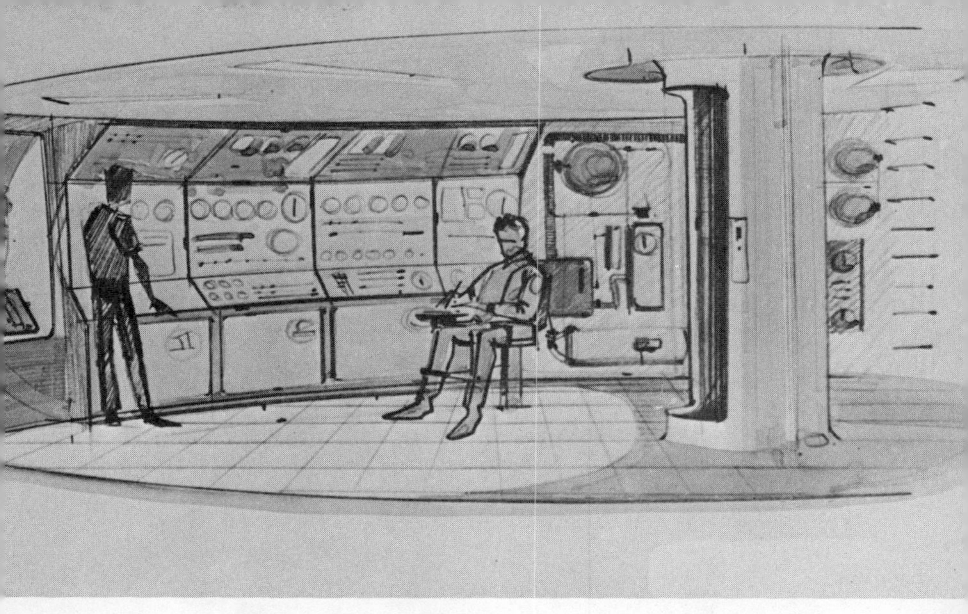

Figure 11-13. Space Station Control Room.

Figure 11-14. Space Station Core Modules.

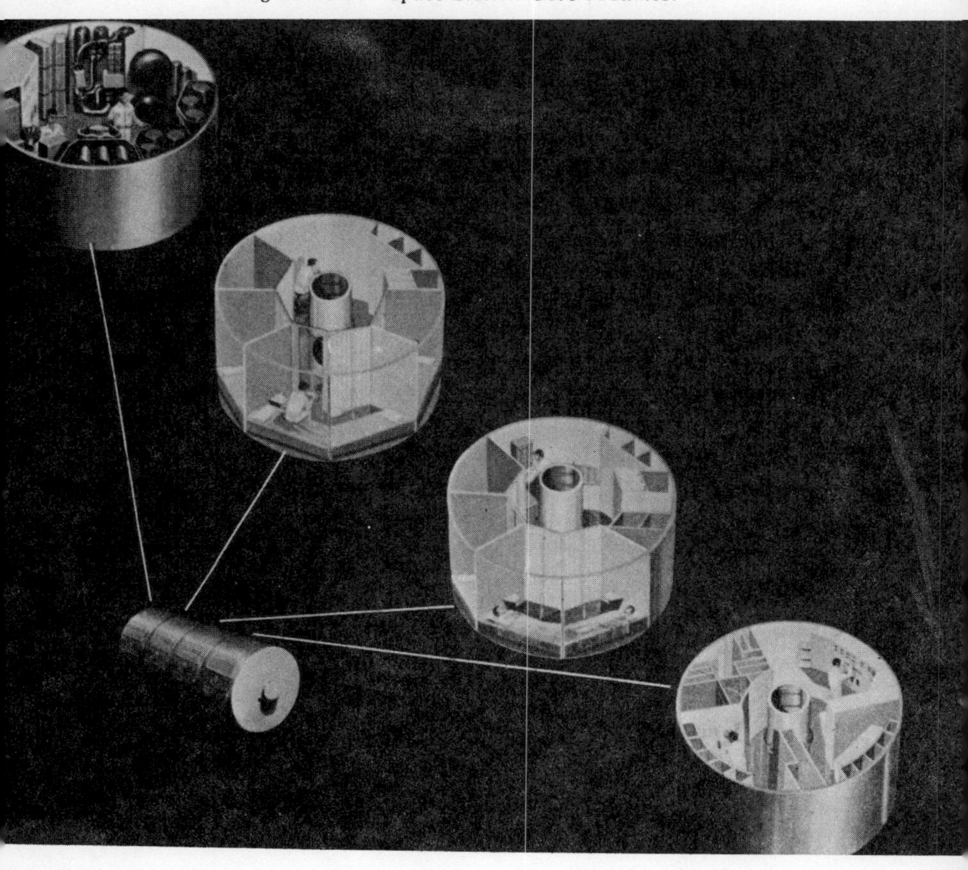

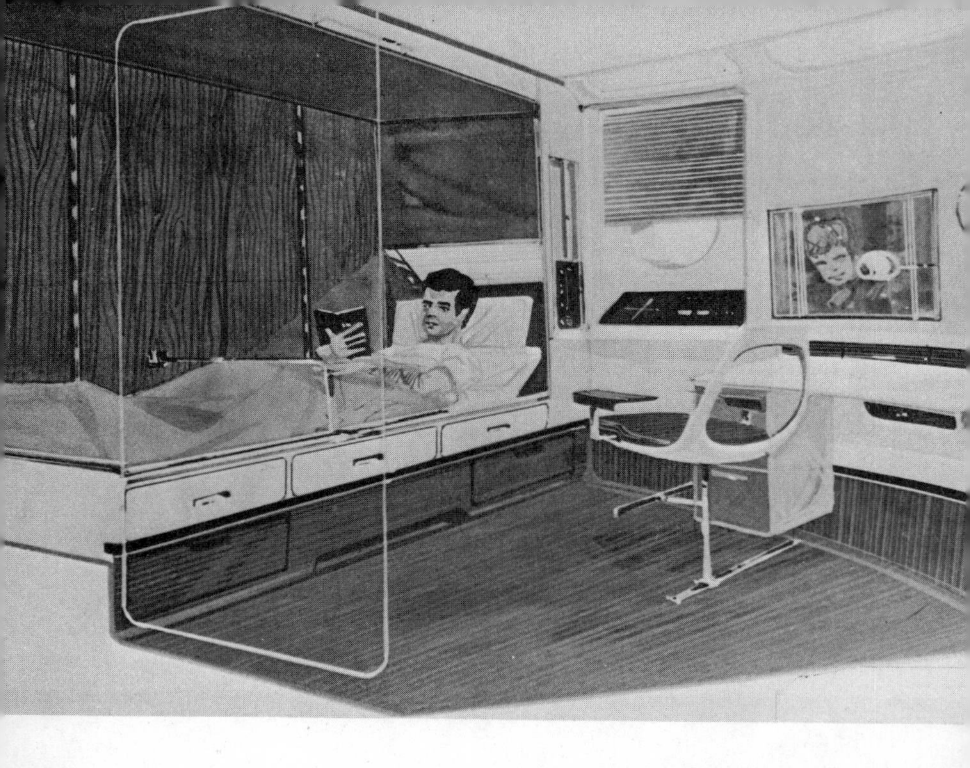

Figure 11-15. Space Station State Room.

The space station is expected to be operated in circular orbits inclined 55° at approximately 270 nautical miles altitude in order to accommodate the wide variety of experiments identified so far, including Earth viewing. In addition, the space station will also be designed for polar and slightly retrograde orbits at 200 nautical miles. For this second design mission, special equipment for artificial gravity assessment will not be required. The original space station may serve as a module of a space base which is being studied as a later programme. There may be more than one growth version of the space station module. However, the space station will also be designed as an independent spacecraft.

One concept of a space station has core stations which may exceed 100,000 pounds in weight. Equipment of the station includes the essential for crew habitability and protection, command control and data management, experiments, utility services and docking.

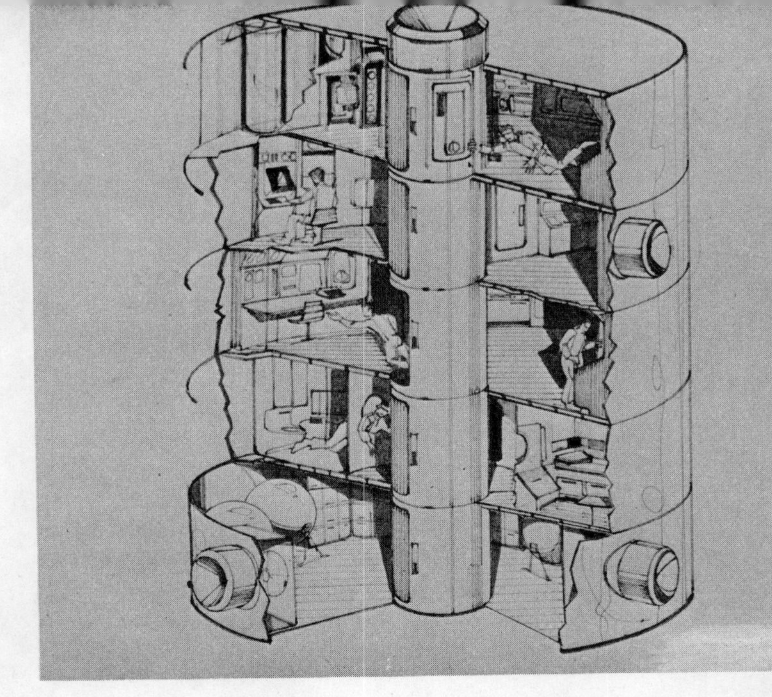

Figure 11-16. Space Station Concept Cutaway.

Autonomy of operation is enhanced by the provision of flexible on-board data management and checkout systems, television, multiple voice channels, broad band experiment data transmission, extensive telephone inter-communications and real-time continuous transmission paths to the ground via data relay satellites. This capability provides the mission flexibility required for the wide range of applications planned for this multi-purpose module.

This concept has five decks. Two will be devoted to living, eating, sleeping and controlling the space station. One deck will receive and house supplies. One will house the subsystems such as electrical and environmental control. The fifth deck would be a laboratory equipped for the conduct of a wide variety of experiments. This would be supplemented by other modules containing specialized experiments. Electric power would be supplied by large solar cell arrays or by a nuclear power system (*Figure 11-16*).

The station will provide living accommodations for a crew of 12, some responsible for the maintenance and operation of the space station — others for the conduct of experiments and

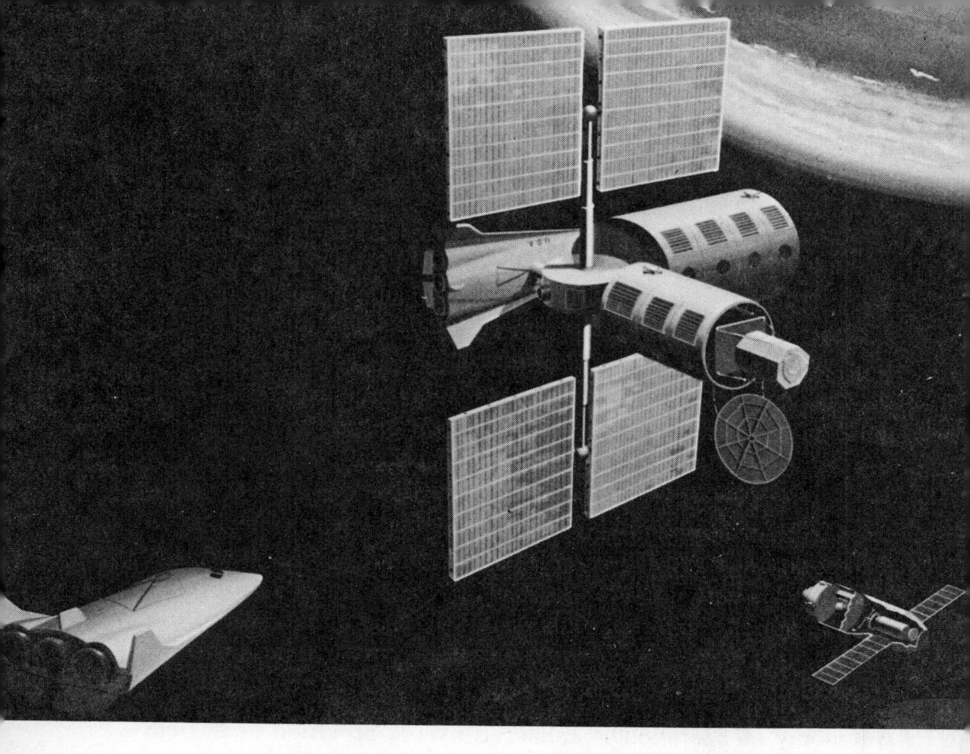

Figure 11-17. Manned Earth Orbital Space Station.

observations. To keep crew morale high and to stimulate creative and effective research operations, a great deal of care will be taken to selection of the every-day living facilities. Food served will be as near as possible to that served on Earth. Quiet, private areas or staterooms will be provided for the crewmen. Adequate personal hygiene facilities will be provided. A wardroom area will serve both for dining and as a conference room.

The space station and the space shuttle can be operational in the late 1970s if the national decision to go ahead is made. This capability will signal the merging of the manned and unmanned activities in space. It will also initiate a new era in scientific exploration as investigators in all regimes make use of the unique characteristics in space to add to our store of knowledge about the Earth, the Moon, the solar system and the universe (*Figure 11-17*).

CHAPTER TWELVE

Planning for the 1970s and 1980s

By L. B. James

During the first decade of space exploration, the programme of the United States progressed from the 31-pound Explorer I satellite in Earth orbit to Apollo spacecraft weighing 50 tons sent to the Moon. From manned flights of a few thousand miles and 15-minute duration, the 500,000 mile round trip, eight-day mission which landed men on the Moon and returned them safely to Earth became a reality. The rudimentary data of the Vanguard satellite paved the way to development of the sophisticated Nimbus and Tiros operational weather satellites now serving 50 nations. A commercially successful communications satellite system relays television around the world as a result of experiments with the early Echo balloon.

Yet these achievements are only the beginning of the long-term exploration and use of space by man. It is now apparent that we can seriously consider unmanned space flight missions to the farthest corners of our solar system; manned stations in Earth orbit or on the Moon; reusable space vehicles that can shuttle passengers, scientific gear and supplies to an orbital base at a tiny fraction of today's space travel cost; nuclear space vehicles for logistics supply of a lunar base or for planetary operations; a manned expedition to the planet Mars; large astronomical observatories in Earth orbit; and Earth resources satellites for beneficial services to man here on Earth.

To make the best possible selection from the broad spectrum of space possibilities available to the United States, President Richard Nixon appointed a Space Task Group in early 1969 and charged the group with making definitive recommendations on the direction which the American space programme should take in the post-Apollo period. In September, 1969, the Space Task Group presented to the President a co-ordinated space programme for the future. This programme contained the recommendation that a future national space programme should emphasize balance and unlike the single-minded objective of the 1960s to land a man on the Moon, the post-Apollo programme should pursue a steady rate of progress over the entire spectrum of space possibilities.

Specifically, the Space Task Group recommended emphasis on the following objectives:

—increase the use of space capabilities for services to man;

—enhance the defence posture of the United States and thereby support the broad objectives of peace and security for the world by exploiting space techniques for military missions;

—increase man's knowledge of the universe by a continuing strong programme of lunar and planetary exploration, astronomy, physics, the earth and life sciences;

—develop new systems and technology for economical space operations that emphasize multiple use of the same space vehicles for different tasks and reuse of space vehicles in repeated space flights;

—promote a sense of world community through a programme which provides opportunity for broad international participation and co-operation.

In March, 1970, President Nixon announced the United States' future goals in space which were based on the recommendations of the Space Task Group. He stated that the following specific objectives would be pursued:

1. **Continued exploration of the Moon.** Future Apollo manned lunar landings will be spaced so as to maximize the scientific return from each mission. Decisions about manned and unmanned lunar voyages beyond the Apollo programme will be based on the results of these missions.

2. **Bold exploration of the planets and the universe.** In the next few years, scientific satellites of many types will be launched into Earth orbit to bring new information about the universe, the solar system and our planet. During the next decade,

unmanned spacecraft will be launched to all the planets of our solar system, including an unmanned vehicle which will be sent to land on Mars and to investigate its surface. In the late 1970s, the "Grand Tour" missions will study the mysterious outer planets of the solar system—Jupiter, Saturn, Uranus, Neptune and Pluto. One major but longer range goal is to eventually send men to explore the planet of Mars.

3. **The cost of space operations should be reduced substantially.** Less costly and less complicated ways of transporting payloads into space must be devised. The feasibility of reusable space shuttles as one way of achieving this objective is currently being examined.
4. **Man's capability to live and work in space should be extended.** The Skylab Programme will be an important part of this effort. It is expected that men will be working in space for months at a time during the coming decade. On the basis of the experience gained in the Skylab Programme, a decision will be made on when and how to develop longer-lived space stations. Flexible, long-lived space station modules could provide a multi-purpose space platform for the longer-range future, ultimately becoming a building block for manned interplanetary travel.
5. **The practical applications of space technology should be hastened and expanded.** The development of Earth Resources Satellites—platforms which can help in such varied tasks as surveying crops, locating mineral deposits and measuring water resources—will enable the assessment of our environment and use our resources more effectively. Application of space-related technology in a wide variety of fields, including meteorology, communications, navigation, air traffic control, education and national defence should be continued.
6. **Greater international co-operation in space should be encouraged.** Progress will be faster and accomplishments will be greater if nations join together in the space effort.

This set of space objectives reflected not only the views of the Space Task Group, but those held by virtually all groups favouring a continuous, vigorous space programme. The question that remains is the pace at which these objectives are to be pursued in the coming years.

With these broad objectives in mind, let us now turn to the plans that the National Aeronautics and Space Administration has mapped for the future of America in space.

Unmanned Spaceflight

In order to fully understand the comprehensiveness of space exploration and applications, it is necessary to consider the future plans and possibilities of both manned and unmanned missions and how they complement each other.

Unmanned space efforts for the decade of the 1970s will include flights to all the planets in the solar system, development of Earth Resources Satellites, and continued development and utilization of satellites for communications, navigation, air traffic control, education, and national defence.

In the important field of Earth applications, plans have been laid to launch two Applications Technology Satellites (ATS) in the first half of this decade. The design characteristics of these satellites will provide basic capabilities for air traffic control, satellite laser communications, satellite-to-satellite tracking and data relay, propagation in the millimetre wave region, community broadcasting for instruction and education, and thermal mapping of the Earth from geostationary orbit.

One important utilization of the ATS to be launched in 1973 is the Indian Instructional Television Experiment. The experiment provides for NASA to position the ATS spacecraft in a location visible to India and retransmit instructional television material received from the Indian communications satellite station at Ahmedabad. These retransmissions will be received with augmented television receivers located in 5000 villages selected by the Indian Government, which will also be responsible for the programme material and the Ahmedabad station.

The Earth Resources Survey Programme is progressing toward the launch of the first Earth Resources Technology Satellite (ERTS) in 1972 followed by a second satellite launch in 1973. This programme will be a major step in establishing a capability for responsible management of the Earth resources and human environment. These satellites offer the key to world-wide crop prediction and an agricultural educational television service to advise farmers all over the world, in their own language, of vital and profitable information such as choice of suitable seed material, irrigation, fertilization and the like. The same satellites can be used for updating maps and to keep track of the rapid urbanization of mankind.

The field of hydrology will also benefit from these orbital techniques. By obtaining data on snow cover and ice occurrence in rivers and glaciers and by mapping circulation patterns in coastal waters, estuaries and lakes, the hydrologists will be able to provide a sufficient quantity of water to the places of need and minimize the detrimental effect of natural events such as floods and man-made effects such as pollution. Since the oceans affect our lives in many ways, systematic Earth observation from orbit will greatly benefit oceanography. Correlation of the various data collected by satellites will enable us to more efficiently route shipping, predict long-range weather patterns, and, by studying the migration habits of fish, we may some day be able to direct fishing fleets to herring and tuna schools.

Turning from the studies of the planet Earth, the United States will pursue a balanced approach to the studies of the other planets in the solar system. The exploration strategy calls for the detailed exploration of the planet Mars and the broad-based exploration of the other planets. This approach will permit a comparison of the planets which is fundamental to the basic goals of planetary exploration.

We know from studies dating back to Galileo that the planets are different from each other in most respects and similar in some respects. In addition to the large variation in the size of the planets and number of their moons, they vary in terms of the types of atmospheres they possess, their densities, their energy characteristics, and the manner in which they rotate on their axes and how they orbit the Sun. Yet, there are many similarities. For example, both Saturn and Jupiter emit more energy than they receive from the Sun and the planets Earth and Venus are similar in size, density and gravitation force. It is the study of these differences and similarities that will increase our understanding of the origin and evolution of the solar system and the evolution of life in our solar system.

In the 1970s, most emphasis will be placed on the detailed exploration of Mars. This planet, which orbits the Sun some 40 million miles outside the Earth's orbit, has been of interest to scientists for centuries. The flybys of Mariner 6 and 7 in 1969 provided a major step forward in exploring the mysterious red planet. After trips of more than 250 million miles both spacecraft passed within 260 miles of their aim points. Over 200 photographs and more than 5000 ultra-violet and infra-red spectra of the planet's surface and atmosphere were returned by the two spacecraft and excellent results on the atmospheric surface temperature and pressure were obtained. As a result of these missions, we now know that the atmosphere of Mars is composed primarily of carbon dioxide. Nitrogen was not detected and an upper limit of a few per cent has been placed on its abundance.

One of the major surprises was the discovery of two unexpected types of terrain, termed "chaotic" and "featureless". The chaotic terrain is a series of short ridges, slumped valleys, and irregular topography. Some irregular/chaotic terrain exists on the Moon but it is not comparable to that observed on Mars. About 300,000 square miles of chaotic terrain on Mars were observed by Mariner 6.

The "featureless" terrain has neither craters nor other distinguishable characteristics. One such area observed by Mariner 6 was the bright desert region called Hellas, which covers an area of almost 300,000 square miles. Even more intriguing is that while Hellas is now seen as one of the brightest areas on Mars, it was observed as one of the darkest areas in 1954. It is quite possible that there are other featureless plains on Mars which could be detected by spacecraft in the next few years.

The National Aeronautics and Space Administration will attempt for the first time in 1971 to place two spacecraft in orbit around the planet Mars. The designed operational lifetime of the spacecraft while in orbit will be about three months.

The objective of the Mariner Mars 1971 project is to explore Mars from orbit long enough to observe a large portion (about 70 per cent) of the planet's surface from an altitude of about 1000 miles, and to look at selected areas to observe temporal changes of surface markings such as the wave of darkening which has been observed from Earth.

The Mars missions will involve two identically instrumented spacecraft. They will be named Mariners 8 and 9 and will be assigned separate missions when they reach Mars. Each spacecraft will weigh approximately 2200 pounds and will be launched by an Atlas-Centaur launch vehicle from Kennedy Space Centre, Florida. The spacecraft will be based on the design of Mariners 6 and 7, which flew past Mars, but did not orbit, in the summer of 1969. A basic change will be the addition of a retro-engine to insert the spacecraft into Martian orbit. Carrying 970 pounds of propellant, each engine will be capable of five mid-course or trim manoeuvres and can provide a velocity change of 3400 miles-per-hour for orbital insertion (*Figure 12-1*).

Both spacecraft will carry a wide angle television camera for maximum coverage of the surface and a camera with a telephoto lens for more detailed pictures. Other scientific experiments include an infra-red radiometer to measure surface temperatures, an ultra-violet spectrometer to analyze constituents of the atmosphere and an infra-red interferometer spectrometer to analyze the planet's atmosphere and surface. Two other scientific experiments will be performed without special instrumentation: an occultation experiment that measures the effect of the Martian atmosphere on radio signals from the spacecraft to determine atmospheric density and a celestial mechanics experiment that will use tracking data to refine further the mass of Mars, Earth and the Astronomical Unit which is the distance from the Sun to Earth.

Each spacecraft will be assigned a scientific mission to carry out in Martian orbit. The first spacecraft will be launched in May, 1971, and will arrive at Mars in November, 1971, to perform a reconnaissance mission. It will be inserted into an orbit with an apoapsis (high point) of 10,500 miles, a periapsis (low point) of 1000 miles and an orbital period of 12 hours. Using its wide

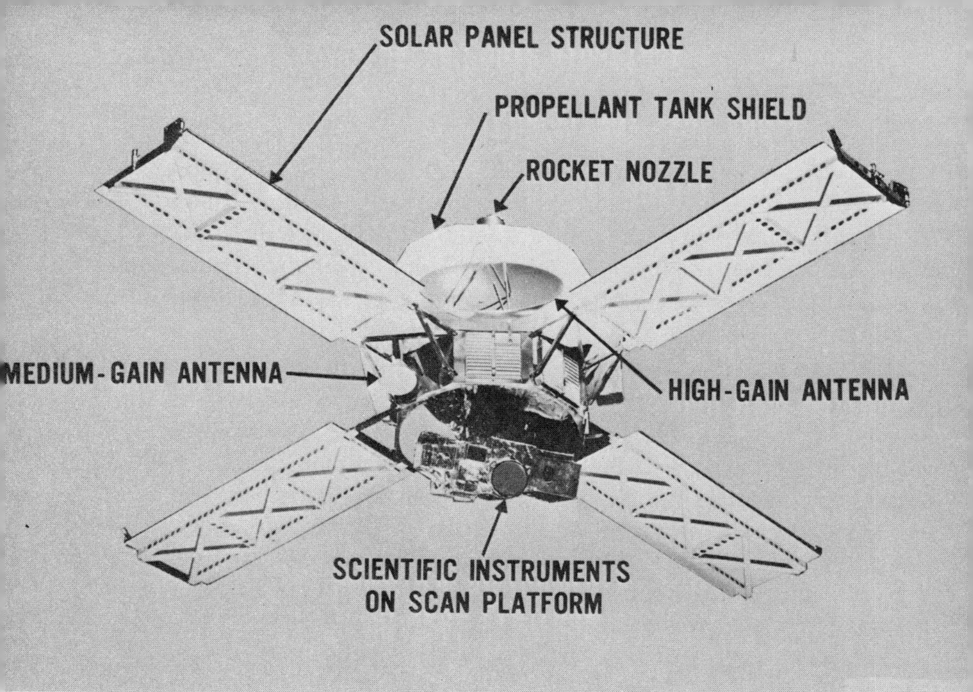

Figure 12-1. Mariner 1971 Mars Orbiter.

angle TV camera with a resolution of about two-thirds of a mile on Mars' surface from an altitude of 1000 miles, the spacecraft will systematically map the surface of Mars from 60 degrees South to 40 degrees North during the first 90 days in orbit. The spacecraft's orbit will be synchronized to the viewing period of the giant 210-foot diameter antenna of the Goldstone station of the Deep Space Network in California's Mojave Desert. The antenna will receive daily transmissions of photographs and other scientific data.

The second spacecraft, launched no earlier than eight days after the first, will arrive at Mars in November, 1971, to perform a variable features mission. It will be inserted into an orbit with an apoapsis of 27,000 miles, periapsis of 1000 miles and an orbital period of 32.8 hours. This orbit will allow the spacecraft to

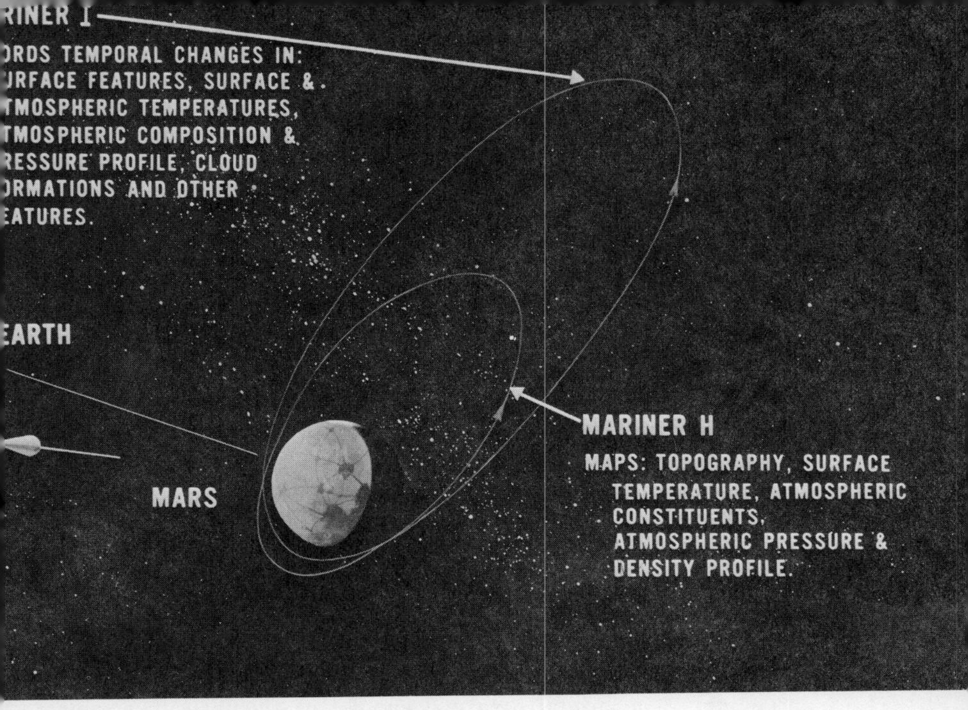

Figure 12-2. Mariner 71 Orbits Mars.

observe the same area on Mars every fourth day. Thus it will repeatedly photograph features of interest to scientists since these features appear to change with the Martian seasons (*Figure 12-2*).

In 1976 the National Aeronautics and Space Administration will land instruments on the surface of Mars as the next exploration step. The Viking Project is directed at the major unknowns concerning the chemical, physical and environmental properties of the Martian surface and the near surface atmosphere. An orbiter will team up with a lander to permit correlation of remote observations with direct measurements in the atmosphere and on the surface.

A series of scientific investigations will be conducted during the 1975-1976 missions. The orbiter which will carry the lander into orbit has a complement of three bore-sighted instruments to help select the best landing site. The lander will begin its investigations during its descent with the determination of the vertical profile of the composition and structure of the Martian atmosphere. After touchdown, the two cameras on the lander will visually characterize the landing site and help in the selection of surface samples. These samples will be analyzed for organic materials, water and biological activity. The chemical analysis will be conducted by a combined gas chromatograph and mass spectrometer. An integrated life detection instrument will look for evidence of life in the form of photosynthesis, respiration, metabolism and/or growth. A meteorological station, a seismometer, and investigations of soil physical and magnetic properties will complement the chemical and biological determinations.

Data obtained by the orbiter overhead will alert the scientists who have investigations on the lander as to the best time to take the samples, such as during the passage of the wave of darkening. The orbital data will enable the scientists to correlate the lander data with local and regional weather activities. When not employed directly with the lander investigations, the orbiter will continue the visual, thermal, and water mapping of the planet and the study of significant dynamic properties.

Now let us turn inward toward the Sun, to the planets Venus and Mercury. Venus is a planet which is very similar to the Earth in terms of size, density and gravitational force, yet it has a considerably different atmosphere. Its surface pressure is over 100 times that of the Earth's atmosphere and its surface temperature approaches 900° Fahrenheit. The atmosphere of this planet has been examined by the Mariner 2 and Mariner 5 spacecraft during their flybys and by three Russian atmospheric probes.

The planet closest to the Sun is tiny Mercury. Its size is somewhat larger than the Earth's moon and it has a very high density. The high density is probably due to either a high concentration of iron or a relatively large loss of volatile material from its interior. It would appear that either Mercury's original formation was different from that of the Earth or else its evolutionary process has been significantly different. Thus far, no atmosphere has been detected on Mercury. However, very little is known about this planet because of its proximity to the Sun, which makes it difficult to study from the Earth using astronomical facilities.

The first spacecraft mission to Mercury will occur in 1973 with the Mariner Venus/Mercury 1973 spacecraft. This will also be the first attempt to perform a gravity-assist dual planet mission of the type which will be performed in the exploration of the outer planets in the late 1970s. The spacecraft will be launched in October, 1973, during an opportunity which will permit the spacecraft to fly by Venus, obtain a gravity-assist deflection, and then move inward toward the Sun and fly by Mercury.

Top priority in this mission is being afforded to the scientific investigation of the planet Mercury. Investigations will be made of the planet's environment, atmosphere, surface and body characteristics. While passing Venus, however, ultra-violet photographs and atmospheric data will be obtained, where possible. In addition, exploration investigations will be performed in the interplanetary medium while the spacecraft is enroute from Earth to Mercury (*Figure 12-3*).

Figure 12-3. Mariner 1973 Venus/Mercury Spacecraft.

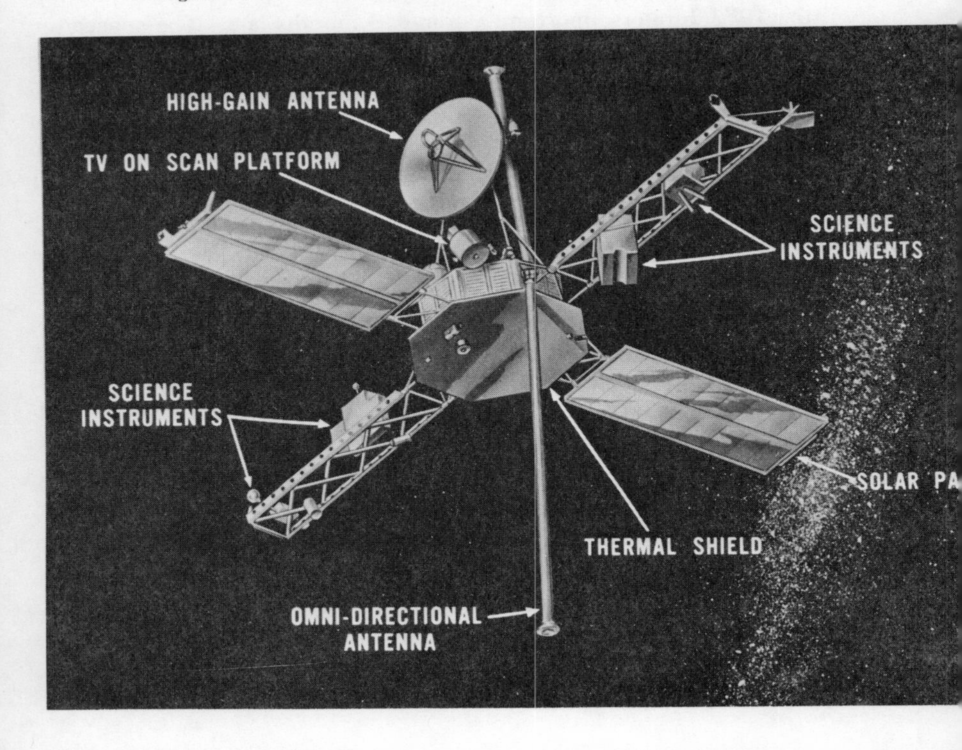

The exploration of the outer solar system will receive considerable attention during the decade of the seventies. For the first time, this area of high potential for new scientific discoveries will be exploited.

Jupiter, one of the most interesting planets in our solar system, will receive greater study in the first half of the 1970s. Jupiter has a mass some 318 times that of the Earth yet rotates very rapidly — its period of rotation being somewhat less than 10 hours. The atmospheric constituents of the Jovian atmosphere have been detected as hydrogen with minor amounts of methane and ammonia. An amount of helium has also been inferred.

Jupiter has 12 satellites and is endowed with multicoloured bands around it, the structure of which changes with time. It is the first planet that was discovered to emit more energy than it absorbs from the Sun. Current measurements indicate that the planet emits about twice as much energy as it absorbs. The precise measurement of Jupiter's temperature in different wavelengths and from the night side of the planet will shed light on this mystery.

The Pioneer F spacecraft will be launched during the Jupiter opportunity in March, 1972, and Pioneer G will be launched during the 1973 Jupiter opportunity, some 13 months later. The spacecraft will take just under two years to reach Jupiter and will cover a flight path of almost 500 million miles. These spacecraft will permit exploratory investigations of the interplanetary medium beyond the orbit of Mars, the nature of the asteroid belt, and the environmental characteristics of the planet Jupiter.

With a complement of 13 scientific investigations on each mission, it will be possible to study spacecraft hazards associated with flights through the asteroid belt, to measure the gradient of the Sun's influence on interplanetary space, and to investigate the penetration of galactic cosmic radiation into the solar system. While the spacecraft are in the vicinity of Jupiter the scientific instruments will measure properties of charged particles, magnetic fields and electromagnetic emissions associated with the planet. The spacecraft will also take a number of pictures of the planet at resolutions considerably better than those possible from the Earth. The data returned by Pioneers F and G will be used in studies of the

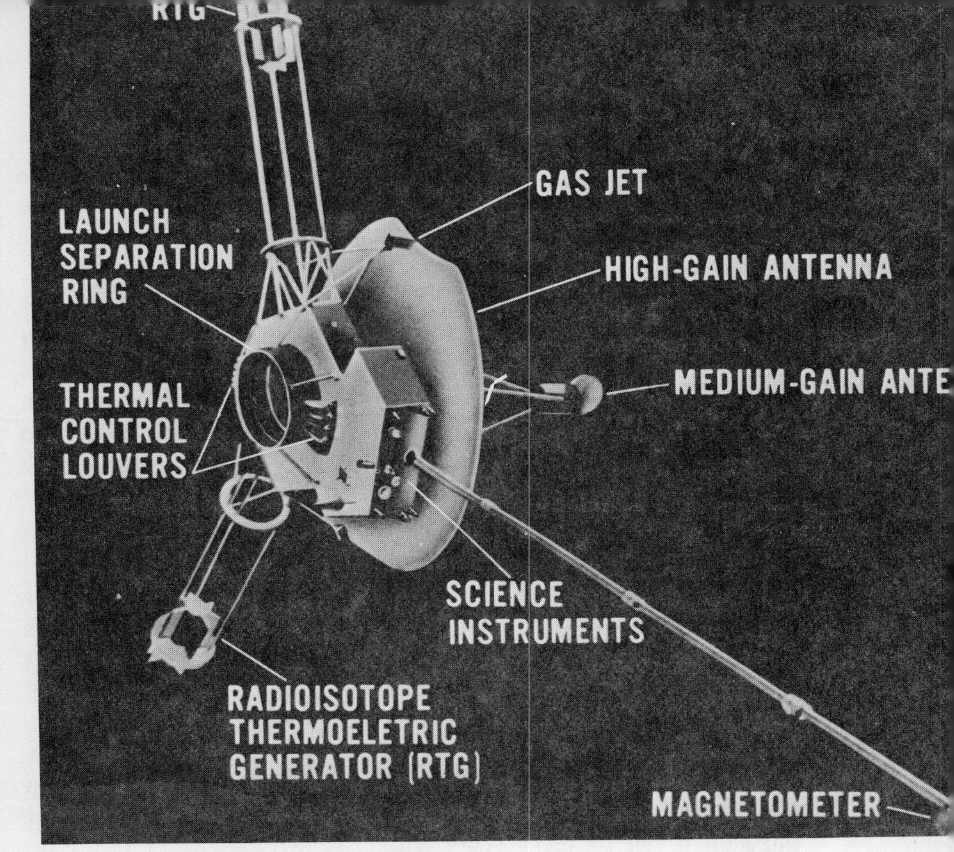

Figure 12-4. Pioneer F/G Spacecraft.

composition and dynamics of the atmosphere surrounding Jupiter, its cloud structure and its interaction with the planetary medium. The infra-red radiometer experiment should make it possible to analyze the thermal balance of Jupiter and the planet's source of energy (*Figure 12-4*).

After Jupiter come Saturn, Uranus, Neptune and Pluto in the most far-reaching space missions yet conceived by man. The best outer planet alignment in 179 years, occurring in the 1976 to 1980 time period, opens the outer planets to exploration in an effective and timely manner. The infrequency of such favourable alignment is due to the slow movement of the outer planets about the Sun.

Plans for two three-planet Grand Tours in the late 1970s are being developed by the National Aeronautics and Space Administration. One such mission would fly by Jupiter, Saturn and Pluto, and the other would go to Jupiter, Uranus and Neptune.

Either conventional or solar-electric propelled spacecraft with a nuclear isotope power source to operate spacecraft equipment will be employed for the planetary tours. From Jupiter on, the spacecraft will use the gravitational attraction of each planet to spin on to the next. This method of exploring the outer solar system is an extension of the "interplanetary billiards" proposal which explains how the heavy mass and strong gravitational fields of Jupiter and the other larger planets make large deflections and speed changes possible for passing spacecraft. As the name implies, this method of space travel has been compared to the ricocheting of billiard balls.

The basic Grand Tour spacecraft will weigh about 1200 pounds and will continue past Neptune or Pluto, escaping the solar system. The trajectories projected for the missions extend into intergalactic space. The planetary carom effect will enable a spacecraft to reach Pluto in seven or eight years where a direct flight to Pluto (at closest 2670 million miles from Earth) would take 41 years. Neptune, the next farthest out, might be reached in 8.9 years instead of the 18-plus years via direct flight.

The technological challenges of the Grand Tour missions are formidable but not insurmountable. The spacecraft will have to be capable of automatically replacing failed equipment because it will take hours for telemetry signals to reach Earth and to return a corrective command to the spacecraft. Such self-repairing computer and data storage systems are presently under development. Since the spacecraft's power will be generated by electrical conversion of heat produced by a nuclear source (plutonium), work is also proceeding on a radioisotope thermoelectric generator. The great distance from which signals must be received at Earth stations would require the spacecraft to have a large, precisely parabolic antenna. It will take four hours for a radio signal transmission from Neptune to Earth.

Although the specific science instrumentation remains speculative, the key mysteries of the outer planets will require photography, atmospheric measurements in ultra-violet and infra-red wavelengths, plus radiation-detecting equipment. Along the way, cometary particles also could be measured in the asteroid belt between Mars and Jupiter. Instrumentation required for the planets is also suitable for measuring interplanetary regions, and even the intergalactic regions beyond the solar system.

These automated space programmes are indicative of the major unmanned projects that the United States hopes and plans to pursue in the future. Now let us turn to the manned space flight objectives for the 1970s and 1980s.

Manned Space Flight

During the past 10 years there has been universal personal identification with the astronauts and a high degree of interest in manned space activities which reached a peak both nationally and internationally with the Apollo programme. Sustained high interest, judged in the light of current experience, however, is related to the availability of new tasks and new challenges for man in space. The presence of man in space, in addition to its effect upon public interest in space activity, can also contribute to mission success by enabling man to exercise his unique capabilities and thereby enhance mission reliability, flexibility, and the ability to react to unpredicted conditions.

The approach used in planning manned space flight for the next two decades is based on the Space Task Group recommendation that manned planetary exploration be a focus for the development of new capabilities. Therefore, the establishment of a foundation for manned planetary exploration becomes a transcendent objective in defining the programme of the future.

The basic Apollo capability will be extended to provide more increasingly important results of scientific knowledge about the Moon. Through 1974, Apollo flights will take teams of two astronauts to different landing sites of particular scientific interest while a third astronaut remains with the Command Module in lunar orbit. Each subsequent mission will offer more involved and extended scientific exploration and a longer stay-time on the lunar surface (eventually up to 54 hours). By the latter part of 1971, there are plans to provide the astronauts with a lunar rover which will enable them to explore areas of the Moon which may be several miles distant from the landing site. These extended Apollo missions will allow more detailed selenological studies at a number of lunar sites and a range of experiments in lunar physics and chemistry. They will also provide the necessary base of experience required for future missions of longer duration on the lunar surface.

The Skylab programme, derived from Apollo hardware, will be placed in low Earth orbit in late 1972. The prime objectives of this programme are to establish physiological and psychological data on man for extended space flight missions and support high resolution solar astronomy at short-wave length that is not directly observable from the surface of the Earth. In addition, the Skylab programme will support a broad spectrum of experimental investigations in other scientific disciplines. For a detailed explanation of this programme see Chapter 10.

The Skylab programme is a first step toward manned utilization of space, but further steps will be taken to realize the full potential of this capability. In the regions of Earth orbit, a permanent, flexible space station supplied by a reusable space shuttle will form the basis of the major space transportation system of the late 1970s. The National Aeronautics and Space Administration is currently involved in the definition phase of the Space Station. The overall objective to this effort is to obtain the technical and managerial information required so that a choice of a single approach to a space station design can be made from the alternate approaches available. The Space Shuttle effort has progressed to the point where the definition phase studies can now proceed. These studies are the next logical step in providing a basis for evaluation of competing designs to the point of preparedness necessary for initiating of development. The Space Shuttle and Space Station are discussed in detail in Chapter 11.

Planning for the 1970s and 1980s

To this point, manned space flight programmes which are being defined or which are in the development stage have been discussed. However, the following discussion is directed toward space systems which could become a reality in the late 1970s and 1980s although no decision has yet been made by the United States to move ahead with definition studies or development.

A new mode of lunar exploration could well begin in the late 1970s with the establishment of a lunar orbit station. The station would be established by placing a space station module (similar to the Earth-orbit module) in polar orbit around the Moon. In addition to performing orbital science, this station would make possible the visitation of any point on the lunar surface every 14 days with the introduction of a new system, the Space Tug.

The Space Tug is one of four major new pieces of equipment which would be required in the balanced space programme of the next two decades. The Space Tug concept is a highly versatile multi-application system that would utilize three major modules — crew, propulsion and cargo — which may be used separately or together with a variety of supplementary kits depending on mission support function required. These kits would contain components such as landing legs, environmental control systems, power, guidance and navigation and manipulator arms. Each module would be recoverable and economically refurbishable.

Significant improvements in lunar exploration would be introduced with the advent of the Space Tug. In delivering payloads to the Moon, the Space Tug would provide improved performance to the Saturn V launch vehicle by adding a fourth stage to the three-stage rocket. An unmanned launch would initiate this operating mode by transporting a Space Station Module into lunar polar orbit at approximately 60 miles altitude. Manned launches would deliver Command Modules, Space Tugs and support cargo to the orbiting station. The Command Module would be used for Earth-to-Moon-and-return crew delivery and would be modified to remain at the station for extended periods. For the lunar landing, the Space Tug propulsion module, crew module, landing legs and other appropriate support kits would descend from the Lunar Orbit Station to the lunar surface for exploration missions of 14 to 28 days. After

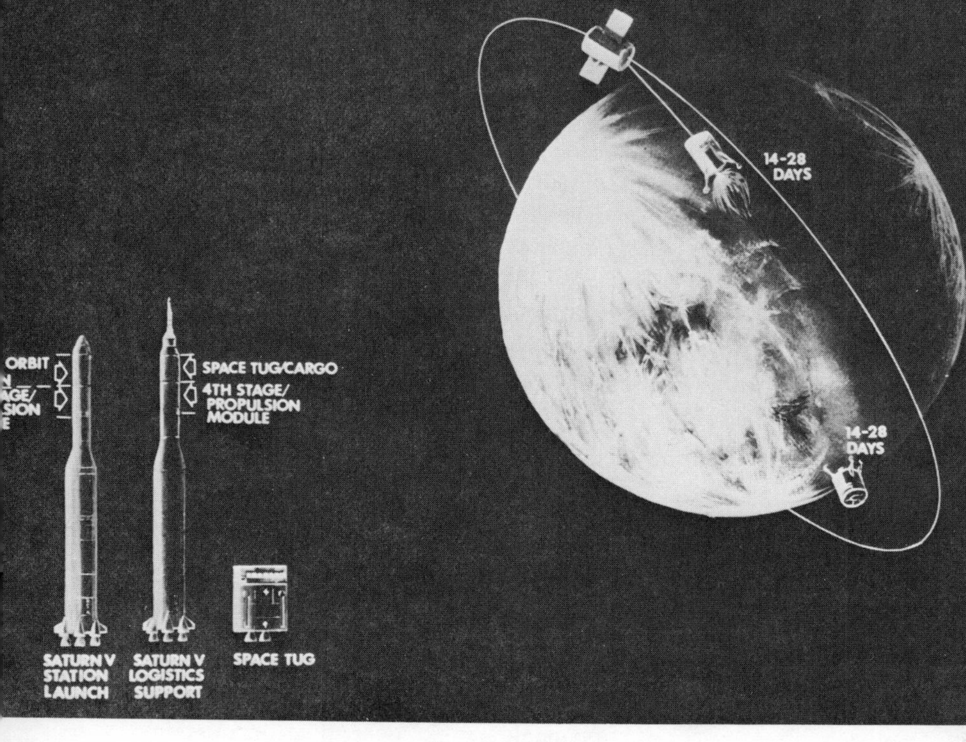

Figure 12-5. Lunar Orbit Station.

Figure 12-6. Space Tug/Lunar Applications.

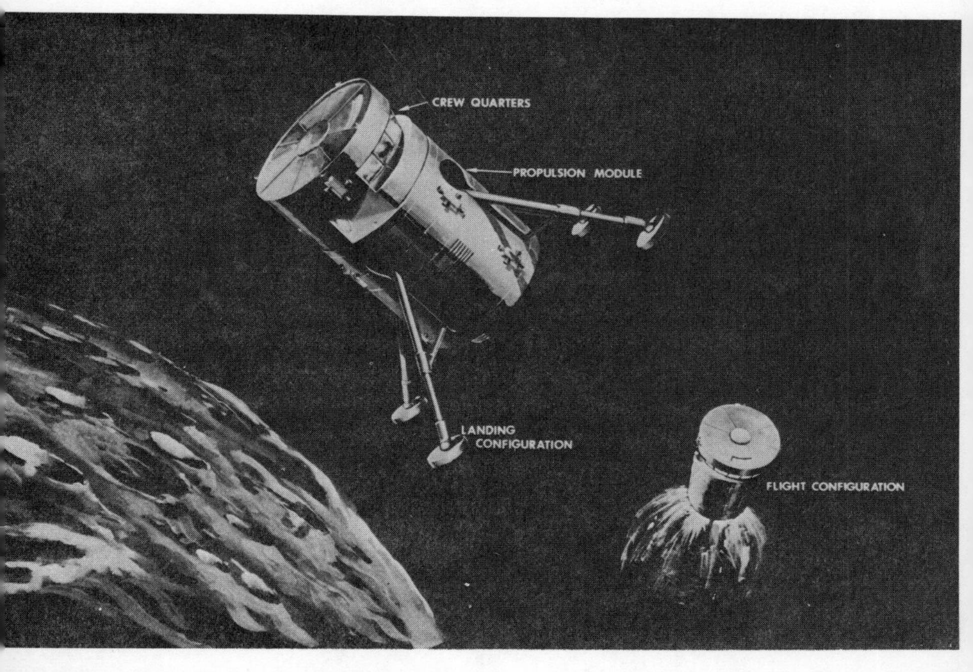

surface mission completion, the tug would ascend to the Orbit Station to refuel and resupply for another surface sortie (*Figures 12-5* and *12-6*).

The Lunar Orbit Station would provide a station from which many activities could be conducted. Surface roving vehicles would be used during the surface excursions to increase mobility on the lunar surface. Selenological samples would be collected from the surface, returned to the station and analyzed in lunar orbit. The first major lunar surface farside radio telescope could be deployed and the polar orbit of the station would make possible the complete mapping and remote sensing of the Moon. At the end of this mode of lunar exploration, enough surface experience would have been accumulated so that a lunar surface base could be implemented if it was deemed desirable.

The versatile Space Tug would be configured for conducting operations and tasks in Earth orbit as well as lunar orbit. Such operations might include the movement of space stations, movement of large payloads in the vicinity of the space station, and satellite placement, retrieval and maintenance services (*Figure 12-7*).

The introduction of a low cost, reusable Earth-orbit to lunar-orbit space transportation system will be required to support long term high energy missions, such as synchronous and lunar surface bases. At this time, it appears that the nuclear shuttle is the most economical system to fulfil these requirements. A prototype of the nuclear engine required for just such a vehicle has already performed a number of highly successful static tests and could be operational by 1978. In addition to cislunar missions, the nuclear shuttle would serve as an economical propulsion system for manned planetary missions in the 1980s (*Figure 12-8*).

The nuclear shuttle, designed for multiple reuses, will be able to operate in either a manned or unmanned mode. It will possess the capability of long term, cryogenic storage in space and will have the ability to station keep in orbit between mission applications.

A two-stage configuration of the Saturn V rocket would launch the nuclear shuttle into orbit. Thereafter, the space shuttle would be used to transport fuel and payload into Earth orbit to support

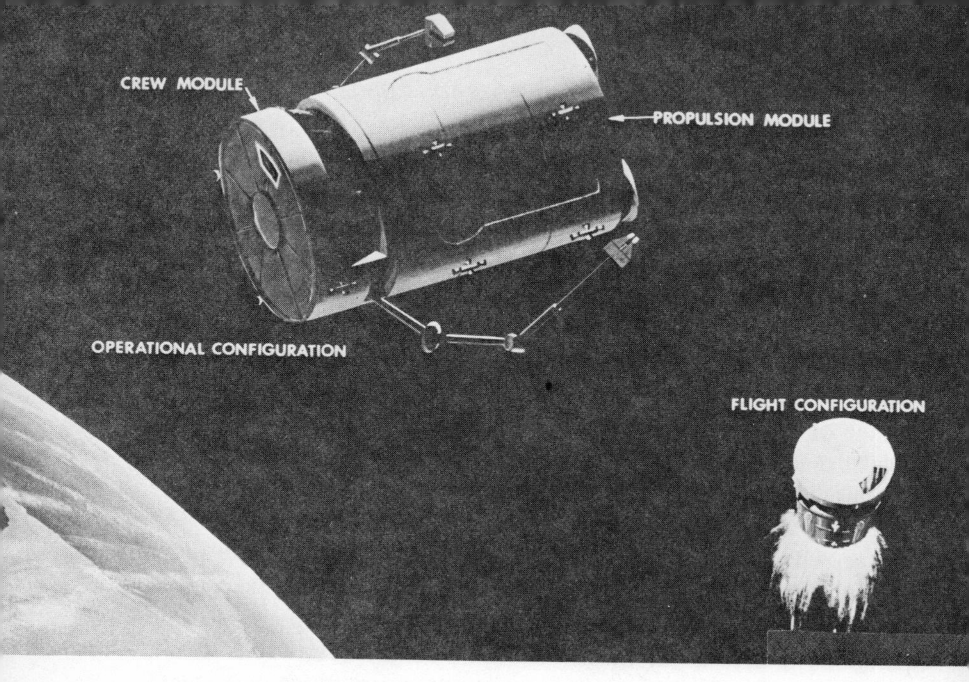

Figure 12-7. Space Tug/Earth Orbit Applications.

Figure 12-8. Nuclear Shuttle.

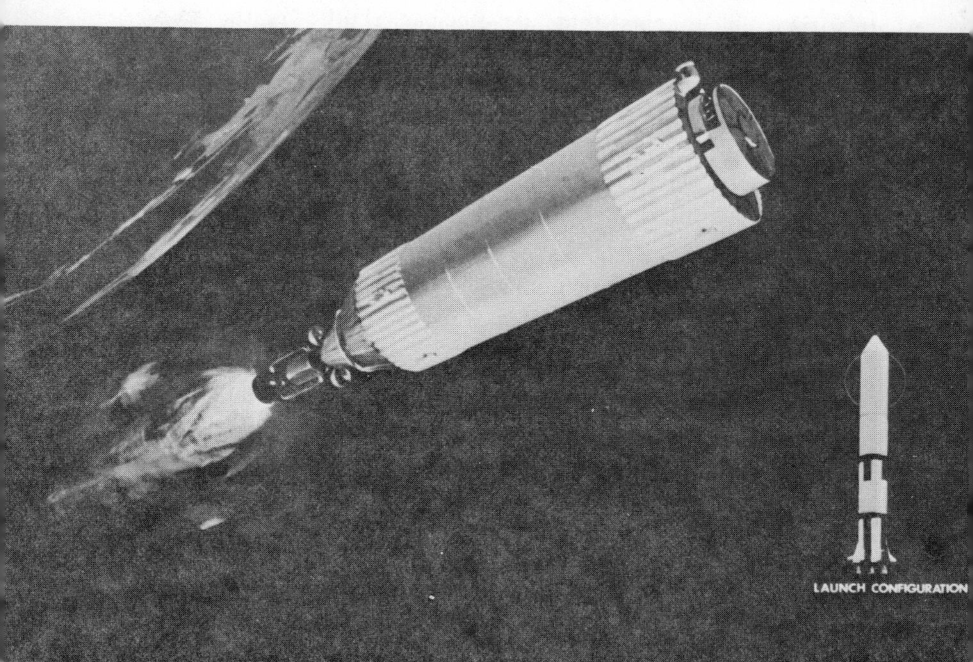

the nuclear shuttle operations (*Figure 12-9*). The nuclear shuttle would have necessary maintenance performed by personnel of the orbiting space station between mission applications.

The nuclear shuttle will initially be used as a logistics vehicle for supporting manned operations in synchronous orbit, in lunar orbit, and on the lunar surface. The shuttle should, therefore, be adaptable to transporting payloads of varying size, weight and configurations. The payload weight that the nuclear shuttle will be able to transport to its destination is a function of the weight it will be expected to return to Earth orbit. For instance, in early phases of the programme, the Space Tugs will be returned to Earth orbit from both synchronous and lunar orbit station to be refuelled and refurbished. Several representative payload configurations for the shuttle are illustrated in Figure 12-10. Configurations of both Cargo Modules and fully fuelled Space Tugs would be able to be transported on a single flight. The space station or experiment modules would also be able to be transported to any point of scientific interest in cislunar space. Other applications of the nuclear shuttle would include rotation of crews between synchronous and lunar orbit stations and transportation of specialized equipment required for lunar surface operations.

Figure 12-11 illustrates the use of the three major space transportation elements just discussed. The sequence shown depicts the cargo transfer from a space shuttle to the nuclear shuttle which is to depart from Earth orbit to lunar orbit. The space shuttle nose is hinged open and an interior expulsion device moves the space shuttle cargo forward and out into the open, exposing the individual cargo modules for removal (Phase I). A Space Tug with a manipulator equipped crew module then extracts the Cargo Module from the space shuttle pallet and docks with it (Phase II). The Space Tug then propels itself and the Cargo Module toward the nuclear shuttle (Phase III), stationed at some distance away, and transfers the Cargo Module to the nuclear shuttle cargo structure by hard docking (Phase IV). This operation and other similar cargo transfers would be a standard operation in Earth orbit for any cargo destined for the Moon.

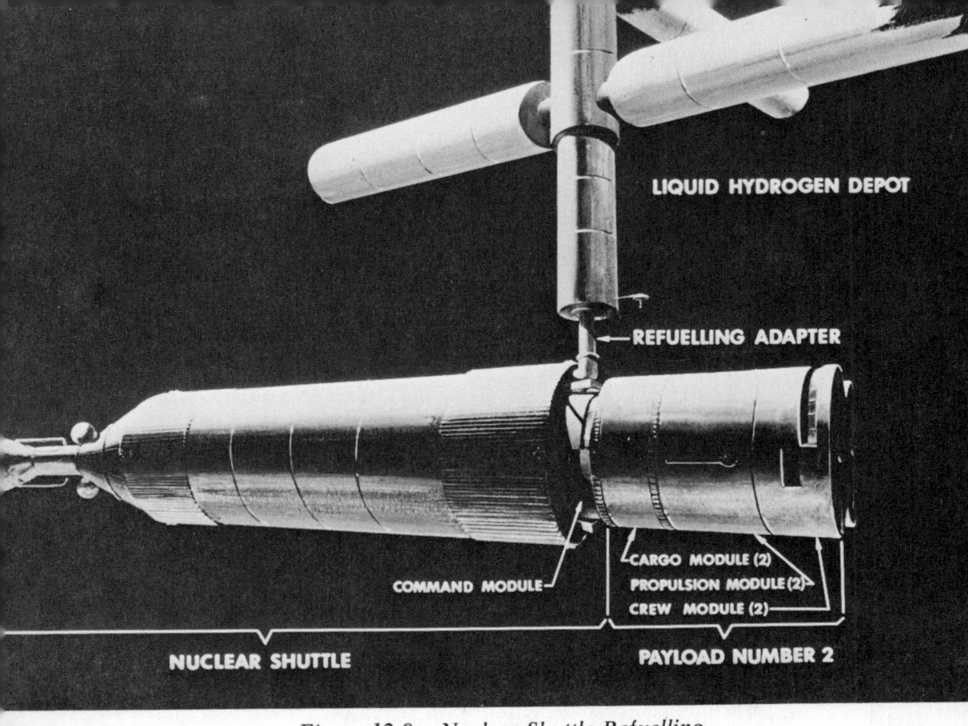

Figure 12-9. Nuclear Shuttle Refuelling.

Figure 12-10. Nuclear Shuttle.

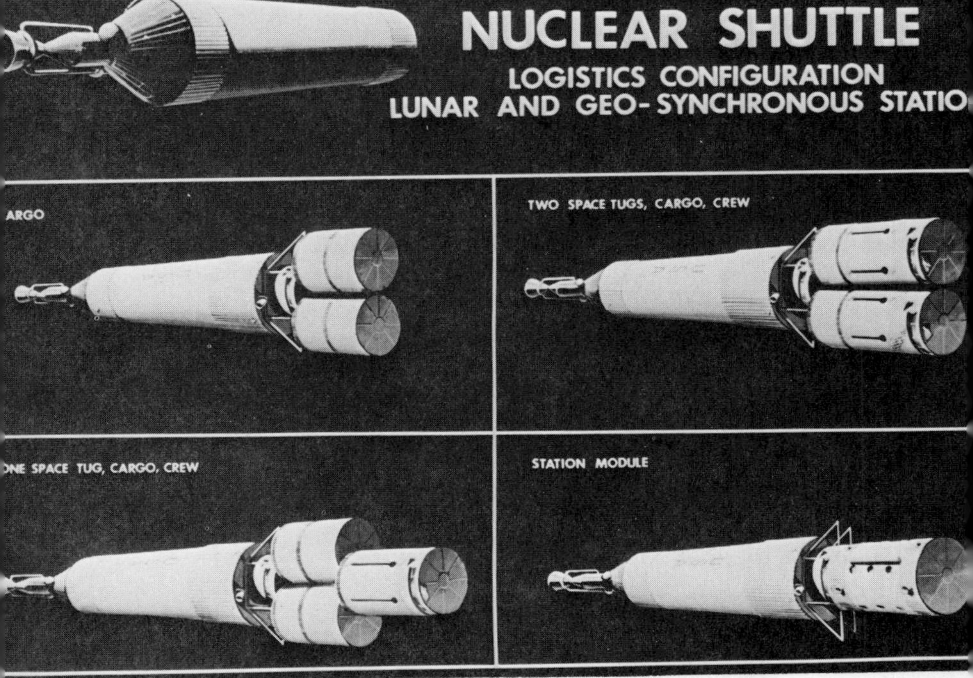

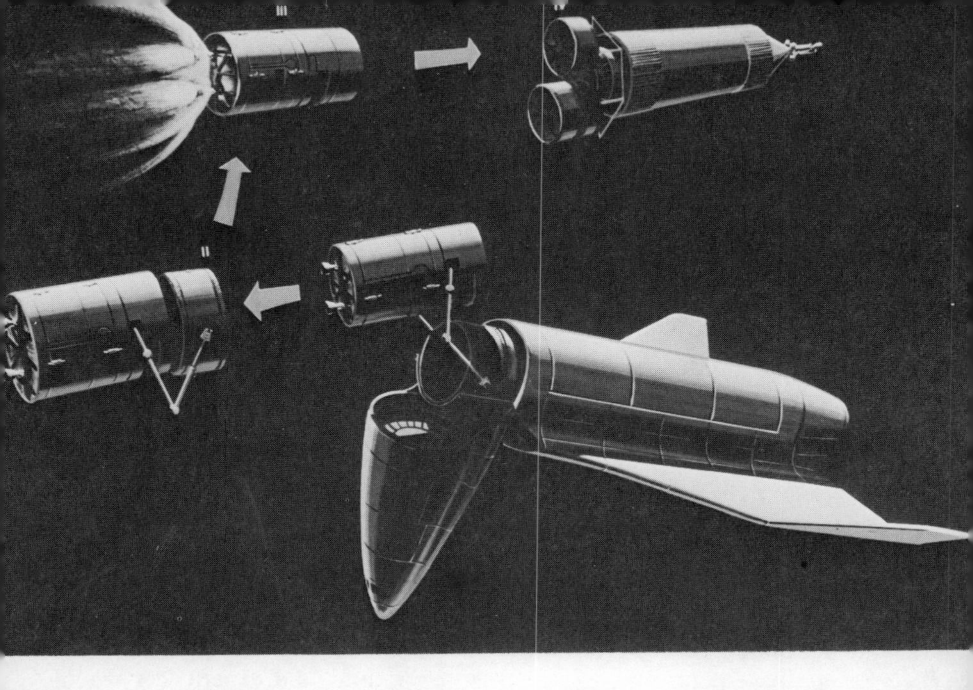

Figure 12-11. Earth Orbit Cargo Transfer. Space Shuttle to Nuclear Shuttle.

With the introduction of the nuclear shuttle for logistics support, the development of a permanent base on the lunar surface then becomes feasible. If lunar resources could be developed to sustain a base, the preparation and operation of the base could provide invaluable data on manned planetary operations in the 1980s and later.

The initial build-up of the base would require that a Space Station Module be placed on the lunar surface from lunar orbit with the propulsion module of the Space Tug. The station would be manned and logistically supplied via the operation of the Space Tug from lunar orbit; at this phase of the lunar programme all equipment, supplies and crew rotations would be supported from Earth orbit with the nuclear shuttle (*Figure 12-12*). In-depth selenological exploration would begin as the base expanded and specialized equipment would be assembled. Drills of several hundred feet capability could be employed to determine if lunar resources could be exploited. Large optical X-ray and gamma-ray telescopes could be erected, and an extended base operation could be developed (*Figure 12-13*).

Successful exploration missions on the lunar surface and the establishment of a lunar base could conceivably lead to discoveries of materials and techniques for utilization of lunar resources by man. Production of propellants, fuel and food on the Moon would greatly reduce logistic requirements from Earth. A permanent colony could result consisting of multiple shelters, various systems for surface transportation, permanent scientific observatories and electro-chemical conversion systems for processing lunar materials. The base would provide not only basic life support and working facilities for the inhabitants, but also essential recreation, entertainment and social functions. It is assumed that by this time in lunar operations, non-astronaut scientists and specialized technicians (including men and women) would make up the largest part of the base personnel. Since these operations and experiences would be fully exploited in manned operations on Mars or other planets, crew rotation would be extended to a year or greater unless emergencies developed.

The space station programme to be initiated in the 1970s will be the first step in an incremental programme which would lead toward a centralized and general purpose Earth orbital laboratory in the 1980s. This centralized facility, or space base, would introduce a new more mature and routine mode of space operations than past programmes. Long term operation with an associated low cost reusable transportation system would enable full exploitation of Earth orbital operations.

Initially, the base would accommodate approximately 50 persons including a small number to perform command, control, service and maintenance functions. Growth to a 100-man capacity could be anticipated by the late 1980s. All personnel working in the space base would be highly trained in specialized disciplines. Since the base would provide an Earth-like environment in addition to large zero gravity facilities, little training, in comparison to present astronaut training, would be required to compensate for the environment.

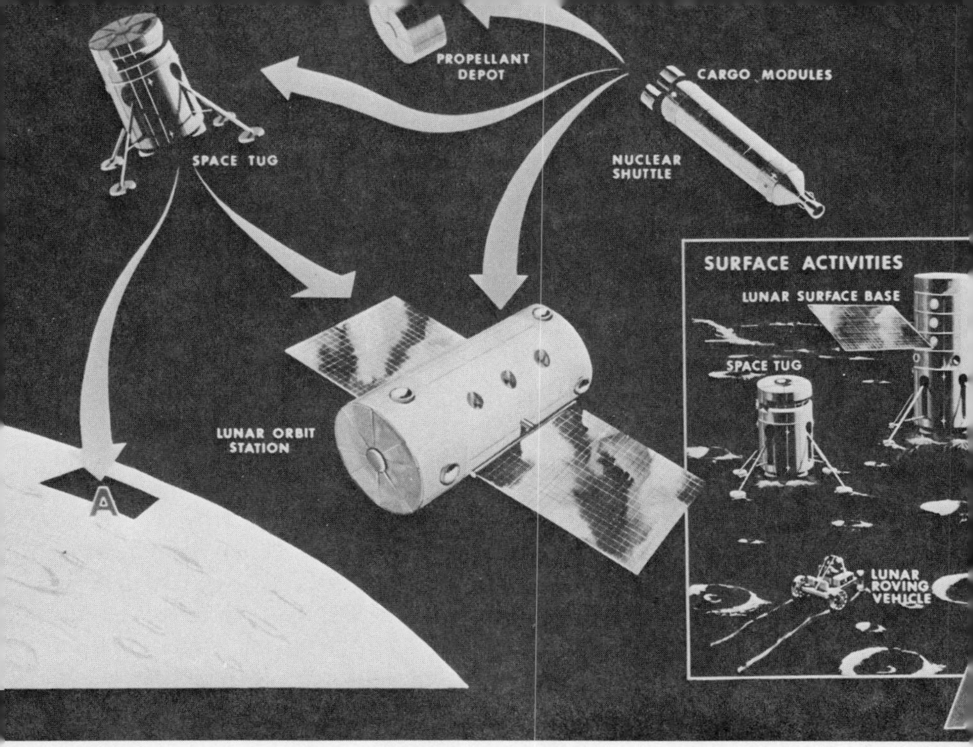

Figure 12-12. Lunar Orbit Operations.

Figure 12-13. Lunar Surface Base.

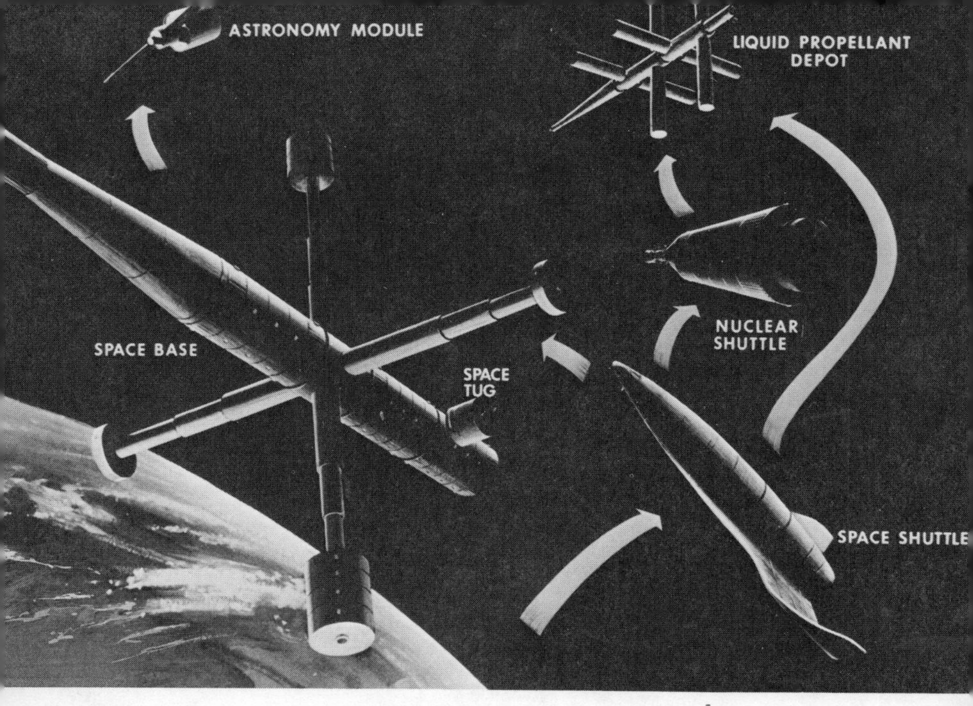

Figure 12-14. Space Base Operations.

One of the space base options being investigated is illustrated in Figure 12-14. The space base would be designed for both zero and artificial gravity. In its final assembled form, portions of the living quarters are conceived to rotate around a central hub of about four revolutions per minute at a radius of some 100 feet from the axis of rotation. Large portions of the base would be counter-rotating to facilitate docking and to support scientific investigations taking place in the weightless environment.

The base would be modular in construction to enable reconfiguration or expansion through in-orbit assembly. It would be powered with a large nuclear power supply and would contain advanced closed loop life support systems. Command post functions would permit highly autonomous mission operations and would reduce mission support ground activities.

Extensive laboratory facilities aboard the base would provide experiment support capability for space astronomy, space physics,

Earth surveys, advanced technology, aerospace medicine, materials processing and engineering operations. Modules which could operate in an attached mode to the base would be launched periodically during the lifetime of the base and docked to one of the many experiment support docking ports provided on the base. Others, which require extremely fine pointing or very low gravity levels not provided by the base, would be operated as free-flying, remote modules which could return and dock to the base periodically for servicing. Some of the astronomy and space biology experiment modules fall into this category.

The space base would provide docking, servicing and recharging functions for the Space Tug which would be utilized in the initial assembly and buildup of the base and which would provide a variety of support functions as a part of the overall base operations. Included in these tug functions would be base exterior inspection and repair; service of the nuclear power supply; support in the unloading of upcoming cargo from the space shuttle to the base; service, inspection and retrieval of remote experiment modules or satellites; and transfer of crew or cargo modules from the space shuttle to a nuclear shuttle for missions to lunar and geosynchronous orbits.

Versatile docking facilities would be provided for the space shuttle at the base to enable accommodation of frequent crew and cargo deliveries. During use of the space shuttle to resupply the nearby propellant storage depot, the base would monitor, support and provide shuttle crew accommodation if needed. Monitoring of propellant transfer operations, servicing and maintenance at the propellant depot would be provided by the Space Base-Space Tug system.

In the 1980s, the precursor activities necessary for manned exploration of the planets will have been accomplished. One of the options available as a logical follow-on in the manned space flight area in the post-Apollo period is a manned mission to Mars. This mission would be based on the information provided by unmanned Mars flights, the Mariner and Viking programmes. To date, no schedule or hardware timetable has been established by the United States to accomplish a manned Mars landing. However, in the interest of planning for the future, the National Aeronautics and Space Administration has conducted a study to determine the feasibility of a Mars mission. The following discussion is based on that study.

The manned Mars mission will be made possible by the new systems developed in the 1970s with the exception of the Mars Excursion Module which would have to be developed. The Space Station Module would be used as a crew compartment and cargo storage area and the Nuclear Shuttle would be used for propulsion. In preparation for the Mars expedition, the crew, cargo, experiments and fuel would be placed into Earth orbit by the Space Shuttle and the Space Station Modules and Nuclear Shuttles necessary for the mission would be placed into orbit by the two-stage Saturn V launch vehicle (*Figure 12-15*).

Although spacecraft may be launched to the Moon approximately once each month, it is the nature of planetary missions that launch windows (the interval of time that a spacecraft may be launched to achieve the proper trajectory) do not occur as frequently. Missions to Mars can be launched only approximately every two years.

The Mars expedition would be made with two space ships, each carrying a crew of six (*Figure 12-16*). In view of the total round-trip flight time of 640 days, the spacecraft would be far more comfortable and roomy than the Apollo Command Module (*Figure 12-17*). The proposal to use two space ships is based on the thought that ship redundancy will be particularly helpful when the crew is so far away from Earth that any idea of help provided from Earth would be completely out of the question. If one ship became incapacitated and unable to return, then its six-man crew could transfer to and return in the other ship. It would be more crowded but acceptable to return all 12 in one ship.

Each spacecraft would be equipped with three nuclear engines. The power manoeuvre of departure from the Earth orbit would be performed by two of these engines. After acceleration from orbital to escape speed to inject the spacecraft into their unpowered flight path to Mars, these two engines would be detached, turned around 180 degrees and fired back to Earth orbit for reuse at a

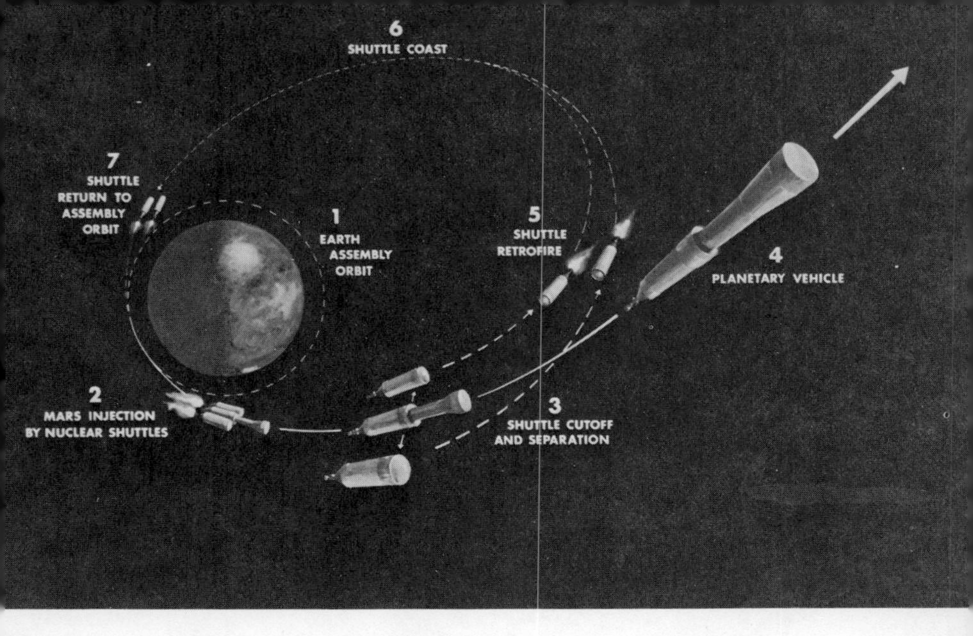

Figure 12-15. Earth Orbit Departure Manoeuvres.

Figure 12-16. Earth Orbit Departure.

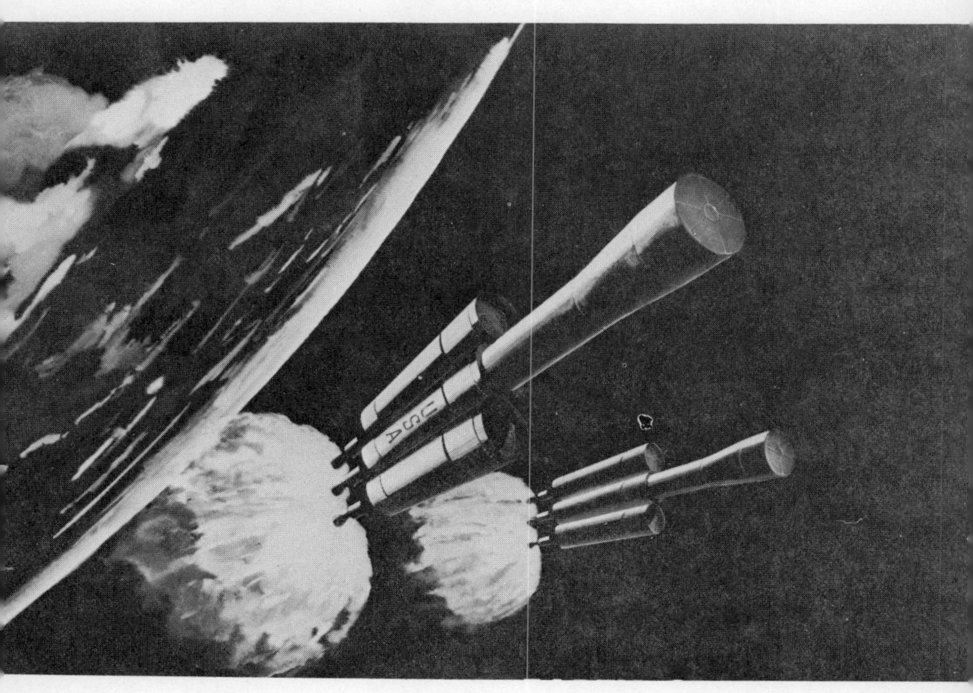

Figure 12-17. En route Spacecraft Configuration.

Figure 12-18. Mars Arrival.

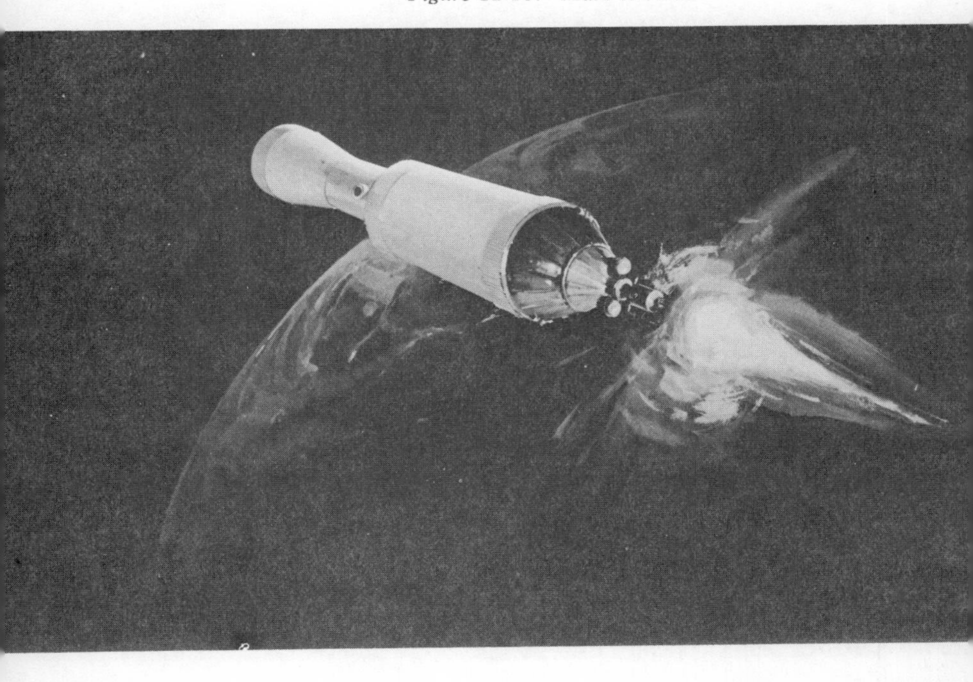

later date. The third nuclear engine would be used for deboosting into the Mars orbit and the return trip to Earth.

Upon arrival at Mars, the spacecraft will be braked by firing the nuclear engine and enter into the Martian gravitational field (*Figure 12-18*). Mars would then pull the craft around and at the lowest point of the approach trajectory, the engine would be fired again to convert what would otherwise be a hyperbolic sweep through the Martian gravitational field into an elliptic orbit of about 24 hours period of revolution around Mars.

After achieving this orbit, unmanned landers will be dispatched to the surface of Mars to determine what the conditions are in particularly interesting, potential landing sites. These vehicles will make an aerodynamic entry, unmanned, into the Mars atmosphere and land at a predetermined spot, each on its own jet power, very much like the Lunar Module. Upon touchdown on the surface, the hatch will open and a small remote controlled vehicle will emerge and begin scooping up surface sample material. The remote controlled vehicle will move 100 to a thousand feet away from the lander so that the samples will be made from an area that is not spoiled by the lander's exhaust jet. The probe will then re-enter the lander, the ascent stage will be fired, and the lander will return to orbit for docking with the spacecraft (*Figure 12-19*).

After the initial samples have been analyzed, the men aboard the spacecraft will enter the Mars Excursion Module (MEM) and descend to the Martian surface. The MEM will be designed to transport the surface exploration crew and their equipment to the Mars surface, provide living accommodations for 30-60 day exploration periods, and transport the crew, scientific data and samples back to the orbiting spacecraft (*Figure 12-20*). The MEM will be an Apollo-shaped vehicle which uses aerodynamic braking to remove most of the velocity of the craft during descent and a terminal propulsion system for the final braking and landing manoeuvres. The descent stage will contain crew living quarters, a scientific laboratory for use during the surface exploration, and a hangar for transporting a small rover vehicle to the surface and storing it during periods of non-use. The descent stage will also

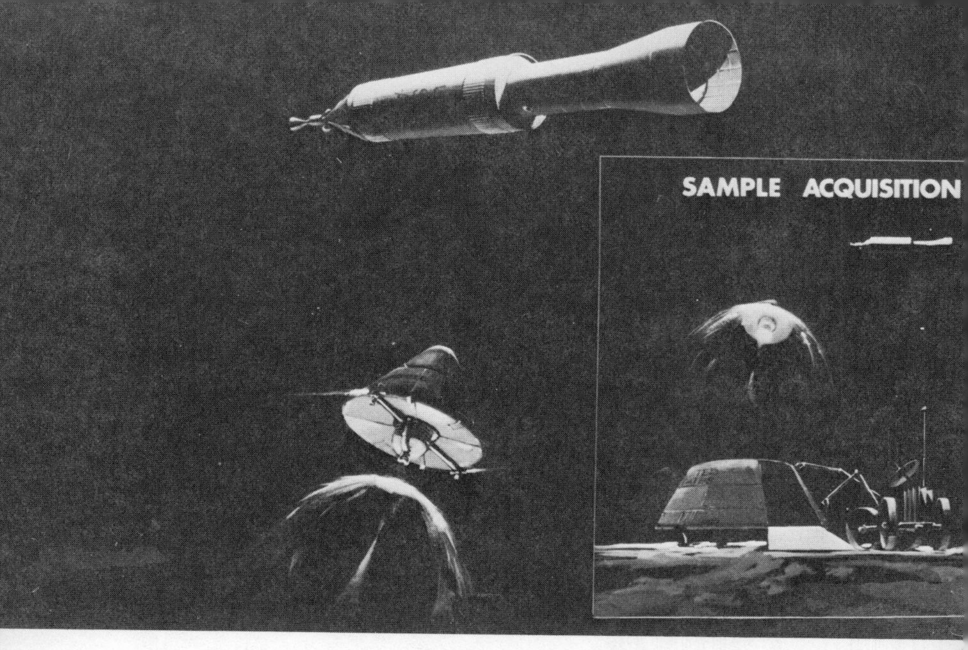

Figure 12-19. Mars Surface Sample Return.

Figure 12-20. Mars Excursion Module Configuration.

Figure 12-21. Mars Initial Landing.

Figure 12-22. Mars Surface Excursion.

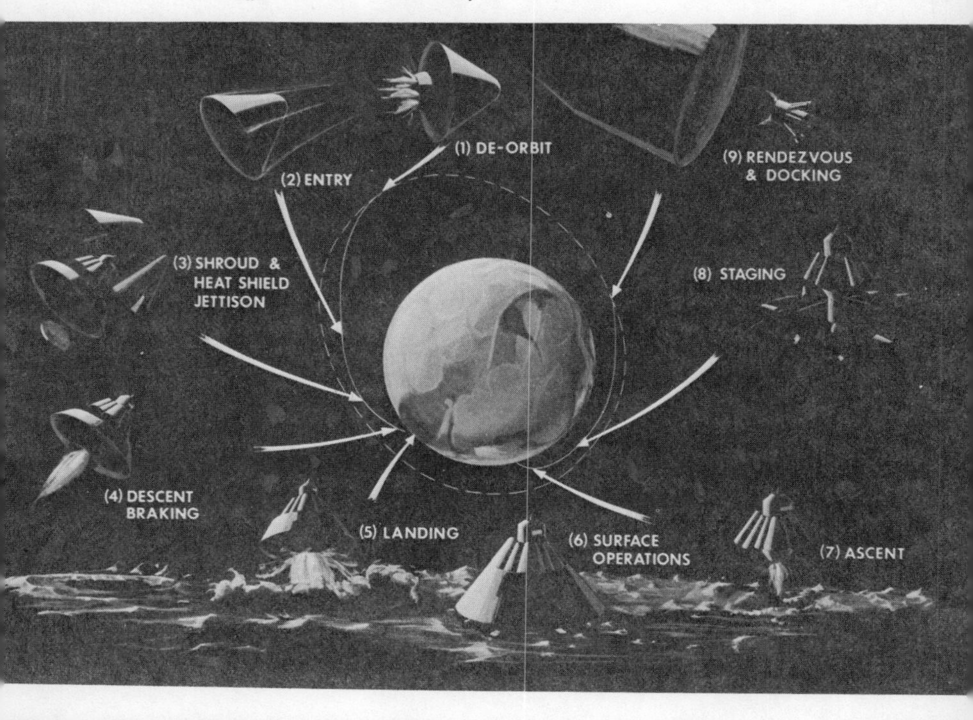

serve as a launch platform for the ascent vehicle. The crew compartment, located above the ascent stage, will be occupied by all crew members during descent and ascent phases of the mission and will serve as a command control centre during the surface exploration.

The scientific objectives of the manned Mars landing mission will be to make geophysical observations including studies of the gravitational field, magnetic field and the internal composition of Mars; collect soil and atmospheric samples; study life forms; study the behaviour of terrestrial life forms in the Mars environment; and search for water and usable natural resources (*Figure 12-21*).

The Mars surface activity on the initial mssion will be similar in many ways to the early lunar exploration activities. Notable, however, is the much longer stay-time of 30 to 60 days per MEM which will permit more extensive observation, experimentation, and execution of scientific objectives. Surface operations will include experiments to be performed in the MEM laboratory as well as the external operations on the Mars surface. The small rover vehicle will permit trips to interesting surface features beyond the immediate landing area.

While the crew on the surface carries out the expedition, that part of the crew which remains in the orbiting spacecraft will conduct experiments, monitor the surface operations, and conduct the necessary spacecraft maintenance.

With the completion of Mars surface activities, the explorers will enter the MEMs, ignite the ascent stages, and return to the mother ship (*Figure 12-22*). The nuclear engine will then be fired again to drive the spacecraft out of the Martian orbit and project itself into a circumsolar ellipse by way of Venus to Earth (*Figure 12-23*).

During the outbound and inbound legs of the mission, experimental activities will be conducted, such as solar and planetary observations, solar wind measurements, biological monitoring of the crew, test plant and animal observations and (during the return flight) analysis of the Mars samples.

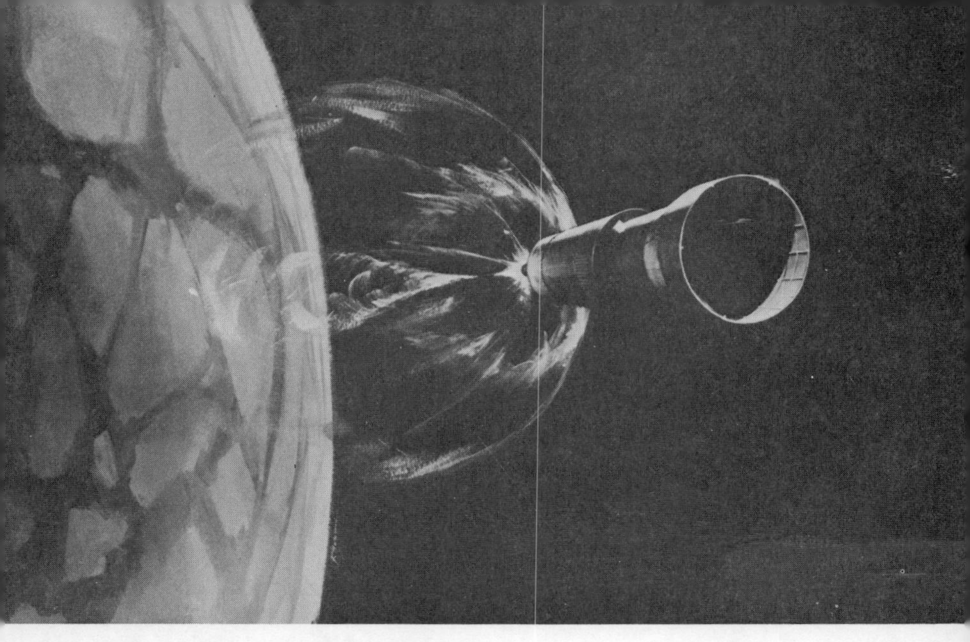

Figure 12-23. Mars Departure.

Figure 12-24. Release of Venus Probe.

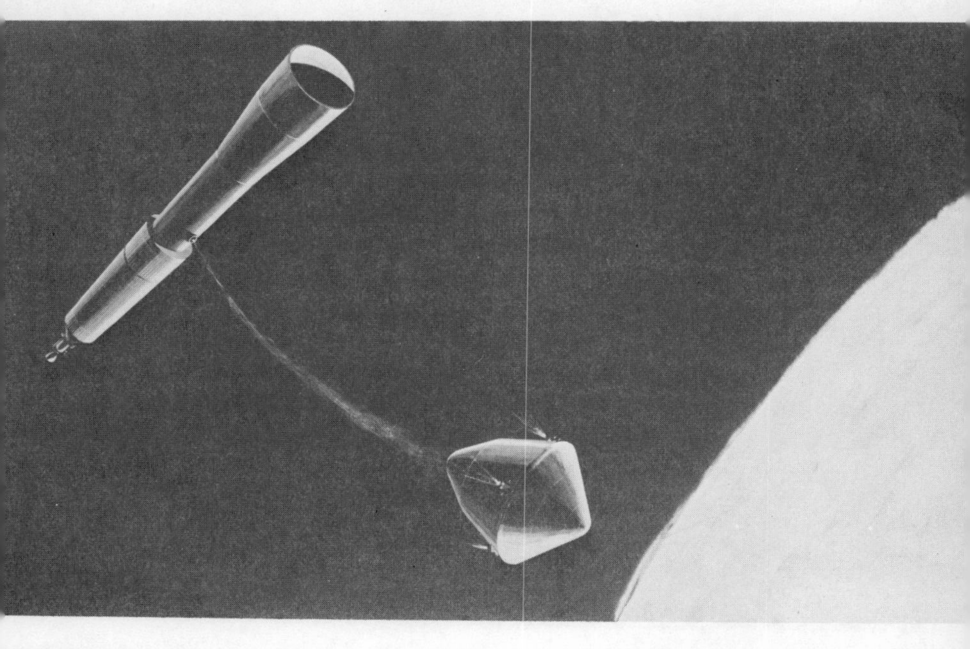

One hundred and twenty-three days after the Mars departure, the spacecraft will fly by the planet Venus. Using the flyby velocity of the spacecraft, it will be planned to inject two small (2000 pound) unmanned probes into the atmosphere of Venus. These probes may land on the Venusian surface or it may be possible to launch them so that they would float in the cooler regions of the Venus atmosphere and make a radar survey of the surface (*Figure 12-24*).

After 640 days or nearly two years, the spacecraft crews would fire the nuclear engines and settle in an orbit around the Earth. From this orbit the crews would then enter another vehicle, most likely the Space Shuttle, and descend to Earth (*Figure 12-25*).

When the findings of the early Mars surface explorations establish the desirability of a more comprehensive exploration of the planet, a temporary base could be established (*Figure 12-26*). This base

Figure 12-25. Return Earth.

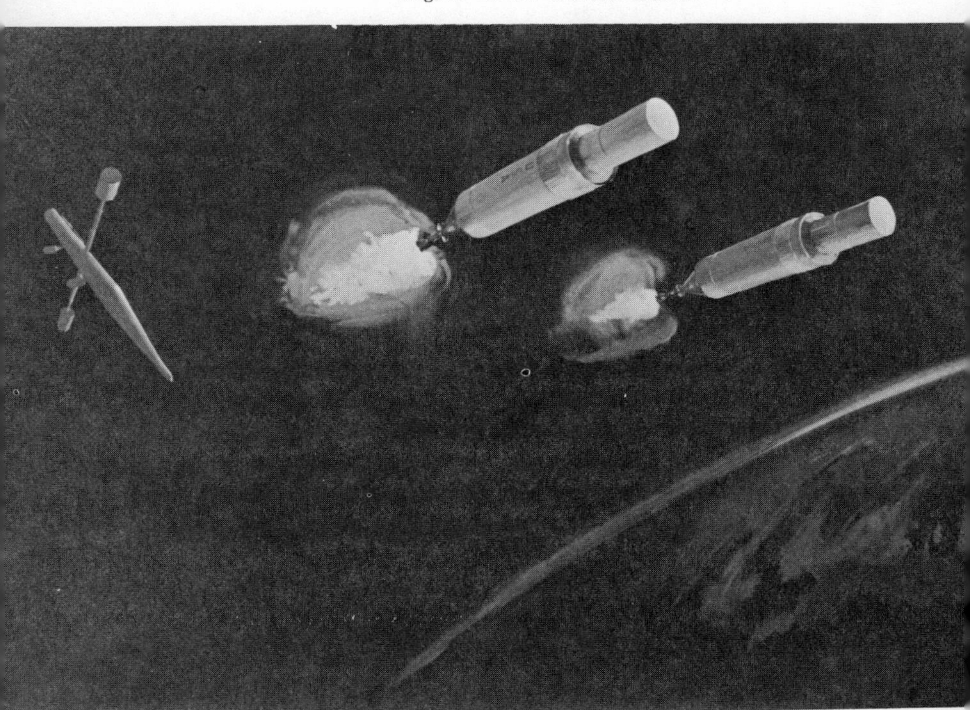

Figure 12-26. Mars Base.

would be used to further the scientific exploration of the planet and to investigate the feasibility of exploiting the planet's natural resources as a means of establishing a more permanent operation. The surface exploration would be complemented by an extensive scientific programme conducted by the crew in the orbiting spacecraft. Of a total mission duration of about 1000 days, approximately 300 days would be spent in Mars orbit and on the Mars surface.

Maximum effectiveness of the temporary Mars base would require an improved surface transportation system that would permit sorties for exploring the planet's surface outside the immediate vicinity of the landing site. This system could be an enlarged and improved version of a similar system used in the lunar programme and earlier Mars missions. Other systems would be common to those used in the first Mars landing mission.

In the far more distant future, a more sophisticated facility would be needed to support Mars operations. The Station Module originally developed for Earth orbit applications and later used as the lunar orbit station, lunar surface base, and planetary mission module could serve as the building block for the surface base. The descent stage of the Mars excursion module would be used for landing the Station Module on the Mars surface and augmenting the Station Module's facilities and services. This facility would provide living quarters, a medical facility, scientific research laboratories and a control centre to monitor operations at remote sites. Redundancy in design and maintainability of systems would assure dependable, long-life operation (*Figure 12-27*).

The facility just described could be developed into a Mars colony capable of exploiting the natural resources of the planet so as to be as nearly self-supporting as possible. It may some day be possible, for example, to establish a plant to manufacture propellants at the colony for use in a logistics spacecraft that shuttles between

Figure 12-27. Temporary Mars Base.

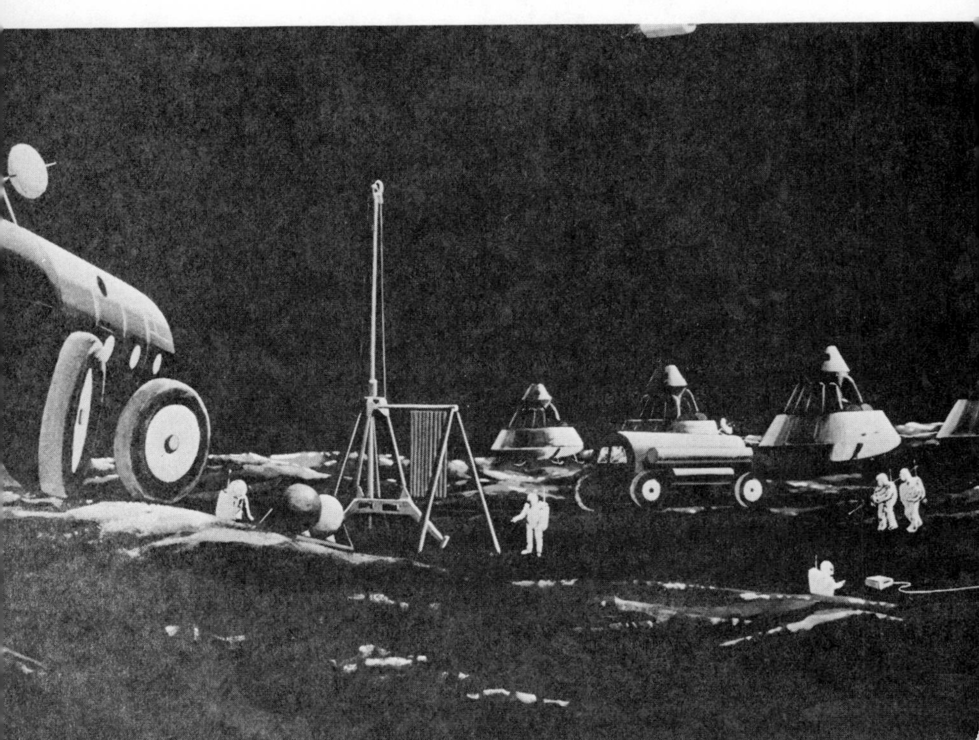

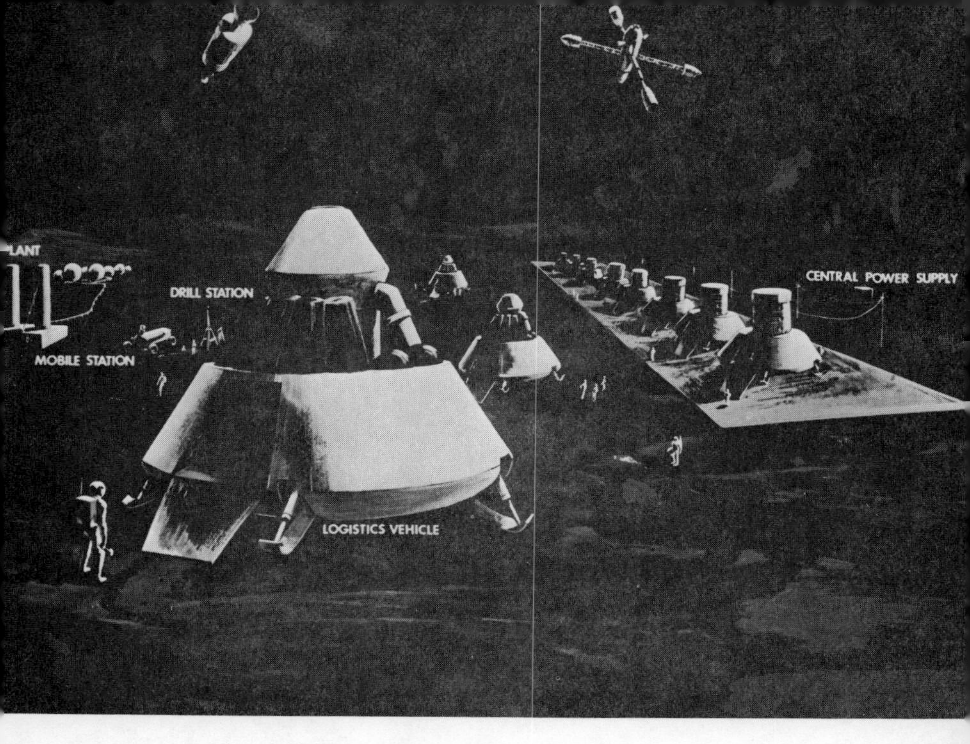

Figure 12-28. Mars Semi-Permanent Base.

the surface of Mars and a station in Mars orbit. It could also be possible to extract the oxygen needed for life support and environmental control systems from minerals (*Figure 12-28*).

The Mars surface colony activities would be accompanied by a significant activity in Mars orbit. An orbiting space base would be maintained in Mars orbit to support the surface operations and to conduct deep space scientific research. It is probable that the tours-of-duty of the Mars exploration crews would include assignments at both the Mars orbiting base and Mars surface colony. The engineering support activities would include assembly of Base Modules from Station Modules and descent propulsion systems transported to Mars orbit by Nuclear Shuttles. The Mars orbiting base may also include a depot where supplies from Earth would be stored until needed by the surface operations and where materials brought up from the Mars surface would be kept until it was convenient to send them to Earth (*Figure 12-29*).

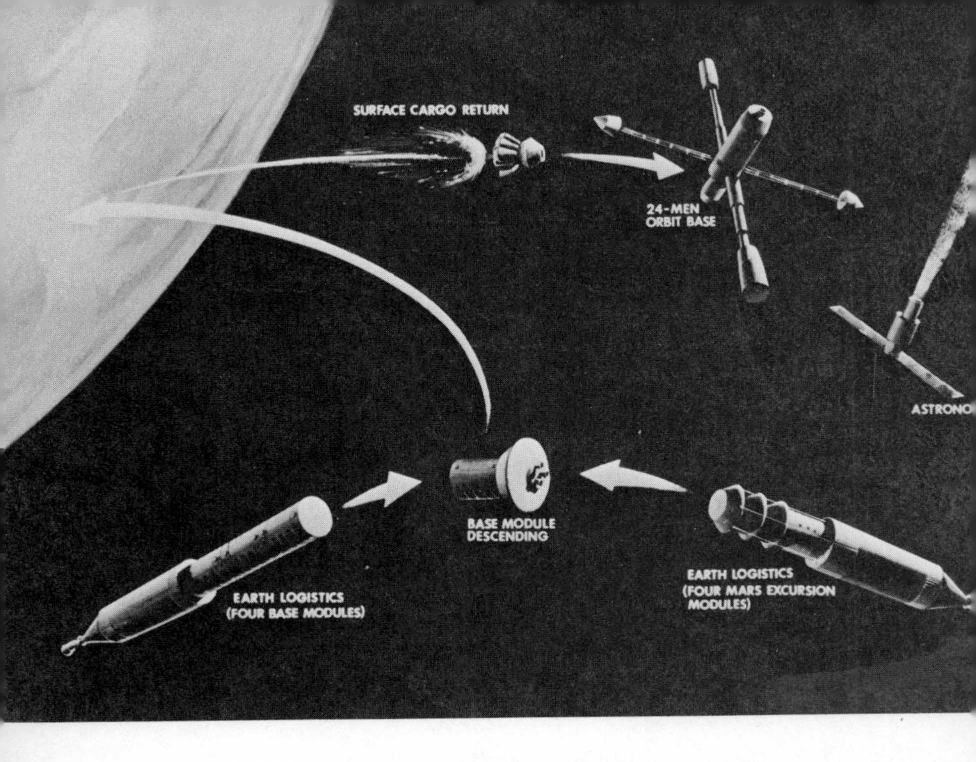

Figure 12-29. Mars Orbital Activity.

Conclusion

Man now has the demonstrated capability to move on to new goals and achievements in space. Space exploration that seemed impossible a few years ago has become today's accomplishment. And, significantly, the challenge of the future is not confined to one or two nations. In the coming years, greater international co-operation in space will develop. The adventures and applications of space missions will be shared by all peoples. Progress will be faster and accomplishments will be greater as nations join together in this effort, both in contributing the resources and in enjoying the benefits.

Our opportunities are great and we have a broad spectrum of choices available. It remains only to chart the course and to set the pace of progress in this new dimension of man.

Science and Mankind

by

M. Oliphant

Sir Mark Oliphant, F.R.S.,
Emeritus Professor, Fellow of the Australian National University, Canberra.

Science and Mankind

1. Science and Technology Now Determine the Course and Nature of Civilization

(a) *The Development of Science and Technology*

Man as we know him today has existed on earth for at least 100,000 years, and possibly for a million years or more. For thousands of years his numbers were small and scattered, so that few remains of the greater part of his primitive history have been found. It was only when he began to practise the simplest forms of technology, chipping stones to make tools and weapons, and making fire, that this most defenceless of all animals could increase substantially in numbers and leave behind durable evidence of his existence. From then on, radioactive carbon-dating of the remains of fires and bones allows a reasonably complete early history to be compiled. It is man's use of tools, that is his development of technology, which has enabled him to become civilized and to dominate the earth. The discovery of metals, which could be hammered or cut into a variety of shapes to suit his needs, and of utensils of clay hardened by fire, some 10,000 years ago, together with the earlier ability to domesticate animals and grow crops, enabled him to live in settled communities and begin the process of civilization.

Language developed enormously, and the written word appeared. The arts, music and literature enriched the lives of mankind. The rules of conduct necessary for orderly life in a community became codified into religion, social behaviour, and the law. Freed from the continuous struggle for existence, the upper classes had leisure to observe nature about them and record their findings. Philosophizing about shapes produced geometry, which was formalized by Euclid. Observation of the heavens showed that some stars, the planets, moved with respect to the fixed pattern of the firmament. Astronomy was born, and given a structure by Ptolemy.

Unfortunately, this early enthusiasm for knowledge of nature did not last, although some astronomical observations continued in the East. In the West, the supremacy of the Church, and the conviction that all philosophy was to be found in the writings of the Greeks, particularly Plato and Aristotle, led to the rejection of any ideas which did not conform with the concepts of nature propounded by these authorities. Revelation, on the one hand, and a formalized theory of a perfect universe, divorced from the imperfections of observation, on the other, virtually prevented progress in science for almost 1,000 years. During this period, skills in fabrication of metals, stone and wood increased, and some of the most beautiful buildings on earth were constructed. But the life of the ordinary man remained almost unchanged. Most lived and worked on the land, producing the food, wood, flax, and so on required by themselves and by the few who associated with the rulers and armies in or around the cities. Travel was difficult and costly, and life centred around the village. Few learned to read and write. Development of the arts, or of any form of intellectual life, depended upon the patronage of the rich. The only universities were centres of theological study, as Cambridge and Oxford in England.

The great discoveries of foreign lands made by intrepid Spanish, Portuguese, Dutch, and above all English navigators in the fifteenth and sixteenth centuries, brought realization that there remained much to learn about the world. The wealth of the East and of the Americas began to pour into Europe, and by the middle of the sixteenth century had created a new class of rich, the merchant adventurers. The demand for education grew, and with it the number of men and women capable of independent thought. The power of the Church was curtailed by the Reformation, and a similar change in the approach to learning accompanied this upheaval.

Slowly, after a turbulent period, the new freedoms and the growth of the entrepreneur gave rise to the Industrial Revolution. The cities grew, factories were built, and the modern age of man commenced. A new era of questioning of all beliefs led to the rapid growth of the experimental method for obtaining information about nature. Mathematics flourished, and with the advent of Newtonian mechanics, in the middle of the seventeenth century, and the foundation of the Royal Society of London for Improvement of Natural Knowledge, the pursuit of science became a recognized and respectable hobby, and even vocation.

At first, the growth of industry was slow. Techniques for producing the basic materials of industry — iron and steel, non-ferrous metals, chemicals, etc. — on a large scale, had to be developed, machinery for fabricating metals devised, and the necessary organization worked out. Until the middle of the nineteenth century science played little part in industrial development. These were the days of the inventor, of Watt and his steam engine, Stephenson and the locomotive, Cartwright and his textile machinery, and so on. The family business, exploiting the creative skill and business acumen of its founders, reigned supreme. Daring men had the courage to build iron ships, and propel them with steam engines. Others, like Telford, used iron and steel to build bridges, towering railway stations, and the Crystal Palace. The need for precision in engineering led to great development of the lathe, of large planing and shaping machines, of forges and equipment for handling heavy loads. The ingenious, practical-minded tradesman was of far greater importance to industry than the scientist.

Probably the most important single factor, which introduced into technology the need for considerable scientific knowledge, was the rise of electrical engineering towards the end of the nineteenth century. The work of Michael Faraday, at the Royal Institution in London, in his endeavours to understand the relationship between electricity and magnetism, had resulted in the invention of the dynamo and electric motor. Swan's development of the carbon filament electric lamp led to an increasing demand for the new, convenient form of lighting, and rapid improvements in the dynamo and motor made this mode of power distribution of growing importance in industry.

During this period of rapid industrial development, science made great strides. Understanding of mechanics, of the properties of gases and liquids, of light, electricity and magnetism, together with corresponding advances in chemistry, made of the physical sciences an awesome edifice of knowledge. Maxwell's theories of electricity and magnetism, the kinetic theory of gases, and beautiful concepts of chemical structure, added new dimensions to understanding. Darwin's theory of evolution brought life itself into this mechanistic picture of existence. An understandable but regrettable arrogance grew among many men of science, who began to believe that they had the key to all understanding. Marx developed his ideas of the ideal organization of society from this malestrom of materialism. By the end of the nineteenth century, it was clear that science had much to offer industry, and engineering became based upon a sound knowledge of fundamental science. Kelvin, in Glasgow, showed that a physicist could both advance his subject itself, and at the same time undertake or suggest developments of importance to technology. The departments of physics and chemistry, with the older mathematics and the newer disciplines of applied science, rapidly assumed a great importance in the universities, old and new.

The certainty of many men of science that they knew all the answers was rudely shattered by three momentous discoveries, all made about 1895. J. J. Thomson, in Cambridge, demonstrated the existence of an elementary particle, the negatively charged electron, which was a constituent of all atoms. Roentgen, in Germany, shrewdly following up a chance observation that penetrating radiations originated in an electric discharge through a gas, discovered X-rays. Bequerel, in France, showed that uranium and thorium emitted radiations spontaneously and continuously, independent of chemical combination or physical state, thus discovering radioactivity.

These three discoveries revolutionized the physical sciences, and provided tools of enormous importance in the development of all science. Over the years, they have had an incalculable effect upon technology. Probably they have influenced our lives as have no other discoveries made in the history of science. From the discovery of the electron, there has flowed understanding of chemical binding and a new approach to the "design" of molecules; the modern picture of the metallic state, of electrical conduction and of magnetism; and the vast technology of electronics. It has given us powerful instruments of progress in most branches of science, like the electron microscope and the computer, and techniques of entertainment such as sound-cinema and TV. Medicine without X-rays would be hard to imagine today, but this same radiation has innumerable other uses. The Braggs showed that it could reveal the positions and spacings of the atoms in a crystal lattice, and it has now provided information enabling unravelling of the structure of complex organic molecules, including that of those carrying genetic information. Rutherford showed that radioactivity was a process of spontaneous atomic transmutation, from uranium or thorium, through successive emissions of α-particles, or helium atoms, and negative electrons or β-radiation, to stable lead. He went on to use the α-particles to probe the interior of atoms, discovering that they were open structures like the solar system, with a tiny, heavy nucleus at the centre, bearing a positive charge of electricity, surrounded by electrons in number sufficient to make the whole atom neutral. There followed experiments showing that the nuclei of light elements could be transmuted, the first observation being the transmutation of nitrogen into oxygen. Thus, he founded nuclear physics, and laid the foundations of the increasing knowledge of nuclear reactions and nuclear structure which led to the release of atomic energy, with all its implications for mankind.

In this rapid, and necessarily incomplete story of the rise of modern science and its increasing importance in technology, I have omitted to mention some developments in basic science which have influenced enormously our picture of matter, radiation, and the universe in which they exist. This is because Einstein's theory of relativity, and Planck's quantum theory, important though they are in basic science, and with a fascinating beauty of their own, have not influenced directly the development of technology. Some applications of the quantum theory, as in chemistry and in the understanding of metals and semiconductors, have contributed indirectly to technological progress. Recognition of the relationship between mass and energy postulated by Einstein, has been important in the development of nuclear accelerators and in nuclear physics. However, their impact on the development of technology as a whole has not been as great as the other discoveries I have mentioned.

Enough has now been said to indicate clearly that modern science is the father of technological development, and that the rate at which each is growing is increasing exponentially. We recognize that science includes a whole spectrum of activities, ranging from the pursuit of natural knowledge for its own sake, through development, to industrial or other applications. It is very important to recognize clearly that each part of this spectrum is critically dependent upon every other. Thus, the unravelling of the genetic code by Crick and Watson would have been impossible without the existence of commercially available X-ray equipment, electron microscopes and computers. Our present-day knowledge of the fundamental particles of matter, the very frontier of physics, depends upon sophisticated engineering of particle accelerators, and complex instrumentation provided by industry. Discovery of quazars and pulsars depended upon the availability of the many techniques used in radio-communications and radar equipment. Nuclear physics has benefited enormously from the great array of instruments developed in the progress towards industrial nuclear power. Also, each part of the spectrum of scientific activity provides its own intense excitement, its own satisfaction, and its own rewards. The idea, sometimes expressed by the misinformed, that basic or "pure" science is a higher activity of man than is development or applied science, is clearly nonsense. The one end of the spectrum contributes to natural philosophy, and hence to the stature of mankind; the other to man's well-being, and indirectly to basic science.

The rate of increase in scientific activity throughout the world is well illustrated by a recent estimate that there are more scientists at work today than in all time before the Second World War, and by far the greater part is financed by governments. It will be appreciated that governments give little financial assistance to activities which will not contribute, in a relatively short period, to an increase in productivity. They support only in a minor way such prestige activities as art, literature, theatre or sport. Governments are therefore convinced that science pays dividends.

Following the 20-year honeymoon period after the war, there is now some decrease in the rate at which expenditure on science rises, but the rate of increase remains substantial.

(b) *The Interaction between Science and Society*

Let us now examine how the practical fruit of science, which is technology, dominates every part of the life of man, determines the course of development of any country, and hence the policies of governments, provides our entertainment, and is even invading the arts and literature.

Advancing technology has made the whole world one, physically speaking. The jet aircraft has brought any two places on earth within a few hours of travel from one another. The globe can now be circled in 24 hours, so that when travelling in an easterly direction local time remains unchanged. The speed of communications is now so great, using coaxial cables, radio and telecommunications satellites, that news of any event is known all over the earth virtually instantaneously. One can telephone anywhere in the world, at will. Improvements in roadways and motor vehicles make travel throughout the continents an everyday experience, as many young Australians have shown. The barriers of language and social customs are fast disappearing as English becomes the recognized international speech, and knowledge of the ways other people live is spread by radio and TV. Every nation is now virtually dependent upon international trade, and none can be self-sufficient. Happenings within any one country are now the concern of all men, and claims of absolute internal sovereignty become less tenable every day. Whether we like it or not, Australia is no longer sufficient unto itself, able to determine her own destiny independently of other nations. She is now part of the community of mankind, and the directions in which she develops are determined as much by America, Japan, China, Britain and Europe, as by her own will. Thus, the gigantic iron-ore deposits of Western Australia would remain untouched red-brown hills, if Japan did not need steel; the wheat fields of western N.S.W. would be bare if China did not buy the grain; Australian industry would come to a halt if machinery from America and Europe was not available; our armed services are critically dependent upon aircraft, naval ships, weapons and know-how from abroad; Australian TV would be dismal without imported programs, and athletics and sport would decline without competition with other nations.

Science and technology know no national boundaries. American, Russian, Chinese, British and Australian scientists can communicate easily and directly with one another about their work, and share in one another's achievements. Recently, when man walked for the first time on the surface of the Moon, people throughout the world shared a common pride as they watched the televised pictures of that momentous event, while it was actually happening. Science and technology have become perhaps the most important avenues of co-operation and understanding within the whole human family. The desire of men of science to talk with one another, regardless of frontiers, is looked upon with suspicion by many governments, but any attempt at isolation rebounds on a nation by reducing the effectiveness of its scientific effort.

Technological advance in every industry is very rapid and accelerating. We have all seen the transformation of earth-moving for construction of roadways, airfields, foundations of buildings, dams and reservoirs, where complex machinery has replaced men with shovels and wheelbarrows. Bulk-carrying ships, loaded and unloaded by ingenious devices, are removing one of the most arduous, dangerous and degrading forms of human labour, and similar changes have taken place in mining and quarrying. New techniques for cutting and forming metals are being introduced. Such traditional industries as steel production are undergoing drastic transformation as a result of the development of cheap methods of producing oxygen by the ton from the air. Alloys with improved strength, resistance to corrosion, magnetic properties, and so on are introduced continually. Industrial gases are transported all over the world in liquid form. Chemical technology produces improved plastics which are easily fabricated and which can replace metals for many applications. Since the discovery of the sulpha drugs and antibiotics, the pharmaceutical industry has poured an increasing stream of improved drugs into medical practice.

Probably, the most significant developments, with far-reaching applications and social significance, have occurred in electronics. The transistor, based upon solid state physics, has revolutionized almost all applications of electronics, and has made possible the development of large and growing numbers of computers and computing systems. Recently, a new magnetic material has been produced in U.S.A., which is capable of storing a million or more bits of information in a cubic inch of the substance, and this information can be fed in and extracted at enormous speed. It is claimed that such a magnetic memory will reduce drastically the space required for storage of information, and will increase the speed of a computer so that more and more complex calculations can be carried out in a given time. Thus, it is now possible to store all the information required for automatic operation of every machine in a complex factory, for checking such operation, and for testing automatically the components produced. Automation on this scale will render the machine-minder redundant where production runs are long enough to justify the investment. Already, many individual machines, including steel rolling mills, are completely automated, and the practice is growing rapidly. Under these conditions, the small production staff of a factory would consist of highly skilled technologists and technicians to set up a run and maintain the equipment. Moreover, the computer can take over many parts of the process of design of an item, or system, allowing, for instance, three-dimensional viewing of a design from every aspect, and checking of its compatibility with other components.

The computer can not only control operations in a factory, it can keep stores, check them, and order material required. It can keep accounts and write cheques and receipts. Management can use it to analyse market information and show future trends, to store information about every employee and make it available on demand. In a similar way, most of the work now requiring hordes of clerical staff in stores and commerce, banks and government departments will be done by computers. In a medical clinic, specialists will feed information, much of it gathered by instruments, into a computer, which will then diagnose the illness and specify treatment. The factual information stored now in a multitude of books in libraries, which has to be sought laboriously by the scientist, engineer, scholar or student, is now being placed in the memories of computers, and is then instantly available on demand. Even translation from one language to another may be possible by computer.

Enough has been said of the power and flexibility of computer applications to indicate that most of the tedious tasks now carried out by men and women will be transferred to computers, and a great part of the labour force will need retraining for more skilled jobs, or will become redundant. Certainly, automation will reduce the hours of work for all but specialists, and so increase the amount of leisure which must be filled by other activities. Large-scale social problems of adjustment are inevitable, as is already apparent in some occupations, such as that of the waterside worker. The machine and the computer are producing irreversible changes in the nature and organization of society. We shall see later that while this revolution can improve greatly the quality of life, it embodies grave dangers of misuse.

There is another aspect of the application of scientific knowledge which we have not yet mentioned. This is its use in warfare.

Killing always leads to a callous disregard for life, and the rights of individuals. This is true of all people, regardless of the principles to which they adhere in peace. The dehumanizing effects of war were apparent in the days of the sword, or of bows and arrows. They have been multiplied enormously by the development of nuclear and other weapons of mass destruction, all of which are fruits of scientific endeavour. There was room for bravery and individual skill in ancient times. It is the impersonal killing, by remote control, which makes of modern war the most degrading activity of man.

Science was applied to some military problems long ago, but it was during the First World War that it was found to be of supreme importance. Nevertheless, it played only a minor part in the conflict, which was resolved by foot soldiers fighting a bloody war in the trenches and fields.

The Second World War began where the first left off, with

two important differences. Development of the aircraft made of it a formidable weapon, but more important, it became an agent of terror against cities and civilian populations, and their associated factories and services, rather than a weapon for use in actual battle. The development of radar increased the power of defensive measures against raiding aircraft, but it also increased the efficacy of night bombing and the effective deployment of military aircraft. Every man, woman and child, and not only fighting troops, became the target of attack by both sides. The crews of bombers did not see the slaughter and suffering they caused when their deadly loads were released by pressing a button. None of the natural human reluctance to kill remained in even the gentlest and most religious of these men. And this ability to put aside all that civilization stood for extended to those at home who produced the aircraft and bombs, any twinges of conscience disappearing rapidly as propaganda fostered hatred and revenge for the death and injury of their civilian fellows.

Late in the war the Germans launched against London small pilotless aircraft, and then high speed rockets, each carrying a substantial explosive charge. These ingenious German rockets were the forerunners of the long-range ballistic missiles now deployed by both America and Russia. After the defeat of Hitler, the scientists and engineers who had developed the rockets were captured and removed to the U.S.A. and the U.S.S.R. The demonstration of the practicability of long-range rockets, able to carry large loads, was a major technological development during the war which changed completely the whole nature of warfare, and which ushered in the age of space exploration.

The revolution in armed conflict, which has made of total war an unimaginable disaster, is due to the development of nuclear weapons. Since the first of such nuclear weapons, known as fission bombs, devastated Hiroshima and Nagasaki, the power of these devices has been multiplied 1,000 times by using a small fission explosion to detonate a far more powerful fusion, or thermonuclear weapon, popularly known as the hydrogen bomb. Nuclear weapons now exist in the armouries of U.S.A., U.S.S.R., Britain, China and France, which are equivalent in explosive power to a million or more tons of normal explosive. A single such weapon releases more than the total power of all the explosives used in the Second World War. It can obliterate completely, in one moment, the largest city in the world, killing millions of people, wounding severely as many more, and causing serious radiation danger to survivors on the periphery. Other nations such as Japan, West Germany, Israel, Canada, Australia, and above all India, could develop these diabolic weapons if they wished to do so. A worsening of the international situation, in any one of several areas, could lead to a wholesale proliferation of nuclear armaments throughout the world.

The efficacy of these terrible weapons has been increased enormously by the development of rocket systems which can deliver them with accuracy to targets in any part of the earth, from launching bases in any country. The number of such weapons and delivery systems possessed by the two great nuclear powers is now sufficient to over-destroy all the major cities and industrial centres of Russia, Europe and North America, together with a large fraction of the population. As China becomes fully armed with nuclear weapons, the same potential fate will threaten also the whole of Asia.

While these developments have been taking place, medical science has been distorted to breed virulent strains of bacteria and viruses which could kill millions of people if released over any country. Chemistry has developed insidious chemicals which, in very small quantities, can kill, maim or render mad all human beings with which they come in contact. These agents of death can be delivered and spread efficiently by rockets similar to those used for nuclear weapons. They will kill men, animals and vegetation, but they do not destroy buildings as do nuclear weapons.

Nations which indulge in these perversions of medical science and chemistry claim, as they do with nuclear weapons, that they would be used only in retaliation against similar forms of attack. But, like nuclear weapons, they are poised ready for use. Knowledge that use of any one of these weapons of mass destruction would inevitably bring retaliation in kind, produces at best an uneasy peace, in which accident, or deliberate provocation by a madman, could lead to all-out war. Such a war would destroy civilization as we know it, and could do irreparable genetic harm to the remnants of the human race.

Mankind therefore finds himself in a situation where his increasing knowledge of nature, and his technological ingenuity have been misused on a colossal scale to create the greatest threat which he has ever faced. World war has become unthinkable, but he stands on the brink of a precipice which he himself has created, frightened to withdraw and take those steps which would eliminate this terrible prostitution of knowledge, and enable it to be used instead to bring prosperity and happiness to all men. The inherent instability of the present situation calls for heroic measures if all the grave consequences of world war are to be avoided. What are the possible solutions to this dilemma?

The non-proliferation treaty, which Russia, America and Britain have jointly sponsored, would be a useful exercise in international control, involving some degree of inspection, and for this reason is worthy of adoption. However, it cannot be claimed that it confers on the world any real security against nuclear warfare. A nation hard-pressed in a war with conventional weapons would almost certainly use nuclear weapons if they were available, and their use appeared to offer some advantage, for in the atmosphere of war, all restraints disappear. With the widespread use of nuclear power stations, almost every nation will soon be able to manufacture plutonium, and under the clause in the treaty allowing withdrawal of a nation from its obligations upon giving six months' notice, any threat to security, real or imagined, could lead to rapid development of nuclear weapons. Recent experience with relatively small wars, fought with ordinary weapons, as in Africa, the Middle East and Vietnam, has shown that these can continue for many years without victory to either side. They breed just that kind of final desperation which ignores the plight of people. Biafran leaders would have been sorely tempted to use any weapon which they could obtain.

A ban on the use of chemical or biological weapons of mass destruction would be likely to suffer the same fate. While conventional wars are tolerated, the ultimate use of such weapons is not only possible, but probable.

At a meeting in New York in 1965, Adlai Stevenson said:

> "The central question is whether the wonderfully diverse and gifted assemblage of human beings on this earth really knows how to run a civilization. Survival is still an open question, not because of environmental hazards, but because of the workings of the human mind. And day by day the problem grows more complex. It was recognized clearly and with compassion by Pope John; to him the human race was not a cold abstraction. Underlying his messages and encyclicals was this simple thought: that the human race is a family, that men are brothers, all wars are civil wars, and all killing is fratricidal."

The only rational steps for the salvation of mankind are those designed to eliminate war as the final arbiter of disagreements between nations, and to substitute the rule of law and order throughout the world. Utopian though this goal may appear, it is the only possible objective. It cannot be achieved overnight, but every step taken should be planned as an intermediate stage in movement towards that end. Complete and general disarmament will be acceptable only when the nations feel safe from revival of armies and armaments by a criminal or irresponsible nation. This means that methods must be devised to detect breaches of the agreement to disarm at a very early stage, before peace is threatened. Science and technology can devise such methods. Much work has been done already, and though more is required, it is becoming clear that effective safeguards are possible. In the final analysis, however, the elimination of the habit of violence from the human race requires much more fundamental changes, which are not so easily amenable to scientific investigation, at any rate at the present time.

There are other dangers which arise from applications of science.

The rapid spread of technology throughout the world is creating unprecedented demands for raw materials, some of which are wasted on a colossal scale. Thus tin may now be classed as a semi-precious metal, in short supply, yet it is almost all wasted in the coating on steel cans for foodstuffs, industrial products like paint, etc., and for beer! Nickel, so important for making steel alloys and munitions of war, is in such demand that it also is becoming semi-precious. Similarly, the known world supplies of phosphate rock, essential for fertilizer production, are being exhausted rapidly, and soon the equally essential material, sulphur, will no longer be available in free form. It is becoming clear that such materials, sources of which are exhaustible, are the property of all men, and not only of the nations on whose territory they occur. Some sort of international caretaking seems essential if they are to be enjoyed by coming generations, and this means international action.

Over-grazing and over-cropping of large areas of the land surface of the earth have resulted in the creation of deserts, the erosion of soils, and serious decreases in fertility, and this goes on despite the knowledge which agricultural science has made available. Problems of pollution of the air, or rivers and lakes and even the sea, by effluents from factories and motor vehicles, sewage works and runoff from fertilized farm lands, some of which are already irreversible, have arisen in many areas. The largest area of fresh water in the world, the Great Lakes of North America, is suffering from this man-made pollution, which is rapidly changing its whole ecology. Recently, animals, birds and fish in the arctic regions of the earth have been shown to have increasing quantities of D.D.T. and other insecticides in their bodies. The extent of such pollution of the Baltic Sea is so great that some remedial action is being taken by countries around it. Again, this is a problem for all mankind.

The surface of the earth is being changed totally by man. We have mentioned already the inroads made annually by the expansion of cities, growth of factories, provision of roadways and airfields. Earth-moving equipment and explosives enable man to shape the land to suit his needs. The natural forests of the earth are disappearing, and in many regions no longer provide a habitat for native animals or birds. Rivers are diverted from their courses and enormous dams are built so that fish are unable to move freely in them. Much of the wild life of the continents is in danger of extinction.

Rapidly, man is gaining complete control of biological evolution on the earth, including his own. He allows only those trees, plants and grasses to continue to exist which he regards as useful to him at present. He chooses which animals, birds and insects may live on the planet. With the aid of modern medicine, he keeps alive the unfit, who would not have survived under the natural conditions where man evolved from lower forms of life, and allows them to breed, increasing the undesirable elements in the gene-pool of his kind. This increasing control of all life throws an enormous responsibility upon the human race. These deleterious effects of the application of technology are seldom deliberate. They arise from ignorance of the overall effects of new processes and materials, and of large-scale activities now possible, upon the natural ecological order and the overall environment. Only increasing knowledge, which scientific investigation alone can bring, can save mankind from the deleterious effect of his own activities.

We have already noted that the development of modern methods of communication has made the whole world one, physically speaking. Yet men are divided as never before. Extreme nationalism is rampant, breeding fear and suspicion. The gap between the standards of living of the advanced and the developing nations grows greater every year. Trivial border disputes and political differences lead to the closing of frontiers and reduced freedom of movement. Countries like U.S.A., U.S.S.R. and China, which have known revolution and benefited from social upheaval, now condemn strenuously any further changes. Religion has created the artificial division of the Indian subcontinent and of Ireland, while military power and politics have divided Europe. The explosive situation in the Middle East, where the Moslem nations are united only in their confrontation of Israel, has closed the Suez Canal and threatens world oil supplies. The great hopes created by the United Nations Charter have crumbled as it becomes an assembly of disunited nations which pretends that 750 million Chinese do not exist. There are gluts of wheat and other grains in North America and Australia, while much of the world goes hungry, and some of the factories of the advanced nations work at reduced capacity, while the poverty-stricken two-thirds of humanity is in desperate need of their products.

There is not time to consider the many other ways in which scientific knowledge has been misused, or used carelessly. I mention only the unfortunate use of the drug thalidomide on pregnant women, resulting in the birth of many deformed babies, and the recent discovery that the defoliants, in use on an extensive scale in the war in Vietnam, can produce similar deformations in children whose mothers have been exposed to them.

2. The Responsibilities of the Scientist

We have seen that science has created modern civilization, so that today our lives, from conception to death, in the home, the factory, office, or any other job, throughout the nation and internationally, are governed by its applications. Existing knowledge, properly applied, can solve quite easily all the problems of material existence, and create a new era of prosperity and happiness for every inhabitant of the Earth. But this same progress in knowledge of nature has created new and serious problems for mankind. The most important of these problems are:

1. Those resulting from the development of weapons of mass destruction, nuclear, chemical and biological, and of means of delivery against which there can be no effective defence. The rapidly increasing armaments race disturbs the economies of the rich nations, and cripples the developing nations, so that the gap between the "haves" and the "have nots" grows steadily.
2. The population explosion resulting from the application of medical science, hygiene and sanitation. About 73 million more people are added to the world's population every year, and these must be fed, clothed, housed, educated and provided with the manifold services necessary for life. The population of India grows by 13 million each year, more than the total population of Australia.
3. The ecology of the earth — the relationships between it and its plants, animals and man — is changing rapidly as a result of advancing technology and increasing population, and there is no real knowledge of the ultimate results of this interference with the balance of nature.

4. The de-humanizing and de-personalizing results of push-button warfare and of control of the individual in the computer age, and the grave dangers of passive acceptance of this loss of freedom.
5. Application of new drugs and new techniques in medicine, of food additives and methods of processing foods, of pesticides, fungicides, and so on, without exhaustive examination of their overall effects on health and well-being.
6. The effects of technological advance upon the organization of industry and government. Are those applicable in the past necessarily the best in the age of science and technology?
7. How can the whole process of education, which because of the rapidity of change must now cover a lifetime, be fitted to the needs and responsibilities of the new era?

These questions concern every member of society, but because they arise as a result of advance in natural knowledge, the scientist has a special responsibility to endeavour to contribute to their solution.

Many suggestions have been made. It has been proposed that there should be a moratorium on science and technology, a halt in further development, until existing knowledge and its applications have been fully digested by mankind. This is clearly impossible in a competitive world, in which a nation accepting such a pause in the garnering of new knowledge and technologies would fall behind in the race. Another suggestion is that men of science should refuse to work upon warlike or other obviously harmful applications of their knowledge. But scientists are ordinary citizens working in a particular field, and will respond to a call for patriotism or of financial gain as will any other cross-section of the people. A proposal that a scientist should withhold any discovery he makes, if in his opinion it can be misused in any way, has little meaning, for neither he nor his fellow-scientists can foresee its consequences. Who can say whether man will devise some terrible use of Professor McCusker's quark, if its existence is confirmed, though that probability appears now as remote as did the atomic bomb when Rutherford discovered the nuclear atom.

Those who have thought deeply about this question conclude that the scientist can fulfil his obligation to society in one way only. This is by endeavouring, through every means in his power, to create awareness of the situation among the public, which can then make sensible decisions through the ballot box, and in other ways. Through public discussion, radio and TV talks, and by raising these questions with his fellows, he can help create a climate of opinion in which solutions will be sought, and found. He is no more capable of resolving political or economic issues than are others, but he can see that questions involving science and technology, and their consequences, are not left to uninformed politicians, or financiers and industrialists, whose sole objective is gain.

Some men of science are inarticulate, and unable to contribute personally to this recognition of their responsibilities, though they can help indirectly. Others are self-centred, and determined not to allow involvement in public debate to interfere with what they regard as their only duty, the search for knowledge. Many will fear that participation in any activity smelling remotely of politics or protest will prejudice their careers. It is inevitable that there will be extremists who will use this activity to push their own political or sociological beliefs, and a few who just see it as an opportunity for personal publicity. However, the majority are now recognizing their responsibilities, and are doing their best to aid the solution of these pressing problems.

Finally, there can be little doubt that you who are young, whose future as men and women of science, and as citizens of the world, lies before you, are more likely to bring about the necessary reformation than are those already immersed in the system. It is the young who make the important discoveries in mathematics and science. It is they who have the uncommitted minds and the courage to look at all problems without preconceptions or embarrassment. I hope that you will prove as successful in these important directions as you undoubtedly will in your learning and research.

NOTES

NOTES

NOTES

NOTES

NOTES

NOTES